T0342296

Structure from Diffraction Methods

Inorganic Materials Series

Series Editors:

Professor Duncan W. Bruce
Department of Chemistry, University of York, UK

Professor Dermot O'Hare
Chemistry Research Laboratory, University of Oxford, UK

Professor Richard I. Walton
Department of Chemistry, University of Warwick, UK

Series Titles

Functional Oxides
Molecular Materials
Porous Materials
Low-Dimensional Solids
Energy Materials
Local Structural Characterisation
Multi Length-Scale Characterisation
Structure from Diffraction Methods

Structure from Diffraction Methods

Edited by

Duncan W. Bruce
University of York, UK

Dermot O'Hare
University of Oxford, UK

Richard I. Walton
University of Warwick, UK

Contents

Inorganic Materials Series Preface

Back in 1992, two of us (DWB and DO'H) edited the first edition of *Inorganic Materials* in response to the growing emphasis and interest in materials chemistry. The second edition, which contained updated chapters, appeared in 1996 and was reprinted in paperback. The aim had always been to provide the reader with chapters that while not necessarily comprehensive, nonetheless gave a first-rate and well-referenced introduction to the subject for the first-time reader. As such, the target audience was from first-year postgraduate student upwards. Authors were carefully selected who were experts in their field and actively researching their topic, so were able to provide an up-to-date review of key aspects of a particular subject, whilst providing some historical perspective. In these two editions, we believe our authors achieved this admirably.

In the intervening years, materials chemistry has grown hugely and now finds itself central to many of the major challenges that face global society. We felt, therefore, that there was a need for more extensive coverage of the area and so Richard Walton joined the team and, with Wiley, we set about a new and larger project. *The Inorganic Materials Series* is the result and our aim is to provide chapters with a similar pedagogical flavour but now with much wider subject coverage. As such, the work will be contained in several themed volumes. Many of the early volumes concentrate on materials derived from continuous inorganic solids, but later volumes will also emphasise molecular and soft matter systems as we aim for a much more comprehensive coverage of the area than was possible with *Inorganic Materials*.

We approached a completely new set of authors for the new project with the same philosophy in choosing actively researching experts, but also with the aim of providing an international perspective, so to reflect the diversity and interdisciplinarity of the now very broad area of inorganic materials chemistry. We are delighted with the calibre of authors who have agreed to write for us and we thank them all for their efforts and cooperation. We believe they have done a splendid job and that their work will make these volumes a valuable reference and teaching resource.

DWB, York
DO'H, Oxford
RIW, Warwick

Preface

Inorganic materials show a diverse range of important properties that are desirable for many contemporary, real-world applications. Good examples include recyclable battery cathode materials for energy storage and transport, porous solids for capture and storage of gases and molecular complexes for use in electronic devices. Some of these families of materials, and many others, were reviewed in earlier volumes of the *Inorganic Materials Series*. When considering the property-driven research in this large field, it is immediately apparent that methods for structural characterisation must be applied routinely in order to understand the function of materials so that their behaviour can be optimised for real applications. Thus 'structure–property relationships' are an important part of research in this area. In order to determine structure effectively, advances in methodology are important: the aim is often rapidly to examine increasingly complex materials so as to gain knowledge of structure over length scales ranging from local atomic order, through crystalline long-range order to the meso- and macroscopic scales.

No single technique can examine all levels of structural order simultaneously and the chapters presented in this volume deal with recent advances in the key techniques that allow investigation of the structure of inorganic materials that are ordered over distances significantly greater than atomic length scales, *i.e.* crystalline materials. Crystalline materials are substances built from regularly repeating 'structural motifs', which may be atoms, ions or molecules, either individually or as groups. All crystals have a defined 'unit cell' that is repeated to form a translationally invariant tiling of space.

Most of the techniques employed to study the identity of the crystal unit cell and its three-dimensional periodicity are based on the elastic scattering of radiation from the material. However, tuning of the wavelength of the scattered radiation allows the periodic order to be probed in subtly different ways and over a range of length scales. So although some of the diffraction methods discussed in this volume may be familiar to the reader (such as single-crystal and powder diffraction), recent advances have both broadened their applicability (*e.g.* study of much smaller single crystals is now possible, and *ab initio* structure solution is feasible from polycrystalline powder) and made them available more routinely. It is therefore timely to provide up-to-date overviews of their use.

Also included are techniques that can probe the details of the three-dimensional arrangements of atoms in nanocrystalline solids, which allow aspects of disorder in otherwise crystalline materials to be studied. Electron diffraction and total scattering techniques have thus developed so rapidly in recent years that separate chapters on these techniques are warranted. Small-angle scattering is a technique we were keen to include as it is often overlooked as one that can probe the ordered structure of materials below the length scale of powder diffraction methods.

We approached an international set of expert authors to write the chapters in this volume with the brief to provide an introduction to the principles of their technique, to describe recent developments in the field and then to select examples from the literature to illustrate the method under discussion. We believe they have done an excellent job in all respects and hope that the chapters provide a valuable set of references for those who wish to learn the principles of contemporary diffraction methods in the study of Inorganic Materials.

DWB, York
DO'H, Oxford
RIW, Warwick
September 2013

List of Contributors

William Clegg School of Chemistry, Newcastle University, Newcastle upon Tyne, UK

Lu Han School of Chemistry and Chemical Technology, State Key Laboratory of Composite Materials, Shanghai Jiao Tong University, Shanghai, China

Kenneth D. M. Harris School of Chemistry, Cardiff University, Cardiff, UK

Keiichi Miyasaka Graduate School of EEWS, WCU Energy Science & Engineering, KAIST, Daejeon, Republic of Korea; Department of Applied Quantum Physics and Nuclear Engineering, Graduate School of Engineering, Kyusyu University, Fukuoka, Japan

Theyencheri Narayanan European Synchrotron Radiation Facility, Grenoble, France

Reinhard B. Neder Crystallography, Department of Physics, University of Erlangen, Erlangen, Germany

Osamu Terasaki Graduate School of EEWS, WCU Energy Science & Engineering, KAIST, Daejeon, Republic of Korea; Department of Materials & Environmental Chemistry, EXSELENT, Stockholm University, Stockholm, Sweden

P. Andrew Williams School of Chemistry, Cardiff University, Cardiff, UK

1

Powder Diffraction

Kenneth D. M. Harris and P. Andrew Williams
School of Chemistry, Cardiff University, Cardiff, UK

1.1 INTRODUCTION

As discussed in Chapter 2, **single-crystal X-ray diffraction**[1−3] (**XRD**) is the most widely used and the most powerful technique for determining crystal structures, and this technique led to many monumental scientific discoveries in the 20th century. The wide-ranging scope and the routine application of single-crystal XRD in the modern day have arisen both through advances in instrumentation and through the development of powerful strategies for data analysis, such that crystal structures can now be determined rapidly and straightforwardly in all but the most challenging cases. The central importance of single-crystal XRD in the physical, biological and materials sciences will continue to be further developed and exploited in the years to come. Thus, provided a single crystal of sufficient size and quality is available for the material of interest, successful structure determination by analysis of single-crystal XRD data is nowadays very routine.

However, the requirement to prepare a suitable single crystal specimen for single-crystal XRD experiments represents a major limitation of this technique. As a consequence, the crystal structures of many important crystalline materials remain unknown simply because the material cannot be prepared as a crystal of appropriate size and quality for single-crystal XRD studies. In such cases, however, the material can usually be prepared as a microcrystalline powder, and therefore it is still

Structure from Diffraction Methods, First Edition. Edited by Duncan W. Bruce, Dermot O'Hare and Richard I. Walton.
© 2014 John Wiley & Sons, Ltd. Published 2014 by John Wiley & Sons, Ltd.

feasible to record **powder XRD** data. The question that immediately arises is whether it is feasible to determine the crystal structure of a material from powder XRD data using techniques analogous to those employed with single-crystal XRD data. Furthermore, are there any aspects of structural science that might actually be more readily investigated by powder XRD than single-crystal XRD?

With the aim of addressing these types of question, the present chapter provides an overview of the current state of the art in the application of powder XRD within chemical and materials sciences, focusing in particular on contemporary opportunities for determining crystal structures directly from powder XRD data. Fundamental aspects of the techniques used to carry out crystal structure determination from powder XRD data are described, and several illustrative examples of the application of these techniques in determining the structural properties of materials across a wide range of areas of chemistry are highlighted. In addition, we discuss the wide-ranging utility of powder XRD in other aspects of the characterisation of solid materials, from routine applications in the identification ('fingerprinting') of crystalline phases to more advanced applications in which *in situ* powder XRD studies are exploited to investigate structural transformations associated with phase transitions, solid-state chemical reactions, crystallisation processes and materials synthesis. While the chapter is focused primarily on powder XRD, the complementary opportunities offered by powder neutron diffraction are also discussed.

With the exception of some brief mention of certain specific aspects of experimental techniques for the measurement of powder XRD data, details of the instrumentation used to record powder XRD data lie outside the scope of this chapter, which is focused primarily on the application of powder XRD to determine structural information in chemical contexts rather than on the technical details of experimental techniques. Descriptions of the variety of experimental set-ups that may be used to record powder XRD data and comparisons of their relative merits (*e.g.* transmission mode *versus* reflection mode, Debye–Scherrer *versus* Bragg–Brentano, angle dispersive *versus* energy dispersive, point detectors *versus* position-sensitive detectors) may be found in more detailed monographs on instrumentation.[4–7]

1.2   THE SIMILARITIES AND DIFFERENCES BETWEEN SINGLE-CRYSTAL XRD AND POWDER XRD

As discussed above, the form of the sample studied in single-crystal and powder XRD is intrinsically different: a large, individual crystal in the former case and a powder comprising a huge number of small, randomly oriented crystallites in the latter case. However, while the nature of the sample is different, the physical phenomenon underlying both techniques is the same. In each case, X-ray radiation (usually monochromatic radiation) is incident on the sample. As the wavelength of the X-rays is comparable to the periodic repeat distances within the crystalline material (*i.e.* within the single crystal in the single-crystal XRD experiment and within each crystallite present in the powder sample in the powder XRD experiment), coherent/elastic scattering of the X-rays by the sample gives rise to an 'XRD pattern' (Figure 1.1) in which the radiation is scattered with significant intensity only in certain specific directions, while in all other directions the intensity of scattered radiation is zero.

Because the underlying physical phenomenon is the same, single-crystal and powder XRD patterns contain essentially the same information. However, as a result of the different nature of the sample used in each case, the form of the XRD pattern and the way in which it can be measured are different (Figure 1.1). Thus, in the single-crystal XRD pattern, the intense diffraction 'peaks' are well separated from each other in three-dimensional (3D) space ('reciprocal space') and both the scattering direction and the intensity of each individual intensity

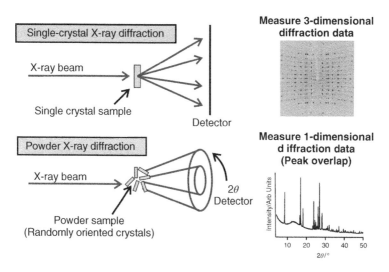

Figure 1.1 Comparison of single-crystal and powder XRD measurements. In powder XRD, the diffraction phenomenon for each individual crystallite in the powder is the same as the diffraction phenomenon in single-crystal XRD. However, the powder comprises a large collection of crystallites with (in principle) a random distribution of crystallite orientations. As a consequence, the three-dimensional diffraction data are effectively compressed into one dimension (intensity *versus* diffraction angle 2θ) in the powder XRD measurement.

maximum can be measured very accurately. In the case of powder XRD, on the other hand, although each individual crystallite in the powder behaves in a similar manner to the single crystal sample in single-crystal XRD, the fact that the powder sample comprises a huge number of randomly oriented crystallites means that only the collective X-ray scattering from the whole sample can be measured. Because of the randomly oriented nature of the crystallites within the powder, the collective X-ray scattering from the whole powder sample comprises a set of coaxial cones of scattered radiation (in contrast to the sharp beams of scattered radiation that arise in single-crystal XRD). The semi-angle of each cone is the diffraction angle, 2θ. As shown in Figure 1.1, the powder XRD pattern comprises the measured diffracted intensity as a function of a single spatial variable, the diffraction angle 2θ. Effectively, in making the powder XRD measurement, the 3D information contained in the diffraction data is 'compressed' into one dimension, as the diffraction intensity is measured as a function of only one spatial variable.

The peaks in the powder XRD pattern arise at specific values of 2θ that satisfy Bragg's law:

$$2\theta_{hkl} = 2\sin^{-1}[\lambda/(2d_{hkl})] \tag{1.1}$$

where λ is the wavelength of the X-rays, the indices h, k and l are three integers (the Miller indices) that uniquely label each intensity maximum ('peak') in the diffraction pattern and d_{hkl} is the interplanar spacing of a specific set of lattice planes (also uniquely labelled with the same Miller indices, h, k and l) in the crystal structure (the set of interplanar spacings, d_{hkl}, for a crystal structure depends on the dimensions of the unit cell – see below).

As a consequence of the fact that the diffraction data are 'compressed' into one dimension in the powder XRD measurement, there is usually considerable overlap of peaks in the powder XRD pattern. Such peak overlap serves to obscure information on the position (*i.e.* the 2θ value) and the intensity of each peak in the powder XRD pattern, and the difficulty of obtaining reliable and accurate information on the peak positions and intensities can impede (or, in severe cases, prohibit) the process of carrying out crystal structure determination from powder XRD data. For materials with large unit cells and low symmetry (such as most molecular solids), there is a very high density of peaks in the powder XRD pattern (especially at high values of 2θ) and the problem of peak overlap can be particularly severe (see Figure 1.2). The 'problem' of peak overlap presents specific challenges in several

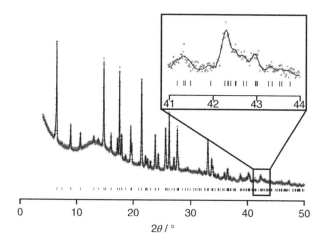

Figure 1.2 Powder XRD pattern of L-phenylalanine hemihydrate. The tick marks at the bottom indicate the 2θ values of individual 'peaks'. The inset highlights a typical region with substantial peak overlap. Note that the density of peaks (*i.e.* the number of peaks within a given 2θ range) increases significantly at higher 2θ values and the overall intensity decreases at higher 2θ values.

aspects of the analysis of powder XRD data, particularly in the context of structure determination. Clearly, the extent of peak overlap may be reduced by recording the data under experimental conditions that give rise to narrow peaks in the powder XRD pattern (*i.e.* 'high-resolution' powder XRD data). As discussed in Section 1.6.4, the widths of peaks in a powder XRD pattern depend both on features of the instrumentation used to record the data and on features of the powder sample, and higher resolution can be achieved (at least in principle) by optimisation of these features. However, while peak overlap may be alleviated by appropriate optimisation of experimental conditions, it will never be eliminated (especially at high diffraction angles).

As a consequence of the problems outlined above, the process of carrying out crystal structure determination from powder XRD data is substantially more challenging than from single-crystal XRD data. Nevertheless, advances in recent years in the techniques used to carry out structure determination from powder XRD data[5,8−26] are such that the crystal structures of many materials can now be determined successfully from powder XRD data, opening up the opportunity to elucidate the structural properties of a wide range of materials that are unsuitable for investigation by single-crystal XRD.

Finally, as a summary comparison between single-crystal XRD and powder XRD, the following generalisations may be made:

(i) *Sample preparation*: Preparation of a single crystal of suitable size and quality for single-crystal XRD can be straightforward, challenging or impossible, depending on the material of interest, whereas preparation of a suitable powder sample for powder XRD is usually straightforward.

(ii) *Data collection*: The time required to record a complete set of single-crystal XRD data is generally longer than the time required to record a good-quality powder XRD pattern for the same material (although the development of area detectors for single-crystal XRD has improved considerably the rapidity of data collection).

(iii) *Structure determination*: Structure determination from single-crystal XRD data is generally very rapid and routine, whereas structure determination from powder XRD data is substantially more time-consuming and challenging; nevertheless, the methodology for analysis of powder XRD data has advanced significantly in recent years and the prospects for achieving successful structure determination from powder XRD data are continually improving.

1.3   QUALITATIVE ASPECTS OF POWDER XRD: 'FINGERPRINTING' OF CRYSTALLINE PHASES

A crucially important, yet very straightforward, application of powder XRD is the identification ('fingerprinting') of crystalline phases, based on the fact that different crystal structures give rise to distinct powder XRD patterns. This type of qualitative characterisation of crystalline materials is exploited in all areas of materials preparation and is also widely utilised in industrial protocols, including quality control, polymorph screening and the characterisation of products from rapid-throughput crystallisation experiments.[27,28]

In utilising powder XRD for 'fingerprinting' of crystalline phases, the aim is to compare the experimental powder XRD pattern of a sample prepared by materials synthesis with those of known materials (either experimental powder XRD patterns recorded for materials prepared by other synthetic routes or powder XRD patterns simulated from known crystal structures that have been determined previously). An example of this type of comparison is given in Figure 1.3. In some cases, the

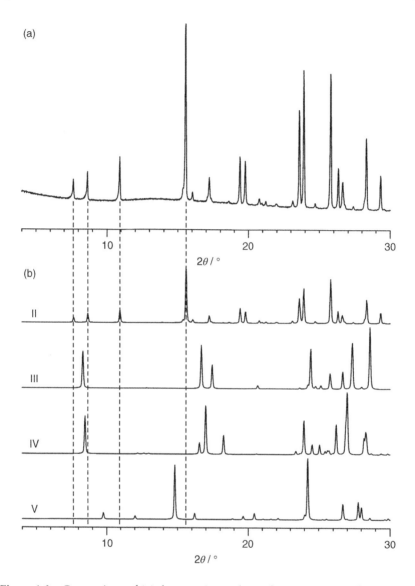

Figure 1.3 Comparison of (a) the experimental powder XRD pattern for a sample of *m*-aminobenzoic acid and (b) the simulated powder XRD patterns for all polymorphs of *m*-aminobenzoic acid with known crystal structures (Forms II, III, IV and V). From this comparison (see the matching of the vertical dashed lines), the sample of *m*-aminobenzoic acid with the powder XRD pattern shown in (a) is readily identified as Form II.

required comparison can be achieved simply by visual assessment of the similarities and/or differences between the powder XRD patterns, leading to straightforward conclusions as to whether the synthesised material either matches a known material or is different from all potential candidate materials of known structure. However, in many cases simple visual inspection of powder XRD patterns does not lead to a definitive conclusion, and deeper scrutiny is required to resolve ambiguities in phase identification. This issue is discussed in more detail in Section 1.9.

1.4   QUANTITATIVE ASPECTS OF POWDER XRD: SOME PRELIMINARIES RELEVANT TO CRYSTAL STRUCTURE DETERMINATION

In this section, we consider some preliminary aspects that are important for understanding the procedures used to determine crystal structures from powder XRD data. Details of the techniques used for structure determination are covered in Section 1.5.

1.4.1   Relationship between a Crystal Structure and its Diffraction Pattern

In the diffraction pattern from a crystalline solid, the **positions** of the diffraction maxima depend on the periodicity of the structure (*i.e.* the dimensions of the unit cell, as defined by the lattice parameters $\{a, b, c, \alpha, \beta, \gamma\}$). On the other hand, the **relative intensities** of the diffraction maxima depend on the distribution of **scattering matter** (*i.e.* the atoms or molecules) within the unit cell. As X-rays are scattered by their interaction with the electrons in a material, the 'scattering matter' in the case of XRD is the electron density distribution within the unit cell. Each diffraction maximum is characterised by a unique set of integers, h, k and l (the Miller indices), and is defined by a scattering vector, H, in 3D **reciprocal space**, given by $H = ha^* + kb^* + lc^*$. The basis vectors a^*, b^* and c^* are called the reciprocal lattice vectors and depend on the crystal structure. The 3D space defining the crystal structure is called **direct space**.

A given diffraction maximum H is completely defined by the structure factor, $F(H)$, which has amplitude $|F(H)|$ and phase $\alpha(H)$. In XRD, the structure factor, $F(H)$, is related to the electron density, $\rho(r)$, within the unit cell by the following equation:

$$F(H) = |F(H)| \exp[i\alpha(H)] = \int \rho(r) \exp(2\pi iH \cdot r)dr \qquad (1.2)$$

where r is the vector $r = xa + yb + zc$ in direct space (where a, b and c are the lattice vectors that define the periodicity of the crystal structure) and the integration is over all vectors r in the unit cell. By inverse Fourier transformation, it follows from Equation 1.2 that the electron density $\rho(r)$ is given by:

$$\rho(r) = (1/V) \sum_{H} |F(H)| \exp[i\alpha(H) - 2\pi iH \cdot r] \qquad (1.3)$$

where V is the volume of the unit cell and the summation is over all vectors H with integer coefficients h, k and l. If both the amplitude $|F(H)|$ and the phase $\alpha(H)$ of the structure factors $F(H)$ could be measured directly from the experimental XRD pattern, then $\rho(r)$ (*i.e.* the 'crystal structure') could be determined directly from the experimental data using Equation 1.3. However, while the values of the amplitudes $|F(H)|$ can be obtained experimentally from the measured diffraction intensities $I(H)$, the values of the phases $\alpha(H)$ cannot be determined directly from the experimental diffraction pattern, which constitutes the so-called **phase problem** in crystallography. To determine a crystal structure from experimental XRD data by using Equation 1.3, it is necessary to use techniques (*e.g.* direct methods or the Patterson method – see Chapter 2) that provide estimated values of the phases $\alpha(H)$. Using the estimated phases $\alpha(H)$ together with the experimentally determined $|F(H)|$ values in Equation 1.3, an approximate description of the electron density $\rho(r)$ and hence the crystal structure can be derived; the quality of the description of the crystal structure is then improved through iterative procedures that progressively improve the quality of the estimated phases $\alpha(H)$. More details on the techniques for overcoming the phase problem are given elsewhere.[1,2]

Importantly, the reverse procedure of calculating the diffraction pattern (*i.e.* the $|F(H)|$ data) for any given crystal structure using Equation 1.2 is a completely 'automatic' calculation. Thus, the diffraction pattern can be calculated automatically for *any* crystal structure using the positions of the atoms in the crystal structure in Equation 1.2, employing a form of Equation 1.2 in which the electron density $\rho(r)$ is approximated by a function that depends on the positions of the atoms in the unit cell. This type of calculation is the basis of the **direct-space**

strategy for structure solution. In the direct-space strategy, a large number of trial crystal structures are generated by computational procedures, the powder XRD pattern for each trial structure is calculated automatically using Equation 1.2 and these calculated powder XRD patterns are then compared with the experimental powder XRD pattern in order to assess the degree of 'correctness' of each trial structure. More details of the direct-space strategy for structure solution are given in Section 1.5.4.2.

1.4.2   Comparison of Experimental and Calculated Powder XRD Patterns

There are two general approaches for comparing experimental powder XRD data with powder XRD data calculated for a structural model during the process of structure determination: (i) comparison of the complete powder XRD profile and (ii) comparison of integrated peak intensities.

The complete powder XRD profile (for either an experimental pattern or a calculated pattern) is described in terms of the following components:

   (i) the peak positions;
  (ii) the background intensity distribution;
 (iii) the peak widths;
  (iv) the peak shapes;
   (v) the peak intensities.

The peak shape depends on characteristics of both the instrument and the sample (see Section 1.6.4 for more details), and different peak shape functions are appropriate under different circumstances. A common peak shape for powder XRD is the pseudo-Voigt function, which represents a hybrid of Gaussian and Lorentzian character, although several other types of peak shape may be applicable in different situations. These peak shape functions and the types of function commonly used to describe the 2θ-dependence of the peak width are described in detail elsewhere.[29]

In comparing experimental and calculated powder XRD patterns through consideration of the complete powder XRD profile, the entire digitised experimental powder XRD pattern is used directly 'as measured' and a digitised powder XRD profile is calculated from the structural model in order to compare it to the experimental pattern. Construction of the digitised calculated powder XRD pattern for a trial structure requires not only the intensities of all peaks in the powder XRD pattern, which are determined using Equation 1.2, but also information on the peak positions, peak widths, peak shapes and background intensity distribution. Clearly, reliable whole-profile comparison between calculated and experimental powder XRD patterns requires that the variables describing all of these aspects of the calculated pattern accurately reflect those in the experimental pattern. The methodology for determining the values of the variables that describe these features of the experimental powder XRD profile is discussed in Section 1.5.3. After the digitised powder XRD pattern for the structural model has been calculated, it is compared directly with the experimental powder XRD pattern (see Figure 1.4) using an appropriate whole-profile figure of merit, the most common of which are the weighted (R_{wp}) and unweighted (R_p) profile R-factors, defined as:

$$R_{wp} = 100 \times \sqrt{\frac{\sum_i w_i (y_i - y_{ci})^2}{\sum_i w_i y_i^2}} \qquad (1.4)$$

$$R_p = 100 \times \frac{\sum_i |y_i - y_{ci}|}{\sum_i |y_i|} \qquad (1.5)$$

where y_i is the intensity of the ith point in the digitised experimental powder XRD pattern, y_{ci} is the intensity of the ith point in the calculated powder XRD pattern and w_i is a weighting factor for the ith point. A significant advantage of using a whole-profile figure of merit of this type is that it uses the experimental data (*i.e.* the digitised data points, $\{y_i\}$) directly 'as measured', without further manipulation. The figure of

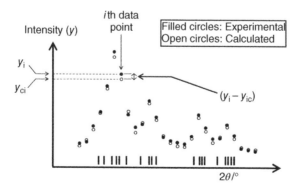

Figure 1.4 Schematic illustration of the comparison between digitised experimental and calculated powder XRD patterns involved in calculating a whole-profile figure of merit such as R_{wp}.

merit R_{wp} is used in Rietveld refinement and in several implementations of direct-space techniques for structure solution (see Sections 1.5.4.2 and 1.5.5).

Consideration of integrated peak intensities, on the other hand, involves analysis of the experimental powder XRD pattern in order to extract a set of integrated peak intensities, $I(H)$ (and hence $|F(H)|$ values), analogous to those measured directly from single-crystal XRD data. Because of the problem of peak overlap in powder XRD data, however, the task of extracting a reliable set of integrated peak intensities from a heavily overlapped powder XRD pattern is non-trivial. The basic techniques for achieving this task are discussed in Section 1.5.3. Methods have been developed to enhance the reliability of the intensity extraction process, by using the extracted intensities in a manner that takes the reliability of the extraction process into account (by making use of the variance–covariance matrix[30]). After extracting a set of integrated peak intensities, the experimental and calculated $I(H)$ data can be compared using the types of figure of merit employed in the analysis of single-crystal XRD data. However, a disadvantage of this approach in the case of powder XRD data is that any errors or uncertainties associated with the process of extracting the integrated peak intensities from the experimental pattern (originating from ambiguities in handling the peak overlap) are inevitably propagated into the structure determination process and may limit the reliability of the derived structural information or even prohibit successful structure determination. As discussed in Section 1.5.4.1, comparison of integrated peak intensities is central to the traditional strategy for structure solution. In addition, some implementations of the direct-space strategy are also based on comparison of integrated peak intensities, with the aim of maximising speed (as such figures of merit are faster to calculate than those based on comparison of the complete powder XRD profile, such as R_{wp}) at the possible expense of reliability.

1.5 STRUCTURE DETERMINATION FROM POWDER XRD DATA

1.5.1 Overview

The three stages of structure determination from diffraction data (single-crystal or powder) are broadly summarised as: (i) unit cell determination and space group assignment, (ii) structure solution and (iii) structure refinement. The aim of **structure solution** is to derive an approximately correct description of the structure, using the unit cell and space group determined in the first stage, but starting from no knowledge of the actual arrangement of atoms or molecules within the unit cell. The methodologies for carrying out structure solution from powder XRD data can be subdivided into two categories, called the **traditional strategy** and the **direct-space strategy**. The main features of these different strategies, and comparisons between them, are discussed in Section 1.5.4. If the structure obtained in the structure solution stage represents a sufficiently good approximation to the true structure, a high-quality description of the structure can then be obtained by **structure refinement**. Nowadays, structure refinement from powder XRD data is virtually always carried out using the **Rietveld refinement** technique.

In the case of structure determination from powder XRD data, there is an additional and crucially important stage, called **profile refinement** (or 'profile fitting'). As depicted in Figure 1.5, which shows a more detailed map of the stages involved in structure determination from powder XRD data, profile fitting is carried out after completing the unit cell determination stage and before commencing the structure solution stage.

In the remainder of this section, we discuss each of the stages involved in the process of structure determination from powder XRD data in turn, describing the entire procedure for the case (so-called '*ab initio* structure determination') in which the structure determination begins 'from scratch', without any prior knowledge of the crystal structure (except, in

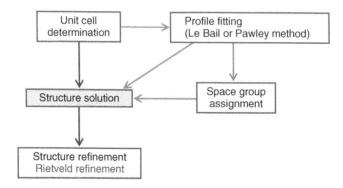

Figure 1.5 Schematic diagram showing the sequence of stages involved in crystal structure determination from powder XRD data.

most cases, knowledge of the chemical composition and, in the case of molecular materials, knowledge of the identity of the molecule).

It is relevant to note that in some cases (particularly in the case of inorganic materials) inspection of the powder XRD data for a new material may allow a particular 'structure type' (*e.g.* perovskite, fluorite, spinel) to be identified. Under such circumstances, following unit cell determination from the powder XRD data, a structural model may be constructed readily on the basis of the identified structure type and then subjected to Rietveld refinement, and hence the structure solution stage of the structure determination process can be bypassed. In the case of molecular materials, this type of scenario, in which a standard structure type can be recognised simply from inspection of experimental powder XRD data, is almost never encountered.

1.5.2   Unit Cell Determination (Indexing)

The first stage of structure determination involves determination of the unit cell $\{a, b, c, \alpha, \beta, \gamma\}$ by analysis of the peak positions in the powder XRD pattern. This task is referred to as 'indexing' the powder diffraction pattern, and involves determination of the Miller indices (h, k, l) corresponding to each observed peak in the experimental powder XRD pattern. Clearly, the structure determination process can progress to the structure solution and structure refinement stages *only* if the correct unit cell is found at the indexing stage, and difficulties encountered in achieving reliable indexing of powder XRD patterns can sometimes be an insurmountable hurdle to successful structure determination.

A wide range of different strategies have been developed for indexing powder XRD data, and are discussed in detail elsewhere.[31] Several widely used indexing programs are available (in particular ITO,[32] TREOR,[33] DICVOL,[34] XCELL[35] and CRYSFIRE[36]), based on a range of different methodologies. In general, these programs require as input data the measured peak positions of about 20 selected peaks, typically at low 2θ values. The existence of significant peak overlap can be particularly problematic in the indexing stage, as certain peaks that may be vital for correct indexing can be obscured or completely unresolved.

We note that a useful check on the validity of the unit cell determined in the indexing stage is to assess whether it corresponds to a plausible density for the material of interest. Assuming that the chemical composition of the material is known, the possible densities may thus be estimated as n times the mass of the formula unit divided by the unit cell volume, where n is any positive integer ($n \geq 1$). If the density of the material of interest is known experimentally (or if it can be estimated from known densities of related materials), this check can provide a useful means of confirming that the unit cell obtained in the indexing stage is at least plausible.

1.5.3 Preparing the Intensity Data for Structure Solution: Profile Fitting

After the unit cell has been determined, the next stage is to prepare the intensity data for space group determination and structure solution using an appropriate profile-fitting technique such as the Pawley method[37] or the Le Bail method.[38] The aim of this stage of the structure determination process is to fit the complete experimental powder XRD profile *via* refinement of variables that describe: (i) the peak positions (the variables that determine the peak positions include the unit cell parameters and the zero-point shift parameter), (ii) the background intensity distribution, (iii) the peak widths, (iv) the peak shapes and (v) the peak intensities.

With regard to (i), the input values of the unit cell parameters used in the profile-fitting procedure are those obtained in the indexing stage, and the refined values that result from the profile-fitting procedure represent a more accurate set of unit cell parameters. It is important to emphasise that no structural model is used in the profile-fitting procedure (except insofar as the unit cell parameters are used to determine the peak positions) and that the intensities in (v) represent a set of intensity variables that are refined to give an optimal fit to the experimental powder XRD pattern without reference to any structural model. Thus, the aim of the profile-fitting procedure is not to determine the crystal structure but rather to obtain reliable values for the variables that describe different features of the powder XRD pattern (*i.e.* (i)–(v) above) in preparation for the subsequent stages of the structure determination process. As now discussed, it is also important to note that different strategies for structure solution make use of different combinations of the variables in (i)–(v) as input information.

The 'traditional strategy' for structure solution requires, as input data, the integrated peak intensities, $I(H)$, extracted from the experimental powder XRD pattern: *i.e.* the fitted values of the integrated peak intensities obtained from the profile-fitting procedure (*i.e.* (v)). In addition, some implementations of the direct-space strategy for structure solution are based on comparison of integrated peak intensities and make use of the set of integrated peak intensities (v) obtained at the profile-fitting stage. After the integrated peak intensities (v) have been extracted, these approaches for structure solution do not make further use of the experimental powder XRD pattern during the structure solution calculations (although the experimental powder XRD pattern is used again in the Rietveld refinement stage).

Alternatively, several other implementations of the direct-space strategy for structure solution involve comparison between experimental and calculated data using a whole-profile figure of merit such as R_{wp}. In these cases, the integrated peak intensities (v) extracted from the experimental powder XRD pattern in the profile-fitting procedure are not used in the structure solution stage. Instead, the variables (i)–(iv) determined in the profile-fitting procedure (together with intensities calculated for trial structural models) are used to construct the calculated powder XRD pattern for each trial structure generated in the direct-space structure solution calculation.

Following the profile-fitting procedure, the space group can be assigned by identifying the conditions for systematic absences in the intensity data (v). Furthermore, various methodologies have been developed to assist the process of space group assignment from powder XRD data based on statistical principles.[39,40]

If the space group cannot be assigned uniquely, structure solution calculations should be carried out separately for each of the plausible space groups. Knowledge of the unit cell volume and space group (together with density considerations) allows the contents of the asymmetric unit to be established. Information from other experimental techniques (particularly high-resolution solid-state nuclear magnetic resonance spectroscopy (NMR)) may be particularly helpful in confirming the number of independent molecules in the asymmetric unit and elucidating other structural aspects that may be useful in assisting the structure solution process (see Section 1.8.2).

1.5.4   Structure Solution

Techniques for structure solution from powder XRD data can be subdivided into two categories: the traditional strategy and the direct-space strategy. We now describe the essential features of each strategy and discuss their relative merits under various circumstances.

1.5.4.1 The Traditional Strategy for Structure Solution

The **traditional strategy** follows a very close analogy to the analysis of single-crystal XRD data. The overall procedure is first to extract the integrated peak intensities, $I(H)$, directly from the experimental powder XRD pattern in the profile-fitting stage and then to use these intensities in the types of structure solution calculation (*e.g.* direct methods, Patterson methods or the recently developed charge-flipping methodology) that are used for single-crystal XRD data (see Chapter 2 for a discussion of these techniques). However, as a consequence of peak overlap in the powder XRD pattern, the extracted intensities, $I(H)$, may be unreliable as a result of errors or uncertainties in the intensity extraction process. Clearly, any unreliability in the extracted intensities can lead to difficulties or failure in subsequent attempts to solve the structure. Such problems may be particularly severe in the case of large unit cells and/or low symmetry, for which the extent of peak overlap is extensive. On the other hand, in cases in which the peaks are sparsely distributed in the powder XRD pattern (corresponding to a low extent of peak overlap), the extracted intensity information is very reliable and will almost certainly lead to successful structure solution through use of the traditional strategy. Thus, for many inorganic materials with small unit cells and/or high symmetry, the intensity extraction process may be very straightforward and reliable, due to the peaks in the powder XRD pattern being well separated, with little or no peak overlap.

In many cases, applications of the traditional strategy for structure solution from powder XRD data have employed the same structure solution software used in the analysis of single-crystal XRD data. Alternatively, software specifically optimised for powder XRD data has also been developed.[41−46]

1.5.4.2 The Direct-Space Strategy for Structure Solution

In contrast to the traditional strategy, which follows a close analogy to the analysis of single-crystal XRD data, the **direct-space strategy** for structure solution[11] (not to be confused with 'direct methods') follows a close analogy to global optimisation procedures. In the direct-space strategy, trial crystal structures are generated in direct space, independently of the experimental powder XRD data, and the suitability of each trial structure is assessed by direct comparison between the powder XRD pattern calculated for the trial structure and the experimental powder

XRD pattern. This comparison is quantified using an appropriate figure of merit. Several implementations of the direct-space strategy employ the weighted powder profile R-factor, R_{wp} (the R-factor normally employed in Rietveld refinement), which considers the entire digitised intensity profile point by point, rather than the integrated intensities of individual diffraction maxima (see Section 1.4.2). Thus, R_{wp} takes peak overlap implicitly into consideration. Furthermore, R_{wp} uses the digitised powder XRD data directly as measured, without further manipulation of the type required when individual peak intensities, $I(H)$, are extracted from the experimental powder XRD pattern. As discussed in Section 1.5.3, some implementations of the direct-space strategy employ figures of merit based on integrated peak intensities, $I(H)$, extracted from the experimental powder XRD pattern in the profile-fitting procedure (*i.e.* the intensity data (v) discussed in Section 1.5.3), rather than considering the complete powder XRD profile using R_{wp}.

The aim of the direct-space strategy is to find the trial crystal structure that corresponds to the lowest R-factor, and is equivalent to exploring a hypersurface $R(\Gamma)$ in order to find the global minimum, where Γ represents the set of variables that define the structure (discussed in more detail below). In principle, any technique for global optimisation may be used to find the lowest point on the $R(\Gamma)$ hypersurface, and much success has been achieved in this field using Monte Carlo/simulated annealing[11,47−72] and genetic algorithm[73−94] techniques. In addition, grid search,[95−99] differential evolution[100,101] and other techniques[102] have also been employed.

We now consider the way in which trial structures are defined within the context of direct-space structure solution calculations, focusing on the case of materials composed of discrete molecules. In principle, the set (Γ) of structural variables could be taken as the coordinates $\{x, y, z\}$ of all atoms in the asymmetric unit (thus, for a structure with N atoms in the asymmetric unit, the total number of variables in the set Γ would be $3N$). However, this approach discards any prior knowledge of the atomic connectivity within the molecule and the geometry of the molecule, and corresponds to the maximal number of structural variables that could be used to define the structure. Instead, it is advantageous to utilise all information on molecular geometry that is already known reliably prior to the structure solution calculation (in general, the identity of the molecule is known before the structure solution calculation is started, and if ambiguities remain concerning the atomic connectivity (*e.g.* tautomeric form), other techniques such as solid-state NMR spectroscopy may be useful to resolve these ambiguities before starting the calculation). Thus, assuming that the identity of the molecule is known beforehand, it is common practice to fix the bond lengths and bond angles at standard values in direct-space structure solution calculations and to fix the geometries of well-defined structural units, such as aromatic rings. In general, the only aspects of intramolecular geometry that require to be determined are the values of some (or all) of the torsion angles that define the molecular conformation. Under these circumstances, each trial structure in a direct-space structure solution calculation is defined by a set (Γ) of structural variables that represent, for each molecule in the asymmetric unit, the position of the molecule in the unit cell (defined by the coordinates $\{x, y, z\}$ of the centre of mass or a selected atom), the orientation of the molecule in the unit cell (defined by rotation angles $\{\theta, \phi, \psi\}$ relative to the unit cell axes) and the unknown torsion angles $\{\tau_1, \tau_2, \ldots, \tau_n\}$. Thus, in general, there are $6 + n$ variables, $\Gamma = \{x, y, z, \theta, \phi, \psi, \tau_1, \tau_2, \ldots, \tau_n\}$, for each molecule in the asymmetric unit, where n is the number of torsion-angle variables.

For the example shown in Figure 1.6, the molecular conformation is defined by three torsion-angle variables ($n = 3$), and thus (with one molecule in the asymmetric unit) the total number of structural variables required in a direct-space structure solution calculation would be nine. In contrast, if knowledge of molecular geometry were discarded and if each individual atom was allowed to move independently in the direct-space structure solution calculation, the total number of structural variables would be 36 (*i.e.* three positional variables $\{x, y, z\}$ for each of the 12 non-hydrogen atoms in the molecule).

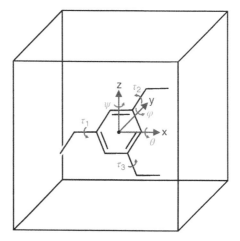

Figure 1.6 Schematic illustration of the structural variables employed in direct-space structure solution: positional variables $\{x, y, z\}$, orientational variables $\{\theta, \varphi, \psi\}$ and torsion-angle variables $\{\tau_1, \tau_2, \tau_3\}$. The bond lengths, bond angles and the geometry of the aromatic ring are fixed at standard values.

The fact that the direct-space strategy takes maximal advantage of reliable information on molecular geometry that is known independently of the powder XRD data prior to the structure solution calculation is an important feature that contributes to the success of this approach. In contrast, the traditional strategy for structure solution does not, in general, utilise prior knowledge of features of molecular geometry (although we note that a few implementations, such as fragment-based Patterson search techniques, do incorporate knowledge of molecular geometry within the framework of the traditional strategy).

The significant advances that have taken place since the early 1990s in the capability to determine the crystal structures of organic molecular solids from powder XRD data have been catalysed by the development of the direct-space strategy for structure solution, particularly because structure solution of molecular materials is very well suited to these techniques. To date, most reported crystal structure determination of molecular solids from powder XRD data has employed the direct-space strategy for structure solution, although there have also been several reports of successful structure solution of such materials using the traditional strategy.

For structure solution by the direct-space strategy, the complexity of the structure solution problem is dictated largely by the dimensionality of the $R(\Gamma)$ hypersurface to be explored (*i.e.* the total number of structural variables in the set Γ) rather than the number of atoms in the asymmetric unit. Thus, the greatest challenges in the application of direct-space techniques arise when the number of structural variables is large. This situation occurs when there is considerable molecular flexibility (*i.e.* when the molecular geometry is defined by a large number of variable torsion angles) and/or when there are several independent molecules in the asymmetric unit.

In summary, for molecular solids or other materials constructed from well-defined modular building units (such as metal–organic framework (MOF) materials), the direct-space strategy is a particularly suitable approach for structure solution, given that a considerable amount of reliable information on the geometries of the molecular building units is already available (*e.g.* from known crystal structures of related molecules or from computational studies of the isolated molecules). For other types of material, for which the peak overlap problem is less severe or for which there is insufficient prior knowledge of the geometry of a suitable structural fragment for use in direct-space calculations, the traditional strategy would generally be the favoured approach for structure solution.

1.5.5 Structure Refinement

Nowadays, the refinement of crystal structures from powder XRD data is virtually always carried out using the Rietveld profile refinement technique.[29,103,104] In Rietveld refinement, the variables defining the powder XRD profile (variables (i)–(iv) in Section 1.5.3) and those defining the structural model (which determine the relative peak intensities in the calculated powder XRD pattern) are adjusted by least squares methods in order to obtain the optimal fit between the experimental and calculated powder XRD patterns. In general, the weighted powder profile R-factor (R_{wp}) is used to assess the fit between experimental and calculated powder XRD patterns. An example of the fit obtained in a typical Rietveld refinement is shown in Figure 1.7. The structural variables in Rietveld refinement are analogous to the variables (*e.g.* atomic coordinates, atomic displacement parameters or site occupancies) that are refined in structure refinement from single-crystal XRD data, and are thus different from the variables used in direct-space structure solution discussed in Section 1.5.4.2. Thus, while bond lengths and bond angles are usually fixed in direct-space structure solution, the bond lengths

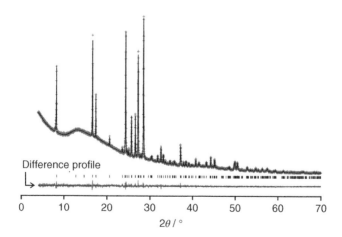

Figure 1.7 Typical result of Rietveld refinement (for Form III of *m*-aminobenzoic acid). The experimental powder XRD pattern is shown as '+' marks and the calculated powder XRD pattern for the structural model is shown as the continuous line overlaid on the experimental data points. The tick marks indicate peak positions. The difference between the experimental and calculated powder XRD profiles is shown as the line at the bottom. Clearly, for a good-quality Rietveld refinement, this 'difference profile' should be as flat as possible, and should ideally reflect only the noise level in the experimental data.

and angles are generally allowed to vary (although often with restraints applied) in the Rietveld refinement stage.

For successful Rietveld refinement, the initial structural model (taken from the structure solution) must be a sufficiently good representation of the correct structure. As Rietveld refinement often suffers from problems of instability, geometric restraints (soft constraints) based on standard molecular geometries generally need to be applied in order to ensure stable convergence of the refinement. Furthermore, it is common to use only isotropic displacement parameters in Rietveld refinement, rather than the anisotropic displacement parameters that are generally refined (except in the case of hydrogen atoms) for single-crystal XRD data. As in structure determination from single-crystal XRD data, the structural model obtained in structure solution is sometimes an incomplete representation of the true structure (particularly when structure solution is carried out using the traditional strategy); in such cases, difference Fourier techniques can be used in conjunction with Rietveld refinement to complete the structural model.

Finally, it is relevant to assess the quality of the structural information that can be obtained by structure determination from powder XRD data *versus* single-crystal XRD data. In general, the final structural parameters obtained from powder XRD data are not as accurate or precise as those that could be determined from analysis of single-crystal XRD data for the same material (assuming that a suitable single crystal is available). Nevertheless, a carefully refined crystal structure from powder XRD data (giving the quality of fit shown in Figure 1.7) provides reliable information on the arrangement of atoms and molecules in the crystal structure and allows an understanding of most aspects of the crystal structure that are of interest to chemists (such as details of the molecular packing arrangement and identification of the intermolecular interactions).

1.6 SOME EXPERIMENTAL CONSIDERATIONS IN POWDER XRD

1.6.1 Synchrotron *versus* Laboratory Powder XRD Data

We now consider the relative merits of using powder XRD data recorded with incident X-rays from a synchrotron radiation source *versus* powder XRD data recorded on a conventional laboratory powder X-ray diffractometer. As elaborated elsewhere,[105,106] we emphasise that X-rays produced from a synchrotron radiation source have the following enhanced properties compared to those produced from a normal laboratory X-ray source: (i) very high intensity, (ii) a high degree of collimation (*i.e.* low beam divergence), (iii) tunability (*i.e.* all X-ray wavelengths are available, and any specific wavelength may be selected by use of an appropriate monochromator), (iv) a high degree of polarisation and (v) the radiation is pulsed. We focus here on the advantages for powder XRD that arise from features (i)–(iii).

As a consequence of the high degree of beam collimation, the intrinsic widths of the peaks (*i.e.* the instrumental contribution to peak broadening; see Section 1.6.4) in a powder XRD pattern recorded using synchrotron radiation are substantially lower than those recorded on a laboratory powder X-ray diffractometer. Thus, in general, the resolution of synchrotron powder XRD data is significantly higher than that of laboratory powder XRD data. As a consequence of the higher resolution, problems arising from peak overlap can be alleviated, at least to some extent, thus increasing the reliability in determining accurate peak positions (which is particularly advantageous in unit cell determination) and increasing the reliability in extracting the intensities of individual peaks from the powder XRD pattern. In this regard, when structure solution is to be carried out using the traditional strategy (or a direct-space technique that uses a figure of merit based on extracted peak intensities), the use of powder XRD data recorded at a synchrotron source can be particularly advantageous. Thus, the success of the traditional strategy for structure solution is generally enhanced by the use of data recorded on an instrument with the highest possible resolution. However, for direct-space structure solution techniques that employ a figure of merit based on a profile R-factor (such as R_{wp}), the most important requirement is not high resolution *per se* but rather that the peak profiles are well defined and described accurately by the peak-shape and peak-width functions used in the structure solution calculation. In such cases, the use of laboratory powder XRD data can be just as effective as the use of synchrotron data, and many examples (including many of those presented in Section 1.10) demonstrate that the use of a good-quality, well-optimised laboratory powder X-ray diffractometer is usually perfectly adequate for research in this field. Within the context of Rietveld refinement, the use of synchrotron data often leads to structural results of greater accuracy, due to the fact that the data in the high 2θ region of the powder XRD pattern are usually of higher quality.

The very high intensity of synchrotron radiation has many advantages in enhancing the scope of powder XRD experiments (provided it does not promote degradation of beam-sensitive materials). Thus, a powder XRD pattern (with a given signal-to-noise level) can be recorded substantially more quickly using a high-intensity synchrotron radiation source than a laboratory X-ray source, which is particularly advantageous for time-resolved studies of structural changes (see Section 1.6.5), as the time resolution in such studies depends on the time required to record a single powder XRD pattern. Rapid data collection is also advantageous in allowing data to be recorded for materials that have short lifetimes, such as unstable materials or intermediate phases that have only a transient existence during structural transformations. The use of a high-intensity synchrotron X-ray source may also be essential for identifying weak diffraction features, such as satellite reflections from incommensurate structures or weak additional reflections that arise for materials exhibiting subtle superstructures.

The tunability of synchrotron radiation is particularly advantageous in 'resonant XRD', in which an XRD experiment is carried out using an incident wavelength that is close to the X-ray absorption edge of an element in the sample. The underlying theory and applications may be found elsewhere.[107−109] In the present context, we note that this technique is particularly valuable for determining the location(s) of the resonant element within the structure and is beneficial when other elements of similar scattering power are present (and therefore difficult to distinguish under non-resonant conditions). The technique is also a powerful way of establishing the distribution of oxidation states of a given element within a crystal structure, recognising that the wavelengths corresponding to the X-ray absorption edges of different oxidation states of the element are different. Examples of the application of resonant powder XRD include the determination of the occupancies of Fe and Ni atoms among the four crystallographically distinct octahedral sites in $FeNi_2(BO_3)O_2$.[110] A seminal demonstration of the application of this technique to distinguish different oxidation states of an element concerns the study of the mixed-valence material α-Fe_2PO_5.[111] In this case, the Fe^{2+} and Fe^{3+} ions were shown to be located on different crystallographic sites by resonant powder XRD studies at the Fe^{2+} and Fe^{3+} absorption edges (the energies of the K-edges of Fe^{2+} and Fe^{3+} differ by ca 3 eV).

1.6.2   Preferred Orientation

In general, structure solution from powder XRD data has a good chance of success only if the experimental powder XRD pattern contains reliable information on the intrinsic relative intensities of the diffraction maxima, which requires that there is no 'preferred orientation' in the powder sample. Preferred orientation arises when the crystallites in the powder sample have a non-random distribution of orientations, and this effect can be particularly severe when the morphology of the crystallites is strongly anisotropic (*e.g.* long needles or flat plates). When a powder sample exhibits preferred orientation, the measured relative peak intensities differ from the intrinsic relative diffraction intensities, limiting the prospects for determining reliable structural information from the powder XRD pattern. In order to circumvent this difficulty, it is recommended that appropriate procedures[112] are carried out to screen powder samples for preferred orientation and that steps are taken to ensure that the sample is free of preferred orientation *before* high-quality powder XRD data are recorded for use in structure determination. For samples that are found to exhibits preferred orientation, a variety of experimental approaches may be used to alleviate its effects, such as using a capillary or end-loading sample holder, mixing the sample with an amorphous material, preparing the sample by spray-drying, or using an appropriate grinding procedure to induce a crystal morphology that is as isotropic as possible. Even if the effects of preferred orientation cannot be eliminated completely from the experimental powder XRD pattern, corrections can be made retrospectively once a sufficiently good structural model is known, *via* refinement of parameters describing the preferred orientation during the Rietveld refinement stage. In general, attempts to carry out structure solution using a powder XRD pattern that is affected *significantly* by preferred orientation are likely to be unsuccessful. It is therefore strongly recommended that time should be devoted to checking the sample for preferred orientation and carrying out careful sample preparation in order to ensure, as far as possible, that a high-quality powder XRD pattern has been recorded for a powder sample with a random distribution of crystallite orientations *before* embarking upon the time-consuming process of carrying out structure solution calculations.

1.6.3   Phase Purity of the Powder Sample

Another issue that can have a profound bearing on the success of structure determination from powder XRD data is the phase purity of the powder sample. So far, the discussion in this chapter has assumed that the powder sample comprises only one crystalline phase. If the powder sample contains a second crystalline phase (*e.g.* a crystalline impurity or a second polymorph of the phase of interest) and is *not known* to contain this second phase, then the structure determination process will almost certainly fail at the stage of unit cell determination (as it will be impossible to find a single unit cell that predicts the positions of all the peaks arising from the two phases in the powder XRD pattern). However, if the existence and identity of an impurity phase or second phase is known beforehand, the peaks arising from this additional phase may be recognised and handled in an appropriate manner that allows the structure of the main phase of interest to be determined successfully. Clearly, the use of other experimental techniques (such as solid-state NMR spectroscopy) may be advantageous for independent confirmation of the phase purity of a powder sample *before* starting the process of structure determination from powder XRD data. In favourable cases, however, careful inspection of the powder XRD data alone may allow the presence of more than one phase to be identified without the need to use information from other techniques, as illustrated by the structure determination of cyclopentadienyl rubidium[113] discussed in Section 1.10.16.

1.6.4   Analysis of Peak Widths in Powder XRD Data

The peak widths in an experimental powder XRD pattern depend on both instrumental factors and features of the powder sample. In many cases, peak widths are dominated by sample-dependent factors, particularly when a high-resolution diffractometer (*e.g.* at a synchrotron radiation source) is used to record the data, and in this situation the peak broadening is described as 'sample-limited'. At the other extreme, peak widths may be dominated by the intrinsic resolution of the powder X-ray diffractometer (*e.g.* for a low-resolution diffractometer and a highly crystalline sample), and in this situation the peak broadening is described as 'instrument-limited'. However, in general, the peak width is a convolution of broadening effects resulting from both the instrument and the sample.

Instrumental factors that influence the peak resolution include the wavelength distribution of the incident radiation (clearly, monochromatic radiation is conducive for recording high-resolution data), the measurement geometry, the degree of collimation of the incident beam and the type of detector. More details of these instrumental factors can be found elsewhere.[4,6] As discussed in Section 1.6.1, the use of

X-rays from a synchrotron radiation source is generally advantageous in reducing the instrumental contribution to peak broadening, primarily as a result of the very high degree of beam collimation. Thus, the highest-resolution powder XRD data are generally recorded at synchrotron radiation sources.

The sample-dependent factors that contribute to peak broadening include crystallite size effects, residual stress and strain and the existence of disorder or stacking faults. With regard to the dependence on crystallite size, it is important to emphasise that the relevant property is the size of the ordered crystalline domains within the crystallites in the powder sample (which is clearly less than or equal to the size of the crystallite particles themselves). For crystalline domain sizes greater than *ca* 0.1–1.0 μm, the peak width does not vary significantly with crystalline domain size. However, for crystalline domain sizes less than this value, the peak width increases significantly as domain size decreases, as represented by the Scherrer equation:

$$W(\theta) = \frac{K\lambda}{\langle D \rangle \cos\theta} \tag{1.6}$$

where W is the peak broadening (in radians), $\langle D \rangle$ is the mean dimension of the crystalline domains, K is the shape factor (usually $K \approx 0.9$), λ is the wavelength and θ is half the diffraction angle (2θ). This equation may be applied to estimate the average crystalline domain size from measurement of peak widths in a powder XRD pattern that is dominated by sample-dependent broadening, although such analysis clearly requires that the instrumental broadening and other sample-dependent contributions to peak broadening are already understood.

The existence of microstrains within the crystallites in a powder sample also gives rise to peak broadening in a powder XRD pattern, resulting from a distribution of tensile and compressive forces, which in turn lead to a distribution of unit cell sizes. Thus, each diffraction peak specified by Miller indices (h, k, l) and characterised by Bragg's law:

$$\theta_{hkl} = \sin^{-1}(\lambda/2d_{hkl}) \tag{1.7}$$

actually arises from a distribution of d_{hkl} values (rather than a single d_{hkl} value in the unstrained case), leading to a distribution of θ_{hkl} values (*i.e.* peak broadening). In this case, the contribution to peak broadening (in radians) is given by:

$$W(\theta) = 4\varepsilon \tan\theta \tag{1.8}$$

where ε is the average residual strain and θ is half the diffraction angle (2θ).

From Equations 1.6 and 1.8, it is clear that peak broadening due to the effects of crystalline domain size and peak broadening due to strain have a different dependence on θ. Thus, studies of the functional dependence of peak broadening on diffraction angle allow the relative contributions of crystalline domain size and strain to be assessed.

We note that the detailed analysis of peak widths and peak shapes in powder XRD data, in order to establish comprehensive insights into stress and strain in materials and to fully characterise their microstructural features, constitutes a major area of activity in materials science.[114]

1.6.5   Applications of Powder XRD for *In Situ* Studies of Structural Transformations and Chemical Processes

From a practical point of view, powder XRD experiments may be carried out fairly straightforwardly to monitor structural changes in a solid as a function of variation of external conditions. In general, such *in situ* studies are more readily carried out by powder XRD techniques than by single-crystal XRD techniques. An important application of powder XRD across several fields of science is the study of structural changes associated with: (i) phase transitions in response to external stimuli such as pressure or temperature, (ii) chemical reactions (*e.g.* thermal or photochemical), (iii) the interaction of solids with external gaseous atmospheres (*e.g.* leading to incorporation of water within the solid due to variation in relative humidity) or (iv) the application of mechanical force (*i.e.* mechanochemistry). In addition, powder XRD may also be exploited for *in situ* studies of the formation of crystalline solids during materials synthesis or crystallisation processes.

As mentioned in Section 1.6.1, a major experimental consideration for this type of study is maximisation of the speed of recording the powder XRD data, as the time resolution of the *in situ* observation depends on the time taken to record each individual powder XRD pattern. Thus, it is clearly advantageous to use both an intense X-ray source (for this reason, *in situ* powder XRD studies are often carried out at synchrotron radiation facilities) and a sensitive detector assembly. Furthermore, recording the powder XRD data in energy-dispersive mode[115] may also be conducive to rapid data collection. In many applications of this type, a crucial challenge is to design an appropriate experimental set-up that allows the powder XRD data to be measured while the required external conditions (*e.g.* high-pressure or mechanical force) are exerted on the powder sample, and at the same time ensuring that the quality of the powder XRD data recorded is not significantly compromised. The design and manufacture of appropriate sample cells for *in situ* studies may present significant technical challenges, and ingenuity in designing appropriate apparatus is a key factor in the successful implementation of these techniques. Within the scope of the present chapter, we do not discuss details of the specific instrumentation designed for applications of this type. Information on the experimental methodology used to carry out powder XRD studies under conditions of high pressure,[116–118] materials synthesis,[119] crystallisation processes[120,121] and solid-state mechanochemistry[122] may be found elsewhere. However, examples of the application of some of these techniques are given in Section 1.10.

A notable advantage of applying powder XRD for *in situ* monitoring of structural changes of the type described above is that phase transitions and chemical reactions in single crystals are often associated with a deterioration in crystal quality, for example through twinning, fracturing or the production of a polycrystalline 'daughter' phase from a single crystal of the 'parent' phase. Under such circumstances, monitoring the transformation of interest using single-crystal XRD is particularly challenging, to the extent that successful structure determination may be impossible, whereas powder XRD data recorded for the same systems can exhibit little or no deterioration in quality throughout the transformation. Thus, the ability to carry out structure determination from powder XRD data is unaffected by the occurrence of twinning in the individual crystallites within the powder, provided each individual twin domain is sufficiently large that the effects of crystalline domain size do not lead to significant peak broadening.

Furthermore, many solid materials (*e.g.* certain clathrate hydrates) do not exist under conditions of ambient temperature and pressure. Such materials require synthesis, handling and characterisation under non-ambient conditions, and in these cases powder XRD generally provides a more straightforward approach for structural characterisation.

In the industrial context, powder XRD allows materials to be investigated directly under the conditions in which they are used in specific applications, which is not necessarily true for single-crystal XRD (by virtue of the requirement to prepare and study a single crystal). For example, solid pharmaceuticals are often administered into the body as compacted powders in the form of tablets. For such materials, it may be important to understand directly (*e.g.* from powder XRD studies of the 'real' tablets) whether they undergo any structural transformation as a consequence of the compaction process. If such transformations occur, the crystal structure of the actual material administered *via* the compacted tablet may be different from the structure of a large single crystal of the same material (as required for single-crystal XRD) obtained *via* a crystal growth process.

Finally, we emphasise that the combined/simultaneous measurement of powder XRD data together with other types of experimental data (*e.g.* spectroscopic data or thermal analysis) on the same sample at the same time is potentially an even more powerful approach for *in situ* characterisation of chemical processes and structural transformations than employing *in situ* powder XRD alone. Examples include simultaneous measurements using the following combined techniques: (i) powder XRD and X-ray absorption fine structure (EXAFS) spectroscopy[123−127] (with the powder XRD data yielding insights into changes in long-range order and the EXAFS data providing information on simultaneous changes in short-range order), (ii) powder XRD and Raman spectroscopy,[128−131] (iii) powder XRD and optical spectroscopy[132] and (iv) powder XRD and differential scanning calorimetry (DSC)[133−137] (this combination of techniques is particularly useful for the study of phase transitions as a function of temperature, as the DSC data pinpoint the transition temperature(s) and the powder XRD data characterise the corresponding structural changes).

1.7   POWDER NEUTRON DIFFRACTION *VERSUS* POWDER XRD

Although the majority of powder diffraction experiments are carried out using incident X-ray radiation, it is important also to mention the complementary opportunities provided by **powder neutron diffraction**. Given that neutron sources (either a nuclear fission reactor or a pulsed spallation source) are available only at centralised national or international facilities, routine day-to-day applications of powder diffraction are necessarily carried out using laboratory-based powder X-ray diffractometers. Nevertheless, powder neutron diffraction is complementary to powder XRD in many aspects and we focus here on highlighting these aspects, which in general may be regarded as ways of enhancing the level of structural understanding of materials beyond the level that can be achieved from powder XRD.

Powder neutron diffraction utilises an incident neutron beam that has a wavelength comparable to those of X-rays. The basis of the diffraction phenomenon arising from the interaction of such neutron beams with crystalline materials is very similar to that for XRD. However, while X-rays are scattered by their interaction with the electrons in a material, it is crucial to emphasise that neutrons are instead scattered by their interaction with the atomic nuclei (so-called 'nuclear scattering'). Furthermore, for magnetic materials, the neutron is also scattered by its interaction with unpaired electron spins (recalling that the neutron has a magnetic moment), giving rise to 'magnetic scattering' (in addition to the nuclear scattering of neutrons from the atomic nuclei). The use of neutron diffraction to establish details of magnetic ordering in materials represents a major area of application of powder neutron diffraction. This area of application has been covered extensively elsewhere[138−141] and is not discussed further in this chapter.

Focusing on the application of powder neutron diffraction to probe the structural properties of non-magnetic materials, the fact that neutrons are scattered by nuclei and X-rays are scattered by electrons means that crystal structure determination from powder neutron diffraction data and crystal structure determination from powder XRD data yield complementary descriptions of the 'crystal structure'. Thus, structure determination from powder neutron diffraction pinpoints the locations of the atomic nuclei within a unit cell, whereas structure determination from powder XRD data leads to knowledge of the electron density distribution, which, by invoking certain assumptions, is interpreted in terms of the positions of the atoms in the unit cell. As a consequence, the accuracy of geometric information (*e.g.* interatomic distances, bond lengths, bond angles) that can be established from powder neutron diffraction data is generally higher than that achievable from powder XRD data. Thus, if specific geometric details must be established with supreme accuracy, powder neutron diffraction would be the technique of choice.

Another important difference is that interference effects cause the X-ray scattering power of an atom to diminish with increasing scattering angle, whereas the scattering of neutrons by an atomic nucleus is essentially independent of scattering angle. As a consequence, a powder neutron diffraction pattern contains significantly more intensity information at high 2θ values than the powder XRD pattern of the same material (as evident from Figure 1.7, powder XRD patterns of organic materials typically contain very little intensity beyond about $2\theta \approx 50°$ for Cu $K_{\alpha 1}$ radiation).

Another important difference between neutron diffraction and XRD is that the relative scattering powers of atoms are significantly different. For XRD, scattering power depends on the number of electrons in the atom and therefore increases monotonically as atomic number increases. As a consequence, light elements (especially hydrogen) are very weak X-ray scatterers and may therefore be difficult to locate accurately in crystal structure determination from XRD data. Furthermore, elements with similar atomic number (*e.g.* Al and Si) have very similar X-ray scattering properties, and may therefore be difficult to distinguish in crystal structure determination from XRD data. In contrast, as neutrons are scattered by the atomic nucleus, the neutron scattering power and properties depend on specific characteristics of the nucleus and also differ between isotopes of the same element. As a consequence, neutron scattering power does not depend on atomic number and varies in an irregular manner on moving across the periodic table. Thus, some light elements are very strong neutron scatterers and certain elements with very similar atomic number can have distinctly different neutron scattering properties. In general, powder neutron diffraction is advantageous over powder XRD for locating accurately the positions of light elements (especially hydrogen) in crystal structures and for distinguishing the positions of elements of similar atomic number (*e.g.* determining the positions of Al and Si atoms in ordered aluminosilicate materials).

There are many situations in structural chemistry in which it is crucial to establish accurate information on the positions of hydrogen atoms in materials, such as determining the structures of metal hydrides and establishing accurate geometric details of hydrogen-bonding arrangements. Following from the above discussion, it is clear that powder neutron diffraction is advantageous over powder XRD in such cases. However, it is important to note that in order to obtain high-quality powder neutron diffraction data for materials containing hydrogen (*e.g.* almost all organic materials) it is highly desirable to record the data for the **deuterated** material, in which the hydrogen is present as the ^{2}H isotope rather than the ^{1}H isotope (in a sample with natural isotopic abundance, 99.985% of the hydrogen nuclei are ^{1}H and 0.015% are ^{2}H). The reason that ^{1}H nuclei in a sample are so undesirable for neutron diffraction is that the scattering of neutrons by ^{1}H nuclei is predominantly incoherent scattering, whereas the scattering of neutrons by ^{2}H nuclei is predominantly coherent scattering. As one of the requirements for observing the phenomenon of diffraction is that the scattering is coherent and elastic (see Section 1.2), it is clear that deuteration of materials is highly desirable in order to maximise the contribution of coherent neutron scattering and hence to record high-quality powder neutron diffraction data. In contrast, neutron scattering from materials containing the ^{1}H isotope of hydrogen is often dominated by incoherent scattering, resulting in very poor-quality neutron diffraction data. However, while sample deuteration is essential for recording high-quality powder neutron diffraction data, the synthetic challenges associated with deuteration of certain materials may be a limiting factor.

Given the complementary characteristics of powder neutron diffraction and powder XRD, the simultaneous use of *both* types of data in 'joint' Rietveld refinements is generally advantageous over the use of powder XRD data alone or powder neutron diffraction data alone. It is therefore common for Rietveld refinement software to allow such joint refinement to be carried out. However, as the opportunity to record powder neutron diffraction data is much less routine (given the requirement to use centralised neutron facilities and, in some cases, to prepare materials containing specific isotopes) than the opportunity to record powder XRD data on a laboratory based diffractometer, a typical strategy is to first tackle structure determination using powder XRD data and then to plan powder neutron diffraction experiments if certain structural issues cannot be adequately understood from analysis of the powder XRD data or if specific aspects of structural knowledge require to be enhanced or established at a higher level of accuracy.

1.8 VALIDATION OF PROCEDURES AND RESULTS IN STRUCTURE DETERMINATION FROM POWDER XRD DATA

1.8.1 Overview

Although software for carrying out each stage of the procedure for structure determination from powder XRD data is now readily accessible and relatively straightforward to use, it is essential that the results from such calculations are subjected to adequate scrutiny to confirm that the structure obtained is definitely correct. Two aspects of validation are relevant: (i) validating the structural information used in structure solution (particularly when the direct-space strategy is employed) and (ii) validating the final structure obtained from Rietveld refinement.

Within such validation processes, it is important to consider information from other experimental and computational techniques, including solid-state NMR spectroscopy, energy calculations (either on individual molecules or on periodic crystal structures), vibrational spectroscopies and techniques of thermal analysis (*e.g.* DSC and thermogravimetric analysis (TGA)). Solid-state NMR spectroscopy can play a particularly important role, as it provides insights into specific structural features independently of the powder XRD data, including: the tautomeric form of the molecule, the number of independent molecules in the asymmetric unit, the question of whether molecules occupy general positions or special positions, direct evidence for the existence of specific interactions (*e.g.* hydrogen bonds), quantitative information on specific interatomic distances and *a priori* insights into the existence of disorder within a crystal structure. Information on several of these structural aspects can be important in setting up the correct structural model for a direct-space structure solution calculation or in validating the final structure obtained from Rietveld refinement.

1.8.2   Validation before Direct-Space Structure Solution

Here we focus on the aspects of validation that may be carried out prior to direct-space structure solution, including: (i) establishing the correct representation of molecular geometry to be used in the calculation and (ii) establishing independent evidence for the correct number of molecules in the asymmetric unit.

In general, the identity of the molecule(s) in the structure and the composition (*e.g.* in the case of a solvate or co-crystal phase) may be established readily by applying a range of analytical techniques, including high-resolution solid-state NMR spectroscopy. Another important issue concerns details of molecular geometry, recognising that many molecules can adopt different tautomeric forms. The structure determination of red fluorescein[49] from powder XRD data provided an early example in which high-resolution solid-state ^{13}C NMR spectroscopy was exploited to distinguish the correct tautomeric form of the molecule prior to direct-space structure solution.

Second, we consider the number of molecules in the asymmetric unit. Following unit cell determination, the number of molecules in the unit cell is generally deduced straightforwardly from density considerations, but such information does not necessarily lead to a unique assignment of the number of molecules in the asymmetric unit, nor to a unique assignment of the space group. In such cases, high-resolution solid-state NMR spectroscopy can often provide independent information on the number of molecules in the asymmetric unit. For example, in the case of an organic material, the high-resolution solid-state ^{13}C NMR spectrum should, in principle, contain one peak for each crystallographically distinguishable carbon atom in the structure (although, in practice, the actual number of *observed* peaks may be less than this number due to accidental peak overlap). Thus, after assigning each peak in the solid-state ^{13}C NMR spectrum to a specific carbon environment within the molecule, it is generally straightforward to assess whether there are one, two or more molecules in the asymmetric unit, or whether there is only a fraction of the molecule (indicating that the molecule is located on a special position).

As an example,[142] before carrying out structure determination of the 1:1 co-crystal of benzoic acid (BA) and pentafluorobenzoic acid (PFBA), the high-resolution solid-state ^{13}C NMR spectrum (Figure 1.8) was found to contain two peaks for the carboxylic acid group of BA and two peaks for the carboxylic acid group of PFBA, leading to the conclusion that there are two molecules of BA and two molecules of PFBA in the asymmetric unit. For the unit cell obtained at the indexing stage, it was deduced that a reasonable density for the material would

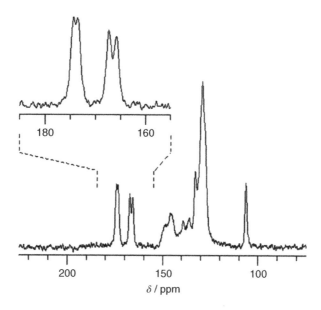

Figure 1.8 High-resolution solid-state ^{13}C NMR spectrum of the 1:1 co-crystal containing benzoic acid (BA) and pentafluorobenzoic acid (PFBA). The two peaks at *ca* 174 ppm represent the carboxylic acid group of PFBA and the two peaks at *ca* 167 ppm represent the carboxylic acid group of BA, from which it is deduced that there are two molecules of BA and two molecules of PFBA in the asymmetric unit. Reprinted with permission from [142] Copyright (2004) John Wiley and Sons Ltd.

correspond to having eight molecules of BA and eight molecules of PFBA in the unit cell, and systematic absences in the powder XRD pattern indicated that the structure has a C-centre and a c-glide plane. This combined knowledge, together with the solid-state ^{13}C NMR data, led to the conclusion that Cc is the correct space group, rather than $C2/c$.

An example in which high-resolution solid-state ^{13}C NMR spectroscopy indicated that a molecule resides at a special position was encountered in the structure determination of an early-generation dendrimeric material,[143] with the solid-state ^{13}C NMR data supporting the assignment that the molecule is located on a crystallographic twofold rotation axis. Structure determination of a new polymorph (β phase) of the latent pigment DPP-Boc[144] from powder XRD data also relied on evidence from high-resolution solid-state ^{13}C NMR spectroscopy to confirm that the asymmetric unit comprises half the DPP-Boc molecule.

1.8.3   Aspects of Validation following Structure Refinement

Aspects of validation following Rietveld refinement are focused on: (i) confirming objectively that the quality of agreement between experimental and calculated powder XRD patterns is sufficient to give confidence that the refined structure is correct, (ii) assessing whether the refined structure is chemically and structurally sensible, (iii) assessing whether there is evidence for disorder in the structure and (iv) assessing whether powder XRD data alone provide an adequate description of the structure or whether complementary techniques (*e.g.* powder neutron diffraction) are required to resolve specific questions (an example of (iv) is given in Section 1.10.8).

In order to assess the quality of the final structure obtained in structure determination, it is important to scrutinise carefully the difference profile obtained in the Rietveld refinement calculation (which represents the difference between the experimental and calculated powder XRD patterns, as shown in Figure 1.7). Clearly, the difference profile should not contain any significant discrepancies, and any minor discrepancies that do exist (*i.e.* discrepancies that are higher than the noise level in the experimental data) must be properly understood before the Rietveld fit can be regarded as acceptable. In this regard, it is important to compare the difference profile obtained in the Rietveld refinement with that obtained *for the same experimental powder XRD pattern* in the profile-fitting stage of the structure determination process. The profile-fitting procedure establishes an upper limit to the quality of fit that can be obtained in a Rietveld refinement calculation for the same experimental powder XRD pattern (and for the same 2θ range). Thus, the Rietveld refinement should aspire to achieve a quality of fit (assessed from the difference profile) that is as close as possible to that obtained in the profile-fitting procedure. If the fit obtained in the Rietveld refinement is significantly worse than that obtained in the profile-fitting procedure, it is probably an indication that the refined structure is incorrect, or at least that some aspect of the true structure is not adequately described in the structural model.

An example that illustrates the importance of applying rigorous scrutiny before accepting the results of Rietveld refinement concerns the structure determination of 3,5-bis(3,4,5-trimethoxybenzyloxy) benzyl alcohol (BTBA) from powder XRD data.[145] Following structure solution by the direct-space genetic algorithm technique, Rietveld refinement gave the fit shown in Figure 1.9a. For this Rietveld refinement ($R_{wp} = 8.25\%$), the difference profile might at first glance appear relatively flat, and it could be quite readily misinterpreted as representing a correct, fully refined crystal structure. However, the discrepancies between experimental and calculated powder XRD patterns in the difference profile obtained in the Rietveld refinement are greater than those in the difference profile obtained in the profile-fitting stage (Figure 1.9b; $R_{wp} = 2.99\%$), suggesting that the structure obtained in the Rietveld refinement, while probably substantially correct, is still not acceptable. Furthermore, it was found that the structure did not contain any $O-H\cdots O$ hydrogen bonding, which was somewhat surprising (although not impossible) for a molecule that has a single hydrogen bond donor (OH group) and several potential hydrogen bond acceptors (oxygen atoms of the $COCH_3$, $COCH_2$ and OH groups). Given these concerns, more detailed scrutiny involving difference Fourier analysis suggested that there was 'missing' electron density in the structure. Following further experimental analysis (primarily by liquid-state ^1H NMR spectroscopy and TGA), it was discovered that the material is actually a monohydrate of BTBA. Further Rietveld refinement was then carried out after adding the water molecule to the structural model, leading to a significant improvement in the quality of fit (Figure 1.9c; $R_{wp} = 4.33\%$), which was considered to be acceptably close to that obtained in the profile-fitting stage (Figure 1.9b). Furthermore, the position of the water molecule in the crystal structure (Figure 1.9d) gives rise to a structurally sensible hydrogen-bonding array (Figure 1.9e) involving the water molecules and the OH groups of the BTBA molecules.

Another aspect of validation concerns the detection of disorder in the crystal structure, recalling that solid-state NMR spectroscopy can provide a means of directly detecting disorder (as well as distinguishing

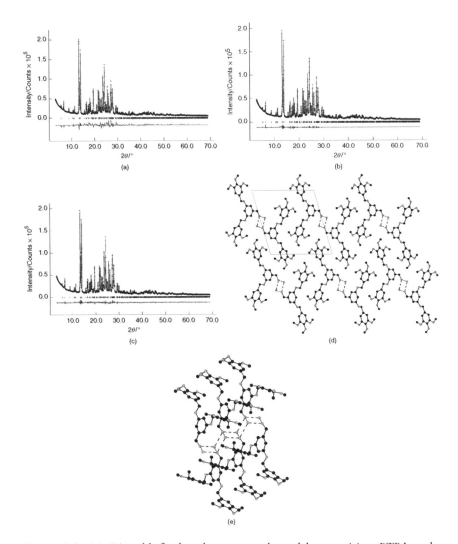

Figure 1.9 (a) Rietveld fit for the structural model comprising BTBA only. (b) Fit obtained in the profile-fitting procedure using the Le Bail technique. (c) Rietveld fit for the structural model comprising BTBA and water. (d) Final refined crystal structure of BTBA monohydrate (dotted lines indicate O–H···O interactions in the extended hydrogen-bonded array, which runs into the page). (e) Crystal structure of BTBA monohydrate viewed in a direction that shows the extended hydrogen-bonded array. Hydrogen atoms are omitted for clarity. Reprinted with permission from [145] Copyright (2005) American Chemical Society.

between dynamic and static disorder) in crystal structures. Cases of severe disorder generally require an appropriate description of the disorder to be incorporated in the structural model in both the structure solution and structure refinement stages. However, if the disorder involves only a localised part of the structure, it may be possible to obtain a reasonable Rietveld fit using an ordered structural model, which can then be improved by incorporating disorder within the structural model. The latter situation was encountered in structure determination of the β polymorph of p-formyl-*trans*-cinnamic acid from powder XRD data[146] (Figure 1.10). The structure was solved and refined as an ordered structure, leading to a good quality of fit in the Rietveld

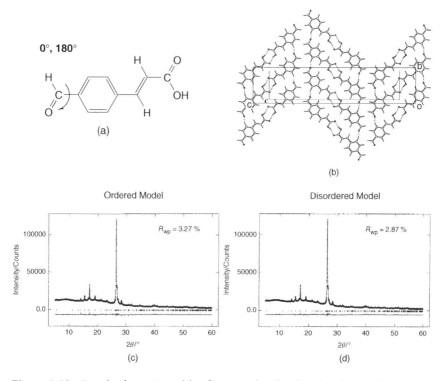

Figure 1.10 Results from Rietveld refinement for the disordered crystal structure of the β polymorph of p-formyl-*trans*-cinnamic acid. The disorder concerns two orientations of the formyl group, as shown in (a). The crystal structure in (b) shows only the disorder component of higher occupancy. The results from Rietveld refinement shown at the bottom are for: (c) an ordered model comprising only the major orientation of the formyl group and (d) the final disordered model. Apart from the description of the order/disorder of the formyl group, all other aspects of these refinement calculations are the same. A slight improvement in the quality of the Rietveld fit for the disordered model is evident. Reprinted with permission from [146] Copyright (2003) John Wiley and Sons Ltd.

refinement ($R_{wp} = 3.27\%$). However, the high-resolution, solid-state ^{13}C NMR spectrum showed evidence of disorder of the formyl group (with the rest of the structure ordered). Subsequent Rietveld refinement considered a model in which the formyl group was disordered between two orientations (with the plane of the formyl group in the same plane as the aromatic ring in each case) and led to an improved fit ($R_{wp} = 2.87\%$) to the powder XRD data.

Energy calculations also have a very important role to play in the validation process, for example to confirm that the crystal structure determined from the powder XRD data is robust and stable under energy minimisation (thus confirming that the structure represents a true minimum on the energy hypersurface for the system of interest). In this regard, periodic density functional theory (DFT) calculations (which take into consideration the space-group symmetry of the crystal structure) provide a convenient and rigorous approach, and are being used increasingly as a validation tool in this field.

1.9   MORE DETAILED CONSIDERATION OF THE APPLICATION OF POWDER XRD AS A 'FINGERPRINT' OF CRYSTALLINE PHASES

In Section 1.3, we introduced the important role of powder XRD as a means of definitively identifying ('fingerprinting') crystalline materials. This qualitative application of powder XRD is used extensively, and is based on comparing the experimental powder XRD pattern of a sample prepared by materials synthesis with the powder XRD patterns of known materials. Usually, the comparison is carried out 'by eye', based on visual assessment of the similarities and/or differences between the powder XRD patterns, rather than subjecting the powder XRD data to more rigorous quantitative analysis. Unfortunately, comparisons based on visual inspection can leave much scope for misinterpretation, with the potential to lead to erroneous structural assignments. In this section, we discuss factors that can cause difficulties in assessing the similarities and differences between powder XRD patterns and suggest more rigorous protocols for carrying out such comparisons.

Examples of the types of ambiguity that can arise in comparing powder XRD patterns by visual inspection include the following scenarios: (i) the synthesised material is the same as a known material, but the powder XRD patterns appear significantly different, such that the match to the known material is not easily identified by visual comparison; (ii) the main features of the powder XRD patterns resemble each other significantly, but with some small differences.

In case (ii), visual comparison tends to focus on the main features of resemblance between the patterns, and the small differences are often overlooked. However, small differences between powder XRD patterns can represent real and significant structural differences. Thus, if the differences are ignored, there is the risk that two materials that have genuinely different structures will be wrongly assigned as the same. In such cases, the two structures may indeed share some structural features in common, but it is nevertheless crucial to establish the reasons for the small differences between their powder XRD patterns. Relevant issues in this regard include superstructures, subtle differences of symmetry, differences in the occupancy of a component within the structure (*e.g.* a solvent molecule) and differing degrees of disorder. An illustrative example[147] is shown in Figure 1.11.

With regard to visual comparison, it is important to emphasise that the appearance of an experimental powder XRD pattern can be influenced by several factors that are not related to the actual crystal structure of the material, but which originate instead from features of the instrumentation and/or the mode of data collection, as well as from features relating to the microstructural characteristics of the powder sample itself (*e.g.* the size, shape and orientational distribution of the crystallites in the powder). Thus, as in case (i) above, powder XRD patterns recorded for two samples with the same crystal structure may actually look quite different as a result of differences in some of these other factors, which include the following:

(i) *Peak widths:* As discussed in Section 1.6.4, the peak shapes and peak widths in a powder XRD pattern depend, *inter alia*, on the crystallinity of the sample, and significant peak broadening can arise when the crystalline domain size is small. Furthermore, peak shapes and peak widths depend on features of the instrumentation and the data-collection procedure. If the powder XRD patterns of two samples with the same crystal structure have significantly different peak widths, the visual appearance may differ substantially, especially in regions of the pattern in which there is significant peak overlap.

(ii) *Peak intensities:* If the powder sample exhibits 'preferred orientation' (*i.e.* a non-random distribution of orientations of the crystallites within a powder; see Section 1.6.2), the relative intensities of

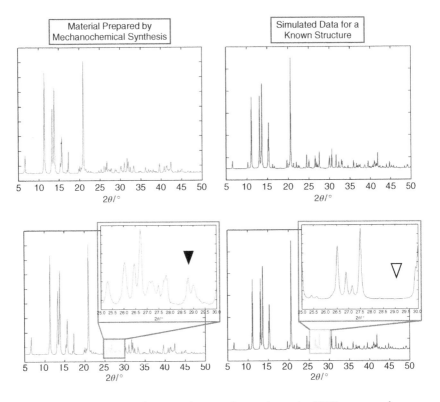

Figure 1.11 Comparison between the experimental powder XRD pattern of a material prepared by mechanochemical synthesis (left side) and the simulated powder XRD pattern of a potential candidate of known structure prepared previously *via* a solvothermal route (right side). Visual comparison (top part) might tend to suggest that the two materials are very similar. However, detailed comparison (bottom part) reveals important differences between the powder XRD patterns. In particular, the difference in the region $2\theta \approx 29°$ leads to the conclusion that the structures of the two materials cannot be identical.

peaks in the powder XRD pattern will deviate from the intrinsic relative intensities characteristic of the crystal structure. Hence, the powder XRD patterns recorded for two samples of the same material but exhibiting different degrees of preferred orientation may look substantially different, as demonstrated in Figure 1.12. This issue is pertinent in comparing an experimental powder XRD pattern with a simulated powder XRD pattern for a known crystal structure, as the simulated pattern implicitly has no effects due to preferred orientation.

(iii) *Peak positions:* Shifts in the peak positions in an experimental powder XRD pattern may arise from a number of instrumental factors. Furthermore, powder XRD patterns recorded at different temperatures may differ significantly in appearance (particularly in regions with substantial peak overlap) as a result of anisotropic thermal expansion/contraction. This issue is particularly relevant when an experimental powder XRD pattern recorded at ambient temperature is compared with the simulated powder XRD pattern for a known crystal structure determined from single-crystal XRD data at low temperature.

(iv) *Phase purity of the sample:* Crystalline impurity phases present in a powder sample (*e.g.* residual amounts of starting materials from a synthetic procedure) contribute additional peaks to the experimental powder XRD pattern. As a result, the powder XRD pattern may look significantly different from that of a pure sample of the main phase. Careful inspection should be carried out to assess the presence of impurity phases in the sample, and other analytical techniques should also be applied in order to identify the presence of chemical impurities in the sample.

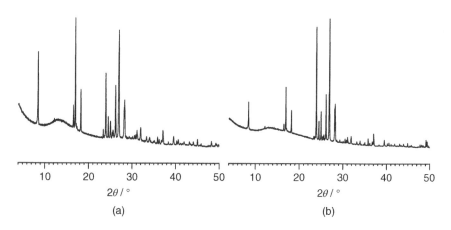

Figure 1.12 Experimental powder XRD patterns recorded with two different sample packing methods for the same sample of Form IV of *m*-aminobenzoic acid. In (a), the sample was loaded within a glass capillary, whereas in (b), the sample was compressed between two pieces of tape (foil-type sample holder). The significant differences in the relative intensities of the peaks in the powder XRD patterns shown in (a) and (b) arise from the preferred orientation in the powder sample.

While the factors listed above can undermine the ability to assess objectively the similarity between two powder XRD patterns by visual inspection, all these factors can be taken into consideration directly by appropriate quantitative analysis of the powder XRD data. Thus, in order to verify whether a synthesised material matches a known material, the recommended protocol is to carry out a Rietveld refinement calculation using the experimental powder XRD data recorded for the synthesised material and taking the crystal structure of the known material as the initial structural model in the refinement. As discussed in Section 1.5.5, Rietveld refinement gives rigorous consideration to fitting the peak profiles, peak positions, peak intensities (including refinement of parameters describing the effects of preferred orientation) and the presence of any known impurity phases (by including such materials as a 'second phase' in the refinement). This protocol allows the question of whether a synthesised material matches or does not match the structure of a known material to be answered with confidence, and provides a substantially more rigorous assessment than visual comparison of powder XRD patterns.

1.10   EXAMPLES OF THE APPLICATION OF POWDER XRD IN CHEMICAL CONTEXTS

1.10.1   Overview

We focus now on illustrative applications of structure determination of materials directly from powder XRD data, demonstrating the current scope of methodologies in this field and highlighting some of the specific issues (and challenges) discussed above. For many materials, conventional crystallisation procedures (*e.g.* growing crystals from solution) do not yield single crystals of suitable size and quality for single-crystal XRD and instead produce only microcrystalline powders. In such cases, structure determination from powder XRD data provides a crucial opportunity for establishing structural understanding of the material of interest. Furthermore, certain solid materials cannot be obtained (even as microcrystalline powders) by crystallisation experiments, but instead can be generated only by other types of preparation procedure. Some preparation processes commonly (or in some cases, inherently) yield microcrystalline products, including: (i) preparation of materials directly from solid-state chemical reactions, (ii) preparation of materials by solid-state desolvation processes, (iii) preparation of materials by solid-state grinding (mechanochemical) processes and (iv) preparation of materials directly by rapid precipitation from solution (as opposed to slow precipitation, which is generally conducive to obtaining large, high-quality crystals). Again, structure determination from powder XRD data may represent the only opportunity for determining the structural properties of new solid phases obtained by such processes. In addition, we highlight examples of the wide-ranging applications of powder XRD in monitoring structural changes in materials through *in situ* studies of chemical processes.

1.10.2 Structure Determination of Zeolites and Other Framework Materials

Zeolites are microporous solids with a wide range of applications in catalytic, adsorption and separation processes. As these important properties are closely related to their crystal structures, knowledge of their structural properties is central to the design and development of applications of these materials. Unfortunately, many synthetic zeolites are available only as microcrystalline powders and cannot be studied using conventional single-crystal XRD techniques. In such cases, a combination of techniques is often used to obtain structural information, such as powder XRD, electron microscopy, high-resolution solid-state NMR spectroscopy and computer simulation.

Synthesis and structure determination of the mesoporous chiral germanosilicate zeolite ITQ-37 have been reported[148] (Figure 1.13) and represent a crystallographic *tour de force*. In the early days of structure determination of zeolites, the structures of several complex zeolites, such as TNU-9, SSZ-74 and IM-5, were determined by a combination of powder XRD and high-resolution transmission electron microscopy (HRTEM).[149−151] However, ITQ-37 was not amenable to this approach as it has extreme electron-beam sensitivity, ruling out the possibility of HRTEM imaging. Instead, a new strategy combining selected-area electron diffraction (SAED) patterns (which require less exposure to the electron beam than HRTEM images) and powder XRD was exploited. The structure was solved using intensity information derived from *both* the SAED data and the powder XRD data, exploiting the charge-flipping algorithm to solve the structure, followed by Rietveld refinement of the powder XRD data.

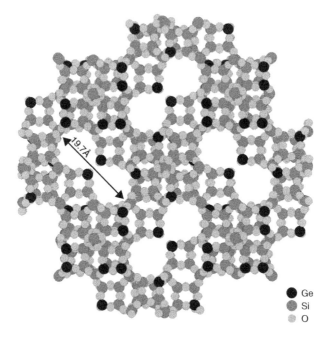

Figure 1.13 The crystal structure of ITQ-37, viewed along the channel direction.

The open-framework structure of ITQ-37, which has the chiral space group $P4_132$ (or $P4_532$), is described in terms of a single tertiary building unit, $T_{44}O_{145}(OH)_7$ (T = Si or Ge), comprising one double 4-ring with one OH group, three *lau* (laumontite) cages and three double 4-rings with two OH groups. The arrangement of the tertiary building units generates a channel system with a gyroidal surface; the channel system comprises two unique large cavities, each connected to three others by windows comprising 30 TO_4 tetrahedra. The pore opening has dimensions of $4.3 \times 19.3\,\text{Å}$. In contrast to previously prepared large-pore zeolitic materials, ITQ-37 is much more stable chemically following removal of the template molecule required for its preparation.

In another recent advance in zeolite structure determination, a 2D charge-flipping method was exploited[152] to solve the structure of the borosilicate zeolite SSZ-82 (Figure 1.14). Electron-density maps generated from the powder XRD data through conventional charge-flipping methods could not be easily interpreted. However, by using 2D projections derived from the powder XRD data, an initial set of structure factor phases could be found (in a similar way to the use of precession electron diffraction), allowing the phases for the full 3D structure to be

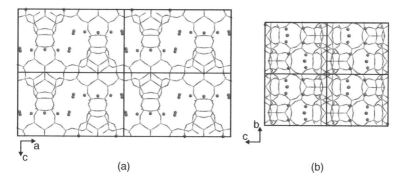

(a)   (b)

Figure 1.14 Crystal structure of zeolite SSZ-82 viewed (a) along the *b*-axis and (b) along the *a*-axis. The zeolite framework is shown as a stick representation. The spheres are oxygen atoms of water molecules.

determined through the charge-flipping algorithm and leading to successful structure solution.

1.10.3 *In Situ* Powder XRD Studies of Materials Synthesis

As discussed in Section 1.6.5, time-resolved *in situ* powder XRD measurements can provide detailed insights into both crystallisation processes and solid-state transformations. A seminal example is the *in situ* powder XRD study[153] of crystallisation of the large-pore oxy-fluorinated gallophosphate ULM-3. Experiments were carried out using both conventional (angular-dispersive) powder XRD and energy-dispersive powder XRD, in each case studying the crystallisation process in a specially designed *in situ* cell (Figure 1.15) and taking advantage of the very high incident X-ray intensity produced by a synchrotron radiation source. Within the sealed cell, the starting materials for the synthesis of ULM-3 (gallium oxide, a phosphorus source, hydrofluoric acid, 1,6-diaminohexane and water) were allowed to react under isothermal hydrothermal conditions, while the powder XRD pattern was repeatedly recorded. Analysis of the results revealed that, when phosphorus pentoxide is used as the phosphorus source, two previously unknown metastable crystalline phases are observed as intermediates during the formation of ULM-3 (Figure 1.16). One of the two intermediate phases was found to be favoured by lowering the pH, whereas the other intermediate phase was found to be favoured by

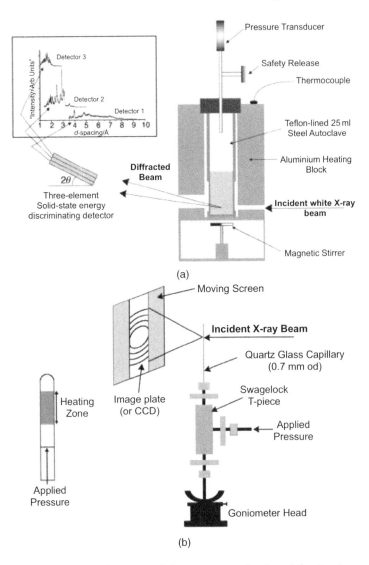

Figure 1.15 Schematic diagrams of the apparatus developed for *in situ* powder XRD studies of the crystallisation process using (a) energy-dispersive powder XRD and (b) conventional (angular-dispersive) powder XRD. Figure kindly provided by Professor D. O'Hare.

increasing the pH. The results of this study support the view that the transformation of the intermediate phases into ULM-3 occurs as a direct solid–solid conversion (however, the possibility that small amounts of amorphous or solution phases may be involved in the transformation cannot be completely discounted).

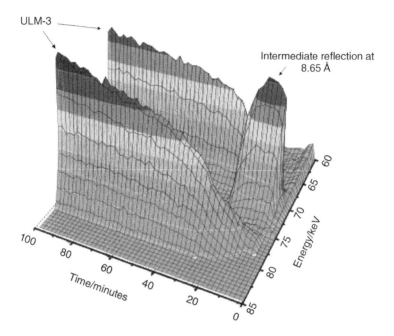

Figure 1.16 Sequence of *in situ* energy-dispersive powder XRD patterns recorded as a function of time during crystallisation of ULM-3 under hydrothermal conditions. Reprinted with permission from [210] Copyright (1999) Cambridge University Press.

1.10.4 Structure Determination of New Materials Produced by Solid-State Mechanochemistry

In addition to conventional solution-phase crystallisation, many novel materials can be prepared only by 'mechanochemical' procedures in which two (or more) solid phases are ground together to generate a new product phase.[154] In some cases, solid-state mechanochemistry represents the only known route for accessing a particular product material (and attempts to recrystallise the mechanochemically obtained product often yield a material with a different structure). However, materials prepared by solid-state mechanochemical procedures are virtually always microcrystalline powders and are not suitable for structure determination by single-crystal XRD.

The first reported[155] application of powder XRD to determine the structure of a molecular co-crystal prepared by solid-state grinding was for the three-component material prepared from racemic *bis*-β-naphthol (BN), benzoquinone (BQ) and anthracene (AN). Grinding together the pure crystalline phases of BN, BQ and AN produced a polycrystalline material with reddish-purple colour (whereas crystallisation of the same components from solution gave a different co-crystal with bluish-black colour). The contents of the asymmetric unit (confirmed by high-resolution solid-state ^{13}C NMR spectroscopy) were established to be one BN molecule, one BQ molecule and one half AN molecule (which resides on a twofold rotation axis). The structure was solved using the direct-space genetic algorithm technique, involving a total of 17 structural variables. The crystal structure (Figure 1.17) is rationalised in terms of three different interaction motifs: edge-to-face interactions between BQ (edge) and AN (face) molecules, face-to-face interactions between BQ and BN molecules and chains of $O-H\cdots O$ hydrogen bonds involving BN and BQ molecules. Hitherto, structural characterisation of co-crystals prepared by solid-state grinding has been limited by the fact that the preparation procedure intrinsically yields microcrystalline powders. Structure determination from powder XRD data clearly has considerable potential in the structural characterisation of new co-crystal phases prepared by such procedures.

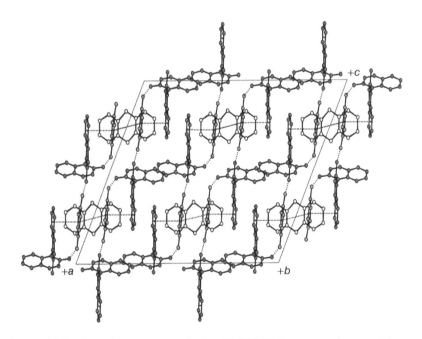

Figure 1.17 Crystal structure of the BN/BQ/AN co-crystal material prepared by solid-state grinding. Dotted lines indicate π-stacking interactions and hydrogen-bonded chains. Reprinted with permission from [155] Copyright (2003) American Chemical Society.

Another example[147] of structure determination of a material prepared by solid-state mechanochemistry concerns a porous interpenetrated mixed-ligand MOF material with composition $Zn_2(fma)_2(bipy)$. The mechanochemical synthesis involved the reagents $Zn(OAc)_2 \cdot 2H_2O$, fumaric acid (H_2fma) and 4,4′-bipyridine (bipy). The crystal structure of this material (Figure 1.18a,b) has some similarity to that of a previously reported[156] DMF solvate material, $Zn_2(fma)_2(bipy)(DMF)_{0.5}$, prepared by a solvothermal route, the crystal structure of which (Figure 1.18c,d) was determined by single-crystal XRD. Nevertheless, there are important structural differences between these materials, including the fact that the bipy ligands in the DMF solvate structure are constrained such that they are planar (they lie in the mirror plane of the C_2/m space group), whereas in the structure of the mechanochemically prepared material

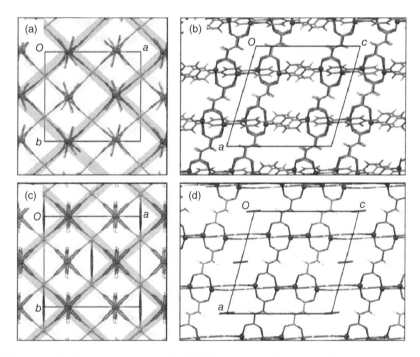

Figure 1.18 Crystal structure of a MOF material, $Zn_2(fma)_2(bipy)$, prepared by mechanochemical synthesis. The structure was determined directly from powder XRD data, and is viewed (a) along the c-axis and (b) along the b-axis. The two (identical) interpenetrated frameworks are indicated by shading. For comparison, (c) and (d) show the corresponding views of the structure of a DMF solvate material, $Zn_2(fma)_2(bipy)(DMF)_{0.5}$, prepared by a solvothermal route. Although there is some similarity between these structures, there are nevertheless important structural differences. Reprinted with permission from [147] Copyright (2010) Royal Society of Chemistry.

(space group $P2_1/a$) the dihedral angle between the two rings of the bipy ligand is 53.2°. Interestingly, desolvation of the DMF solvate yields a material identical to that prepared by mechanochemical synthesis.

Other examples of materials prepared under mechanochemical conditions for which structure determination has been carried out from powder XRD data include: (i) a MOF material with composition $Co(dibenzoylmethanate)_2(nicotinamide)_2$, obtained by thermal desolvation of the corresponding acetone solvate prepared by liquid-assisted grinding (LAG),[157] (ii) a hydrate co-crystal of 5-methyl-2-pyridone and trimesic acid, prepared by grinding a methanol solvate co-crystal of the same components under an ambient atmosphere[158] and (iii) the 1:1 co-crystals of theobromine with trifluoroacetic acid and theobromine with malonic acid, each prepared by LAG.[159]

In another example,[160] a novel acetic acid solvate of the organic light-emitting diode material $[Alq_3]$ (q = 8-hydroxyquinolinate) was prepared by ball milling a mixture of basic aluminium(III) diacetate $[Al(OAc)_2(OH)]$ and 8-hydroxyquinoline with no added solvent. The powder XRD pattern of the resultant microcrystalline powder did not match any known solvate or polymorph of $[Alq_3]$. The combined results of microanalysis, thermogravimetric analysis and solution-state ^1H NMR spectroscopy led to the conclusion that the reaction product is a 1:1 solvate of $[Alq_3]$ and acetic acid. The crystal structure of this material was solved from powder XRD data using the direct-space genetic algorithm technique followed by Rietveld refinement, yielding the structure shown in Figure 1.19.

1.10.5   *In Situ* Powder XRD Studies of Solid-State Mechanochemical Processes

While solid-state mechanochemistry is now applied widely in the preparation of new materials, fundamental understanding of many aspects of mechanochemical processes remains sparse. It is clearly desirable to establish whether such transformations proceed *via* transient intermediate phases, to rationalise the sequence of phases involved and to determine the rate of each step in the reaction sequence. Some insights into these issues may be gained by halting the mechanochemical process at different times and removing the sample for *ex situ* characterisation, for example by powder XRD. In some cases, intermediate phases can be identified, but *ex situ* studies of this type have several potential

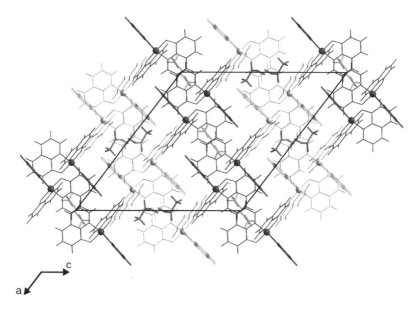

Figure 1.19 Crystal structure of the mechanochemically synthesised acetic acid solvate of [Alq$_3$], determined directly from powder XRD data, containing two independent molecules of [Alq$_3$] and two independent molecules of acetic acid. Reprinted with permission from [160] Copyright (2012) American Chemical Society.

limitations. For example, depending on the time step used for sampling, transient intermediate phases can be missed, and interpretation of the time-dependence of the process can be complicated by the fact that some mechanochemical transformations actually continue after the mechanical force has been removed.

As discussed in Section 1.6.5, *in situ* studies offer significantly better prospects for understanding fundamental underlying principles, but establishing a suitable experimental set-up for *in situ* studies is often challenging. In this context, a requirement for *in situ* studies of solid-state mechanochemical synthesis is that the technique must be able to probe the solid material directly inside the confined environment of the continuously vibrating sample vessel in a ball mill. Recently, however, Friščić and co-workers[122] reported an experimental set-up for *in situ* powder XRD studies of mechanochemical processes in which the sample stage of the powder X-ray diffractometer is replaced by an adapted ball mill. Although the sample vibration is potentially detrimental to data quality, this effect is minimised by optimising the geometry of the set-up. Furthermore, by exploiting the high-intensity incident X-rays from a synchrotron radiation source, the powder XRD data can be recorded rapidly, allowing excellent time-resolution (*ca* 4 seconds) in studying the kinetics of mechanochemical processes.

The *in situ* strategy has been demonstrated by studies of the mechanochemical formation of zeolitic imidazolate framework (ZIF) materials,[161,162] MOFs prepared by milling zinc oxide and derivatives of imidazole. The reactions were carried out under conditions of 'liquid-assisted grinding' (LAG) or 'ion- and liquid-assisted grinding' (ILAG), in which a small amount of liquid or a liquid plus a salt, respectively, is added to the powder reagents in order to accelerate the mechanochemical synthesis. The reaction between zinc oxide and 2-ethylimidazole (Figure 1.20) provides a particularly emphatic demonstration of the new insights revealed by the *in situ* technique. A sequence of crystalline materials with characteristic structure types can be identified (corresponding to the structure types 'RHO', then 'analcime', then the final 'β-quartz' form). Furthermore, under ILAG conditions, the use of different salts leads to distinctly different rates of the individual steps within this sequence of solid phases.

In addition to identifying the sequence of solid structures involved in a reaction pathway, the *in situ* powder XRD data can be used to determine the kinetics of each step, which involves fitting the time-evolution of peak intensities to kinetic models for solid-state reactions. However, some potential pitfalls must be carefully considered. For example, if the powder sample exhibits preferred orientation, establishing the time-dependent changes in the relative amounts of different phases may require more rigorous quantitative fitting of the data by Rietveld analysis, particularly if the degree of preferred orientation changes during the reaction.

1.10.6 *In Situ* Powder XRD Studies of a Polymorphic Transformation

In recent years, there has been increasing interest, both in academia and in several industrial sectors, in the phenomenon of **polymorphism**[163] (in the context of molecular materials, polymorphism arises when a given molecule is able to form two or more different crystal structures). In the pharmaceutical industry in particular, it is often crucial to understand the diversity of polymorphic forms that are available to a given drug molecule, as different polymorphs can have significantly different solid-state properties (and hence different performance in therapeutic applications) as a consequence of their different crystal

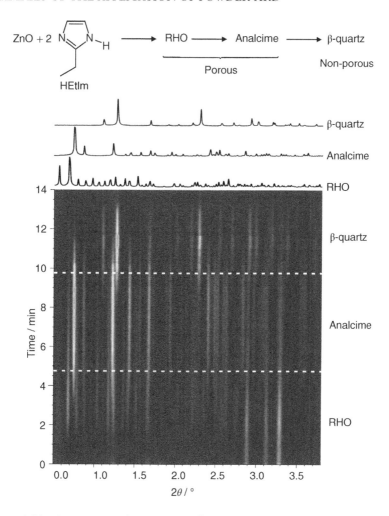

Figure 1.20 *In situ* powder XRD studies monitoring the evolution of the mechanochemical reaction that occurs on milling ZnO and HEtIm together under ILAG conditions (using DMF as the liquid and NH_4NO_3 as the salt additive). Top: the reaction scheme. Middle: characteristic powder XRD patterns of the three structure types observed during the reaction. Bottom: time-resolved *in situ* powder XRD patterns, which allow the evolution of the reaction to be elucidated (bottom). Figure kindly provide by Professor T. Friščić.

structures. Therefore, an important aspect of pharmaceutical research is the discovery, structural characterisation and physicochemical understanding of all polymorphic forms that are accessible to a given drug molecule, as well as understanding the occurrence of transformations between polymorphic forms.

Recently, a new polymorph (Form II) of the non-steroidal anti-inflammatory drug racemic ibuprofen (*rac*-ibuprofen) has been discovered.[164,165] This new polymorph crystallises following quench-cooling of molten *rac*-ibuprofen to a temperature significantly below the glass transition point of super-cooled liquid *rac*-ibuprofen (below at least −105 °C), before annealing at a temperature above the glass transition (−15 °C). However, further work[166] has shown that Form II can actually be crystallised simply by cooling the molten *rac*-ibuprofen to 0 °C and waiting for a sufficient period of time. At ambient temperature, Form II is unstable with respect to the previously known Form I, but Form II is actually produced (as a kinetic product) before Form I in crystallisation from the super-cooled liquid at 0 °C. Thus, on holding the super-cooled liquid *rac*-ibuprofen at 0 °C and recording powder XRD data *in situ* as a function of time, crystallisation of Form II is observed to occur, followed after a period of time by transformation to Form I (the thermodynamic product). In the *in situ* powder XRD results shown in Figure 1.21, crystallisation of Form II commences after the super-cooled liquid has been held at 0 °C for 14.6 hours. The peaks in the powder XRD pattern characteristic of Form II increase in intensity until 17.5 hours (suggesting that crystal growth of Form II continues during this time). After 19.0 hours, transformation of Form II to Form I is seen to commence; peaks due to Form I now appear and increase in intensity,

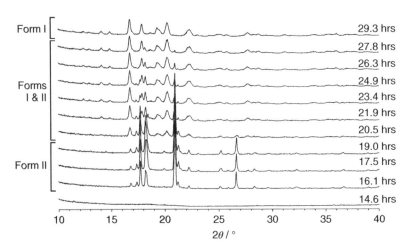

Figure 1.21 *In situ* powder XRD data for the crystallisation of super-cooled liquid ibuprofen at 0 °C, showing powder XRD patterns recorded between 14.6 and 29.3 hours after commencing the experiment. The time taken to record each powder XRD pattern was 88 minutes. Reprinted with permission from [166] Copyright (2012) American Chemical Society.

while the peaks characteristic of Form II decrease in intensity until 29.3 hours, by which time the polymorphic transformation is complete.

1.10.7 *In Situ* Powder XRD Studies of a Solid-State Reaction

In situ powder XRD has been exploited to monitor the formation of $CaTiO_3$, an important dielectric material, from $CaCO_3$ and TiO_2.[167] A mixture of the starting materials was prepared with no prior mechanical activation and was then heated from ambient temperature to 1100 °C, with powder XRD patterns recorded at specific temperature intervals (Figure 1.22). Prior to the start of the *in situ* experiment, it was found that a small amount of $CaTiO_3$ was present (probably due to the short initial calcination process). In the *in situ* powder XRD study, decomposition of $CaCO_3$ to CaO was observed to begin at 600 °C, reaching completion by 800 °C. On further increasing the temperature, formation of $CaTiO_3$ was observed, and on reaching 1100 °C, $CaTiO_3$ was the major product (together with *ca* 10% of the unreacted starting material). In contrast, when the same reaction was carried out using a mixture of the starting materials that had been mechanically activated prior to heating, it was found from the powder XRD data that *ca* 75% of the material at ambient temperature was already present as $CaTiO_3$. During subsequent heating, once all the $CaCO_3$ had decomposed, formation of $CaTiO_3$ proceeded much more quickly than in the non-mechanically activated case and moreover yielded $CaTiO_3$ as a phase-pure product.

1.10.8 Establishing Details of a Hydrogen-Bonding Arrangement by Powder Neutron Diffraction

We consider a case in which structure determination from powder XRD data alone cannot elucidate all aspects of the crystal structure. In particular (see Section 1.7), when details of hydrogen-bonding arrangements are not clearly established from powder XRD data, the advantages of employing powder neutron diffraction must be exploited. In spite of the historical importance of ammonium cyanate $[NH_4^+][OCN^-]$ (first studied by Wöhler over 180 years ago[168]), the crystal structure of this material remained undetermined until

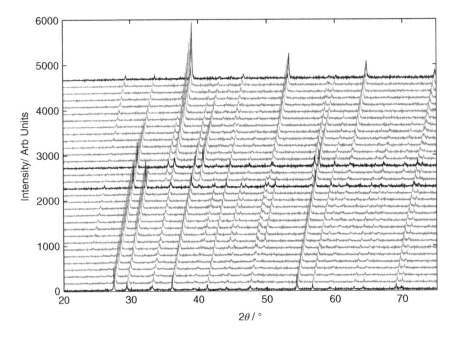

Figure 1.22 *In situ* powder XRD data recorded during the reaction of $CaCO_3$ and TiO_2. Data recorded on increasing temperature are offset vertically (lowest temperature at the bottom). The highlighted powder XRD patterns are those for the starting mixture at 30 °C (lowest), the commencement of $CaCO_3$ decomposition at 600 °C, the end of $CaCO_3$ decomposition at 800 °C and the final product at 1100 °C (highest). Figure kindly provided by Professor J.S.O. Evans and Dr I. Evans.

relatively recently,[169,170] as this material can be prepared only as a microcrystalline powder and is unsuitable for study by single-crystal XRD. Structure determination from powder XRD data[169] established the positions of the non-hydrogen atoms but could not reliably distinguish the correct orientation of the ammonium cation. In the crystal structure, the nitrogen atom of the ammonium cation resides at the centre of a nearly 'cubic' arrangement of O and N atoms (from cyanate anions), which occupy alternate corners of the 'cube'. Two plausible orientations of the ammonium cation may be proposed, in one case forming four N–H · · · O hydrogen bonds and in the other case forming four N–H · · · N hydrogen bonds. However, the locations of the hydrogen atoms cannot be unambiguously determined from the powder XRD data. In order to establish definitively the relative amounts of the N–H · · · O and N–H · · · N hydrogen-bonding situations in the crystal structure of ammonium cyanate, powder neutron diffraction was carried out[170] on the deuterated material $[ND_4^+][OCN^-]$ (actually *ca* 81% D, 19% H). The neutron diffraction results definitively support the structure with $N-D \cdots N$ hydrogen bonding, with no detectable population of $N-D \cdots O$ hydrogen bonding (and hence no disorder between the $N-D \cdots N$ and $N-D \cdots O$ hydrogen-bonding arrangements). Results from solid-state ^{15}N NMR spectroscopy[170] are also consistent with the assignment that the structure has $N-H \cdots N$ rather than $N-H \cdots O$ hydrogen bonding.

1.10.9   Structure Determination of a Material Produced by Rapid Precipitation from Solution

The reaction (Figure 1.23) of $[Co(OH_2)_2\{(OH)_2Co(en)_2\}_2](SO_4)_2$ (denoted **1**; en = ethylenediamine) with NH_4Br to give the chiral complex *cis*-$[CoBr(NH_3)(en)_2]Br_2$ (denoted **2**) has been studied widely in relation to such phenomena as chiral symmetry breaking, spontaneous resolution and chiral amplification. Indeed, this reaction is important historically because **2** was one of the first octahedral metal complexes to be resolved into Δ and Λ stereoisomers, some years after Werner predicted that octahedral ions, $[MXY(en)_2]$, should exist as enantiomeric pairs.

As rapid precipitation of **2** occurs from the solution in which the reaction is carried out, the material obtained directly from the reaction is a microcrystalline powder. Structure determination of this material from powder XRD indicates[171] that it is a new solid phase of **2**. Structure determination was carried out[171] directly from powder XRD data using the direct-space genetic algorithm technique for structure solution, followed by Rietveld refinement. In the crystal structure (Figure 1.23b), the *cis*-$[CoBr(NH_3)(en)_2]^{2+}$ complexes are arranged in two different types of chain, propagated along the *a*-axis and the *b*-axis respectively, with neighbouring complexes in each type of chain linked by $N-H \cdots Br^- \cdots H-N$ interactions. Along the *a*-axis, neighbouring repeat units are related by translation and the chain is relatively straight, whereas along the *b*-axis, neighbouring repeat units are related by the 2_1 symmetry operation and the chain is helical. With regard to chirality, the most important feature of this result is the fact that the structure is racemic (non-chiral space group $P2_1/n$, with one formula unit in the asymmetric unit). Studies of this reaction under a wide range of

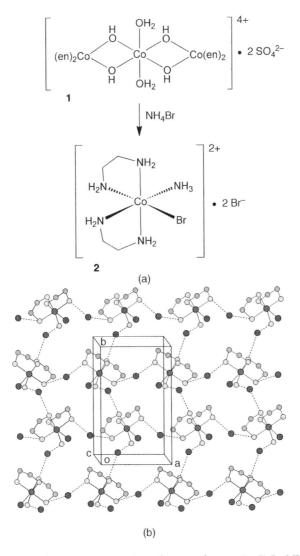

(a)

(b)

Figure 1.23 (a) Solution-state reaction that produces *cis*-[CoBr(NH$_3$)(en)$_2$]Br$_2$ (denoted **2**). (b) Crystal structure of the racemic phase of **2** produced directly from this reaction (the structure is viewed nearly along the *c*-axis, showing the straight and helical chains that run along the *a*-axis and the *b*-axis, respectively; hydrogen atoms are omitted for clarity). Reprinted with permission from [171] Copyright (2006) Royal Society of Chemistry.

experimental conditions led consistently to this new racemic phase of **2**, and conglomerate phases of **2** were never obtained. The implications of this result in relation to previous reports of spontaneous induction of chirality in this system are discussed in detail elsewhere.[171]

1.10.10 Structure Determination of Intermediates in a Solid-State Reaction

An interesting example of the application of structure solution from powder XRD data concerns structural characterisation of the reactants and products of chemical reactions, including structure determination of intermediate phases that are difficult to obtain as pure phases. The decomposition reaction of ammonia trimethylalane (Me_3AlNH_2) to give aluminium nitride (AlN) *via* the intermediates $(Me_2AlNH_2)_x$ and $(MeAlNH)_y$ was first discovered over 50 years ago;[172] the structure of the intermediate $(Me_2AlNH_2)_3$ produced in this reaction has been determined directly from powder XRD data[173] using a direct-space simulated annealing technique for structure solution. Interestingly, this structure is different from that of the trimer $(Me_2AlNH_2)_3$, which was prepared by a different route and solved from single-crystal XRD data.[174] The unit cell of the $(Me_2AlNH_2)_3$ material obtained by indexing the powder XRD pattern was consistent with the asymmetric unit comprising a trimeric unit, which complicated the structure solution process, as this trimer may exist in a variety of boat and twist-boat conformations. The best structure solution was obtained for the boat conformation (Figure 1.24), in contrast to the result for the structure determined from single-crystal XRD data, in which the trimer adopts a twist-boat conformation.

1.10.11 Structure Determination of a Novel Aluminium Methylphosphonate

In general, the traditional strategy for structure solution from powder XRD data has been applied widely for structure determination of inorganic materials, but the direct-space strategy can provide a viable alternative for such materials. Thus, when solving the structure of the γ-phase of $Al_2(CH_3PO_3)_3$ proved difficult with the traditional strategy, the direct-space strategy (using a simulated annealing algorithm) was employed instead.[175] Thermal and spectroscopic data indicated that the material is anhydrous and the asymmetric unit comprises two inequivalent aluminium atoms and three inequivalent phosphonate groups, representing a total of five independent structural fragments in the direct-space structure solution calculation. The structure was

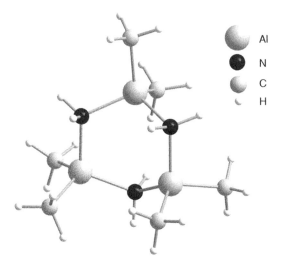

Figure 1.24 Trimeric association of Me_2AlNH_2 units in the polymorph of $(Me_2AlNH_2)_3$ prepared by the decomposition reaction of Me_3AlNH_2, for which the structure was determined from powder XRD data.

found to be lamellar, with the methyl groups protruding from the ends of each sheet (Figure 1.25). Validation of the structure using energy-minimisation calculations based on semi-empirical methods gave results in good agreement with the structure determined from the powder XRD data.

1.10.12   Structure Determination of Materials Prepared by Solid-State Dehydration/Desolvation Processes

For many crystalline materials (especially certain organic compounds), the preparation of a 'pure' (non-solvate) crystalline phase by conventional crystal growth from solution is difficult, due to the competitive formation of solvate crystals. In such cases, a possible route to the 'pure' phase involves carrying out desolvation of a solvate phase at elevated temperature and/or reduced pressure. However, such desolvation processes are often associated with loss of crystal integrity, such that a single crystal of the solvate phase yields a polycrystalline product following desolvation (Figure 1.26). In such cases, powder XRD is essential for determining the structure of the product.

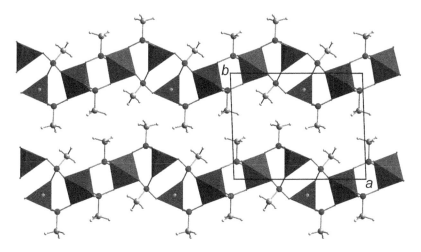

Figure 1.25   Crystal structure of the γ-phase of $Al_2(CH_3PO_3)_3$, determined from powder XRD data, showing that the methyl groups protrude from the ends of each sheet.

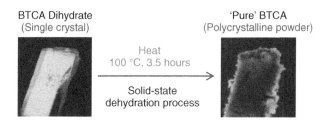

Figure 1.26   Optical micrographs showing that a single crystal of BTCA dihydrate leads to a polycrystalline product ('pure' phase of BTCA) upon dehydration.

Although the chemical properties of benzene-1,2,3-tricarboxylic acid (BTCA) were first studied over 100 years ago, the crystal structure of 'pure' BTCA was not reported until a recent powder XRD study.[176] Like many other organic molecules, BTCA has a strong propensity to form solvate structures when crystallised from solution. The 'pure' (non-solvate) phase of BTCA can be obtained by dehydration of the dihydrate phase of BTCA at elevated temperature, but the material is obtained as a microcrystalline powder (Figure 1.26). The crystal structure of the 'pure' phase of BTCA was solved directly from powder XRD data using the direct-space genetic algorithm technique followed by Rietveld refinement. There are three independent BTCA molecules in the asymmetric unit, and trial structures in the genetic algorithm calculation were defined by a total of 27 structural variables (nine variables are required to define the position, orientation and conformation of each independent molecule). In the crystal structure (Figure 1.27), all carboxylic acid groups are engaged in intermolecular hydrogen bonding to other carboxylic acid groups *via* the double $O-H\cdots O$ hydrogen-bonded motif found in carboxylic acid 'dimers'. The structure of the 'pure' phase of BTCA differs substantially from that of BTCA dihydrate, implying that the solid-state dehydration process is associated with substantial structural reorganisation.

Another example concerns dehydration of the hydrate crystalline phase of chloroquine bis-(dihydrogen phosphate) (denoted $CQ(DHP)_2$) to form anhydrous $CQ(DHP)_2$ as a microcrystalline powder. The structure of the anhydrous phase was determined[177] directly from powder XRD data. The structure exhibits several interesting contrasts with that of the parent hydrate phase, notably concerning the topology of the hydrogen-bonded chains of DHP anions that exist in both structures. Thus, while the hydrogen-bonded chains of DHP anions in the hydrate phase are linear, the corresponding chains in the anhydrous phase have a zigzag topology, resulting from changes in the mode of hydrogen bonding of the DHP anions in the chain. Given the substantial structural reorganisation associated with dehydration, it is not surprising that this process is associated with the formation of a polycrystalline product.

Within the inorganic domain, a new metastable polymorph of strontium oxotellurate(IV) (ε-$SrTeO_3$) has been prepared by dehydration

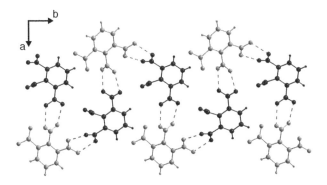

Figure 1.27 Helical hydrogen-bonded chain of molecules, viewed approximately along the c-axis, in the crystal structure of the 'pure' phase of BTCA, determined directly from powder XRD data.

of the corresponding monohydrate,[178] yielding ε-SrTeO$_3$ as a microcrystalline powder. The crystal structure was determined from powder XRD data recorded at ambient temperature and solved using the direct-space strategy. During topotactic dehydration (Figure 1.28), the removal of layers of water molecules caused collapse of the structure along the a-axis, resulting in a structure that shares significant similarity with the monohydrate, although it is now best described as a framework structure, in contrast to the layered nature of the monohydrate.

1.10.13 Structure Determination of the Product Material from a Solid-State Photopolymerisation Reaction

Many crystalline solids undergo chemical transformations, induced for example by incident radiation or heat. An important aspect of such solid-state reactions is to understand the structural properties of the product phase obtained directly from the reaction, and in particular to rationalise the relationships between the structural properties of the product and reactant phases. In many cases, the product phase is amorphous, but for cases in which the product is crystalline, it is usually obtained as a microcrystalline powder that does not contain single crystals of suitable size and quality for single-crystal XRD studies. In such cases, powder XRD data may provide the only opportunity to determine the structural properties of the product phase.

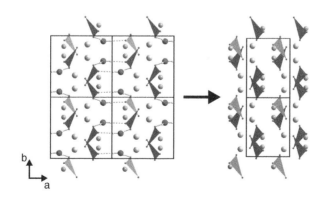

Figure 1.28 Structural change associated with dehydration of SrTeO$_3 \cdot$ H$_2$O (left) to form ε-SrTeO$_3$ (right). In the monohydrate, hydrogen bonds are shown as dashed lines.

An example is the photopolymerisation of 2,5-distyrylpyrazine (DSP) (Figure 1.29), for which polymerisation occurs *via* intermolecular [2 + 2] photocyclisation reactions at each end of the monomer molecule. Although this reaction is a 'classic' solid-state reaction and was studied extensively around 40 years ago,[179,180] structure determination of the polymeric product phase was carried out only recently,[181] with the structure determined directly from powder XRD data using the

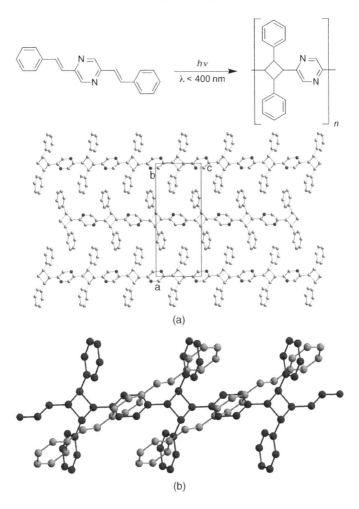

(a)

(b)

Figure 1.29 Solid-state photopolymerisation reaction of DSP (shown at top). (a) Crystal structure of the polymeric product phase obtained directly from the solid-state photopolymerisation reaction of DSP, viewed along the *b*-axis (for clarity, only half of the unit cell is shown along the direction of view). (b) Overlay of the monomer (light grey) and polymer (dark grey) in their crystal structures. Reprinted with permission from [181] Copyright (2008) American Chemical Society.

direct-space strategy for structure solution. In the crystal structure (Figure 1.29a), the polymer chains are aligned along the c-axis and the distance between the centres of adjacent cyclobutane and pyrazine rings corresponds to half the c-axis of the unit cell. From the overlay of the monomer and polymer structures in Figure 1.29b, it is clear that the solid-state reaction is associated with only very small atomic displacements at the site of each [2 + 2] photocyclisation reaction (the displacement of the carbon atoms of the $C=C$ double bonds of the monomer molecules on forming the cyclobutane ring of the polymer is only ca 0.8 Å for one pair of carbon atoms and ca 1.6 Å for the other pair). Such small displacements are completely in accord with the assignment of this solid-state reaction as a topochemical transformation,[182−185] in which the crystal structure of the reactant monomer phase imposes geometric control on the pathway of the polymerisation reaction, thereby controlling the structure and stereochemistry of the polymer product obtained.

Many solid-state reactions (not only polymerisation) intrinsically generate microcrystalline powders as the product phase, and there is clearly considerable potential to exploit powder XRD techniques for structural rationalisation in such cases.

1.10.14 Exploiting Anisotropic Thermal Expansion in Structure Determination

As discussed in Section 1.5.4.1, structure solution from powder XRD data using the traditional strategy relies upon the ability to extract accurate values of the integrated peak intensities in the powder XRD pattern at the profile-fitting stage. The difficulty in obtaining reliable intensities in this way was one of the primary factors that motivated the development of the direct-space strategy for structure solution (as discussed above, many implementations of the direct-space strategy are based on consideration of the whole-profile fit rather than extracted peak intensities). However, it is important to emphasise that structure solution by the traditional strategy may still be successful when the powder XRD pattern suffers from substantial peak overlap, as illustrated by the structure determination of 9-ethylbicyclo[3.3.1]nona-9-ol.[186]

Powder XRD patterns were recorded at several different temperatures. At each temperature, there is substantial overlap of peaks in the powder XRD pattern, and the peak intensities extracted from the data at any individual temperature did not allow the structure to be solved. To overcome this problem, anisotropic thermal expansion[187,188] was exploited as a means of improving the intensity extraction process. As a consequence of anisotropic thermal expansion, different peaks in the powder XRD pattern shift to a different extent as temperature is varied, and hence the nature of the peak overlap changes with temperature. By carrying out a combined analysis of the data recorded at all temperatures studied, a more reliable extraction of the integrated peak intensities can be obtained. Using the accurate set of integrated peak intensities obtained from this multipattern peak extraction process, the crystal structure of 9-ethylbicyclo[3.3.1]nona-9-ol was solved successfully using direct methods (and analysis of difference Fourier maps). The structure has four independent molecules in the asymmetric unit, assembled into a tetrameric unit held together by O–H··· O hydrogen bonds, as shown in Figure 1.30.

1.10.15   Rationalisation of a Solid-State Reaction

In order to rationalise the chemical reactivity of solids,[182,184] it is essential to know the structural properties of the parent phase. Thus, crystal structure determination of reactive materials is a prerequisite for understanding chemical transformations within them. It has been known

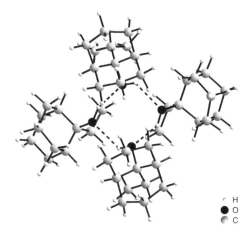

H
O
C

Figure 1.30   Tetrameric association of 9-ethylbicyclo[3.3.1]nona-9-ol molecules in the crystal structure determined from powder XRD data.

since the $1850s^{[189,190]}$ that solid sodium chloroacetate undergoes a polymerisation reaction at high temperature to produce polyglycolide and sodium chloride:

$$n \; ClCH_2COONa \rightarrow n \; NaCl + [CH_2COO]_n \qquad (1.9)$$

Unfortunately, as sodium chloroacetate is a microcrystalline material, it is unsuitable for study by single-crystal XRD. Thus, in the absence of knowledge of the crystal structure, deriving an understanding of the polymerisation reaction and its mechanism were not possible. However, following the development of the direct-space strategy for structure solution from powder XRD data, the structure of sodium chloroacetate was determined.[51] The structure was found to contain rows of chloroacetate anions (Figure 1.31), within which one of the two oxygen atoms of the carboxylate group is ideally positioned to attack the α-carbon atom of a neighbouring chloroacetate anion, expelling a Cl⁻ anion. Propagation of this attack along the row of chloroacetate anions in the crystal structure results in polymerisation to produce polyglycolide. Thus, from knowledge of the crystal structure of the reactant sodium chloroacetate

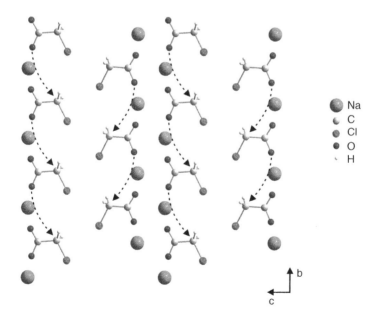

Figure 1.31 Crystal structure of sodium chloroacetate, determined directly from powder XRD data. Within the linear arrays of chloroacetate anions (which run vertically in the figure), the arrows indicate the direction of attack of each anion on its neighbour, leading to the formation of polyglycolide as the reaction product.

phase, the production of polyglycolide may be rationalised directly in terms of a topochemical reaction pathway. The crystal structures of various lithium halogenoacetates have also been determined from powder XRD data,[191] and provide a basis for understanding the chemical reactions in these materials.

1.10.16   Structure Determination of Organometallic Complexes

In structure determination of organometallic complexes from powder XRD data, the presence of a metal atom often simplifies the process of structure solution if it behaves as a dominant X-ray scatterer,[192] although in such cases locating the organic component of the material may prove to be more challenging. An example of successful structure determination in which the traditional strategy was used for structure solution concerns cyclopentadienyl rubidium.[113] Interestingly, the powder sample used in this study was a mixture of two polymorphs of cyclopentadienyl rubidium. Initial attempts to index the powder XRD pattern failed because of the presence of the two phases, but closer inspection revealed that the peaks could be subdivided into a set of narrow peaks (with instrument-limited peak widths) and a set of substantially broader peaks. By subdividing the experimental data in this way, the crystal structure of each polymorph was determined. The use of synchrotron powder XRD data was important in this case as the selective peak broadening could not be distinguished with the lower instrumental resolution of a conventional laboratory powder X-ray diffractometer.

In an example employing the direct-space strategy for structure solution, the crystal structure of silicon phthalocyanine dichloride ($SiPcCl_2$) was determined[55] from powder XRD data. In this case, the Si atom of the molecule was fixed on a crystallographic inversion centre, and as $SiPcCl_2$ is a rigid molecule with no conformational degrees of freedom, the direct-space structure solution calculation required only three structural variables (*i.e.* the variables $\{\theta, \phi, \psi\}$ defining the orientation of the molecule relative to the unit cell axes).

In another application of the direct-space strategy (employing a simulated annealing algorithm) for structure solution, the crystal structure of the malaria pigment β-haematin was determined from powder XRD data.[56] However, more recent work[193] on β-haematin at low temperature (100 K) provided evidence for a second crystalline form, which was

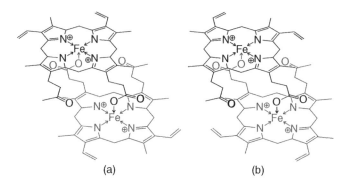

(a)   (b)

Figure 1.32 Centrosymmetric dimers present in (a) the major phase and (b) the minor phase of β-haematin.

present as a minor phase with the known crystalline form as the major phase. Structure determination of the new form was carried out directly from powder XRD data, leading to the realisation that the two forms contain different centrosymmetric dimers (Figure 1.32).

1.10.17 Examples of Structure Determination of Some Polymeric Materials

In spite of the impression of huge molecular size, crystalline polymeric materials are well suited to structure determination from powder XRD data, as the asymmetric unit often comprises a simple monomer unit. However, it can be difficult to prepare polymers of sufficiently high crystallinity to give high-quality powder XRD data. Poor crystallinity is manifested in peak broadening, which often renders the traditional strategy for structure solution unsuitable due to excessive peak overlap. For this reason, early examples of successful structure determination in this area (*e.g.* poly(*p*-phenylene)terephthalate,[194] poly(hydroxybutyrate)[195] and azomethine block copolymers[196]) invoked molecular modelling to generate energetically and sterically favourable models, which were then refined against experimental powder XRD data. Subsequently, the structures of a number of polymeric complexes based on metal diazolates were solved directly from powder XRD data, including [Cu(pz)]$_n$ and [Ag(pz)]$_n$ (Hpz = pyrazole),[197] [Ag(imz)]$_n$ (Himz = imidazole),[198] [NiX$_2$(pydz)]$_n$ (X = Cl or Br; pydz = pyridazine)[199] and [MX$_2$(bipy)]$_n$ (M = Ni or Cu; X = Cl or Br; bipy = 4,4'-bipyridyl).[200] In these cases, structure solution was carried out using either direct methods or a grid search technique[96] to locate the dominant X-ray scatterer, with the positions of the remaining ligand atoms established from simple packing and symmetry considerations.

Several polymer electrolyte complexes comprising salts dissolved in poly(ethylene oxide) (PEO, a high-molecular-weight polymer with formula $(CH_2CH_2O)_n$) have also been characterised structurally by powder XRD. The crystal structures of the $(PEO)_3:NaClO_4$[201] and $(PEO)_3:LiCF_3SO_3$[202] complexes were determined by Rietveld refinement starting from a structural model based on a known structure. However, in other cases, structure solution was tackled directly from powder XRD data. Thus, the structure of $(PEO)_4:NH_4SCN$[203] was solved using the traditional strategy, whereas the structures of $(PEO)_3:LiN(SO_2CF_3)_2$,[52] $(PEO):NaCF_3SO_3$[204] and $(PEO)_6:LiAsF_6$[205] were solved using a direct-space strategy based on a simulated annealing protocol.

1.10.18   Structure Determination of Pigment Materials

By definition, pigments are coloured solid particles that are insoluble in the medium in which they are applied (*e.g.* in paints, plastics and printing inks). Clearly the colouristic properties of pigment materials depend on both molecular and crystal structure. Most pigments can be prepared only as fine powders (poor solubility often prevents the growth of high-quality single crystals from solution) and thus many pigments have eluded structure determination by single-crystal XRD. Furthermore, for good dispersion and optimisation of other pigment properties, real pigment materials are usually prepared and applied as microcrystalline particles. For these reasons, structural characterisation of pigments falls directly within the scope of powder XRD techniques.

Many organic pigment materials contain heterocyclic chromophores such as phthalocyanines, quinacridones and anthraquinones, which constitute rigid structural units of well-defined geometry. For most structures in these classes, structure determination from powder XRD data has focused on direct-space techniques for structure solution or approaches that combine crystal modelling with analysis of powder XRD data. Examples include 6,13-dichloro-triphendioxazine,[206] 1-aminoanthraquinone[207] and the X-form of metal-free phthalocyanine.[208] A crystal modelling approach has also been used to solve the structure of the perinone pigment 2,5-dihydroxy-benzo[*de*]benzo[4,5]imidazo[2,1-*a*]isoquinolin-7-one,[209] for which the structures obtained were 'screened' by comparison with the experimental powder XRD data. A structure giving good agreement between calculated and experimental powder XRD patterns was found (although it was not actually the lowest-energy structure), followed by Rietveld refinement.

The structures of other pigment materials have been solved from powder XRD data using direct-space techniques, including the red phase of fluorescein.[49] Another example is 1,4-diketo-2,5-di-*t*-butoxycarbonyl-3,6-diphenyl-pyrrolo[3,4-*c*]pyrrole (DPP-Boc), an important derivative of the commercial red pigment 1,4-diketo-3,6-diphenyl-pyrrolo[3,4-*c*] pyrrole (DPP). Powder XRD studies revealed a new polymorph (β polymorph) of DPP-Boc, the structure of which was determined[144] from the powder XRD data using the direct-space strategy followed by Rietveld refinement.

1.11   CONCLUDING REMARKS

As demonstrated in this chapter, powder XRD has wide-ranging applications across many different areas of solid-state chemistry and materials science, encompassing routine identification and characterisation of crystalline phases, *in situ* monitoring of structural changes associated with a wide variety of chemical and physical processes and the complete determination of crystal structures. In particular, rapid advances have been made in recent years in the scope and potential for carrying out crystal structure determination directly from powder XRD data, as illustrated by the examples highlighted in Section 1.10. With continual developments in instrumentation for recording powder XRD data of higher quality and with the ongoing evolution of new methodology and more powerful strategies for carrying out crystal structure determination, we may predict with confidence that the opportunities for determining crystal structures from powder XRD data will continue to expand and will make considerable impact in several areas of structural sciences in the years to come.

REFERENCES

[1] J. P. Glusker and K. N. Trueblood, *Crystal Structure Analysis – A Primer*, Oxford University Press, Oxford, 1985.

[2] J. D. Dunitz, *X-Ray Analysis and the Structures of Organic Molecules*, Verlag Helvetica Chimica Acta, Basel, 1995.

[3] G. M. Sheldrick, *Acta Crystallogr. Sect. A*, **64**, 112 (2008).

[4] H. P. Klug and L. E. Alexander, *X-Ray Diffraction Procedures For Polycrystalline and Amorphous Materials*, John Wiley, New York, 1974.

[5] J. I. Langford and D. Louër, *Rep. Prog. Phys.*, **59**, 131 (1996).

[6] R. Jenkins and R. L. Snyder, *Introduction to X-Ray Powder Diffractometry*, John Wiley, New York, 1996.

[7] J. K. Cockcroft and A. N. Fitch, in *Powder Diffraction: Theory and Practice*, edited by R. E. Dinnebier and S. J. L. Billinge, The Royal Society of Chemistry, 2008, p. 20.

[8] A. K. Cheetham and A. P. Wilkinson, *J. Phys. Chem. Solids*, **52**, 1199 (1991).

[9] A. K. Cheetham and A. P. Wilkinson, *Angew. Chem., Int. Ed. Eng.*, **31**, 1557 (1992).

[10] P. R. Rudolf, *Mater. Chem. Phys.*, **35**, 267 (1993).

[11] K. D. M. Harris, M. Tremayne, P. Lightfoot and P. G. Bruce, *J. Am. Chem. Soc.*, **116**, 3543 (1994).

[12] K. D. M. Harris and M. Tremayne, *Chem. Mater.*, **8**, 2554 (1996).

[13] D. M. Poojary and A. Clearfield, *Acc. Chem. Res.*, **30**, 414 (1997).

[14] A. Meden, *Croat. Chem. Acta*, **71**, 615 (1998).

[15] K. D. M. Harris, M. Tremayne and B. M. Kariuki, *Angew. Chem., Int. Ed.*, **40**, 1626 (2001).

[16] V. V. Chernyshev, *Russ. Chem. Bull.*, **50**, 2273 (2001).

[17] W. I. F. David, K. Shankland, L. B. McCusker and C. Baerlocher eds., *Structure Determination from Powder Diffraction Data*, IUCr/OUP, Oxford, 2002.

[18] K. D. M. Harris, *Cryst. Growth Des.*, **3**, 887 (2003).

[19] K. D. M. Harris and E. Y. Cheung, *Chem. Soc. Rev.*, **33**, 526 (2004).

[20] M. Tremayne, *Philos. Trans. R. Soc. Lond. Ser. A*, **362**, 2691 (2004).

[21] A. Altomare, R. Caliandro, M. Camalli, C. Cuocci, C. Giacovazzo, A. G. G. Moliterni, R. Rizzi, R. Spagna and J. Gonzalez-Platas, *Z. Kristallogr.*, **219**, 833 (2004).

[22] R. Černý, *Croat. Chem. Acta*, **79**, 319 (2006).

[23] H. Tsue, M. Horiguchi, R. Tamura, K. Fujii and H. Uekusa, *J. Synth. Org. Chem. Jpn.*, **65**, 1203 (2007).

[24] W. I. F. David and K. Shankland, *Acta Crystallogr. Sect. A*, **64**, 52 (2008).

[25] K. D. M. Harris, *Mater. Manuf. Processes*, **24**, 293 (2009).

[26] K. D. M. Harris, *Top. Curr. Chem.*, **315**, 133 (2012).

[27] M. L. Peterson, S. L. Morissette, C. McNulty, A. Goldsweig, P. Shaw, M. LeQuesne, J. Monagle, N. Encina, J. Marchionna, A. Johnson, J. Gonzalez-Zugasti, A. V. Lemmo, S. J. Ellis, M. J. Cima and O. Almarsson, *J. Am. Chem. Soc.*, **124**, 10958 (2002).

[28] A. J. Florence, B. Baumgartner, C. Weston, N. Shankland, A. R. Kennedy, K. Shankland and W. I. F. David, *J. Pharm. Sci.*, **92**, 1930 (2003).

[29] R. A. Young ed., *The Rietveld Method*, International Union of Crystallography, Oxford, 1993.

[30] W. I. F. David, *J. Appl. Crystallogr.*, **32**, 654 (1999).

[31] A. Altomare, C. Giacovazzo and A. Moliterni, in *Powder Diffraction: Theory and Practice*, edited by R. E. Dinnebier and S. J. L. Billinge, The Royal Society of Chemistry, 2008, p. 206.

[32] J. W. Visser, *J. Appl. Crystallogr.*, **2**, 89 (1969).

[33] P. E. Werner, L. Eriksson and M. Westdahl, *J. Appl. Crystallogr.*, **18**, 367 (1985).

[34] A. Boultif and D. Louër, *J. Appl. Crystallogr.*, **24**, 987 (1991).

[35] M. A. Neumann, *J. Appl. Crystallogr.*, **36**, 356 (2003).

[36] R. A. Shirley, CRYSFIRE Suite of Programs for Indexing Powder Diffraction Patterns, University of Surrey, 1999.

[37] G. S. Pawley, *J. Appl. Crystallogr.*, **14**, 357 (1981).

[38] A. Le Bail, H. Duroy and J. L. Fourquet, *Mater. Res. Bull.*, **23**, 447 (1988).

[39] A. J. Markvardsen, W. I. F. David, J. C. Johnson and K. Shankland, *Acta Crystallogr. Sect. A*, **57**, 47 (2001).

[40] A. Altomare, R. Caliandro, M. Camalli, C. Cuocci, I. da Silva, C. Giacovazzo, A. G. G. Moliterni and R. Spagna, *J. Appl. Crystallogr.*, **37**, 957 (2004).

[41] G. Cascarano, L. Favia and C. Giacovazzo, *J. Appl. Crystallogr.*, **25**, 310 (1992).

[42] J. Jansen, R. Peschar and H. Schenk, *Z. Kristallogr.*, **206**, 33 (1993).

[43] A. Altomare, G. Cascarano, C. Giacovazzo, A. Guagliardi, M. C. Burla, G. Polidori and M. Camalli, *J. Appl. Crystallogr.*, **27**, 435 (1994).

[44] A. Altomare, M. C. Burla, M. Camalli, B. Carrozzini, G. L. Cascarano, C. Giacovazzo, A. Guagliardi, A. G. G. Moliterni, G. Polidori and R. Rizzi, *J. Appl. Crystallogr.*, **32**, 339 (1999).

[45] A. Altomare, R. Caliandro, M. Camalli, C. Cuocci, C. Giacovazzo, A. G. G. Moliterni and R. Rizzi, *J. Appl. Crystallogr.*, **37**, 1025 (2004).

[46] C. Baerlocher, L. B. McCusker and L. Palatinus, *Z. Kristallogr.*, **222**, 47 (2007).

[47] D. Ramprasad, G. P. Pez, B. H. Toby, T. J. Markley and R. M. Pearlstein, *J. Am. Chem. Soc.*, **117**, 10694 (1995).

[48] B. M. Kariuki, D. M. S. Zin, M. Tremayne and K. D. M. Harris, *Chem. Mater.*, **8**, 565 (1996).

[49] M. Tremayne, B. M. Kariuki and K. D. M. Harris, *Angew. Chem., Int. Ed. Eng.*, **36**, 770 (1997).

[50] C. M. Freeman, A. M. Gorman and J. M. Newsam, *Computer Modelling in Inorganic Crystallography*, Academic Press, San Diego, CA, 1997.

[51] L. Elizabé, B. M. Kariuki, K. D. M. Harris, M. Tremayne, M. Epple and J. M. Thomas, *J. Phys. Chem. B*, **101**, 8827 (1997).

[52] Y. G. Andreev, P. Lightfoot and P. G. Bruce, *J. Appl. Crystallogr.*, **30**, 294 (1997).

[53] W. I. F. David, K. Shankland and N. Shankland, *Chem. Commun.*, **931** (1998).

[54] G. E. Engel, S. Wilke, O. König, K. D. M. Harris and F. J. J. Leusen, *J. Appl. Crystallogr.*, **32**, 1169 (1999).

[55] P. Miao, A. W. Robinson, R. E. Palmer, B. M. Kariuki and K. D. M. Harris, *J. Phys. Chem. B*, **104**, 1285 (2000).

[56] S. Pagola, P. W. Stephens, D. S. Bohle, A. D. Kosar and S. K. Madsen, *Nature*, **404**, 307 (2000).

[57] Y. Tanahashi, H. Nakamura, S. Yamazaki, Y. Kojima, H. Saito, T. Ida and H. Toraya, *Acta Crystallogr. Sect. B*, **57**, 184 (2001).

[58] H. P. Hsu, U. H. E. Hansmann and S. C. Lin, *Phys. Rev. E*, **64**, 056707 (2001).

[59] Y. G. Andreev and P. G. Bruce, *J. Phys.: Condens. Matter*, **13**, 8245 (2001).

[60] E. D. L. Smith, R. B. Hammond, M. J. Jones, K. J. Roberts, J. B. O. Mitchell, S. L. Price, R. K. Harris, D. C. Apperley, J. C. Cherryman and R. Docherty, *J. Phys. Chem. B*, **105**, 5818 (2001).

[61] K. Shankland, L. McBride, W. I. F. David, N. Shankland and G. Steele, *J. Appl. Crystallogr.*, **35**, 443 (2002).

[62] V. Brodski, R. Peschar and H. Schenk, *J. Appl. Crystallogr.*, **36**, 239 (2003).

[63] A. Huq and P. W. Stephens, *J. Pharm. Sci.*, **92**, 244 (2003).

[64] V. Favre-Nicolin and R. Černý, *Z. Kristallogr.*, **219**, 847 (2004).

[65] K. Shankland, A. J. Markvardsen and W. I. F. David, *Z. Kristallogr.*, **219**, 857 (2004).

[66] V. Favre-Nicolin and R. Černý, *Mater. Sci. Forum*, **443**, 35 (2004).

[67] V. Brodski, R. Peschar and H. Schenk, *J. Appl. Crystallogr.*, **38**, 688 (2005).

[68] W. I. F. David, K. Shankland, J. van de Streek, E. Pidcock, W. D. S. Motherwell and J. C. Cole, *J. Appl. Crystallogr.*, **39**, 910 (2006).

[69] A. Altomare, R. Caliandro, C. Cuocci, C. Giacovazzo, A. G. G. Moliterni, R. Rizzi and C. Platteau, *J. Appl. Crystallogr.*, **41**, 56 (2008).

[70] S. Pagola and P. W. Stephens, *J. Appl. Crystallogr.*, **43**, 370 (2010).

[71] K. Shankland, A. J. Markvardsen, C. Rowlatt, N. Shankland and W. I. F. David, *J. Appl. Crystallogr.*, **43**, 401 (2010).

[72] X. D. Deng and C. Dong, *J. Appl. Crystallogr.*, **44**, 230 (2011).

[73] B. M. Kariuki, H. Serrano-González, R. L. Johnston and K. D. M. Harris, *Chem. Phys. Lett.*, **280**, 189 (1997).

[74] K. Shankland, W. I. F. David and T. Csoka, *Z. Kristallogr.*, **212**, 550 (1997).

[75] K. D. M. Harris, R. L. Johnston and B. M. Kariuki, *Acta Crystallogr. Sect. A*, **54**, 632 (1998).

[76] B. M. Kariuki, P. Calcagno, K. D. M. Harris, D. Philp and R. L. Johnston, *Angew. Chem. Int. Ed.*, **38**, 831 (1999).

[77] B. M. Kariuki, K. Psallidas, K. D. M. Harris, R. L. Johnston, R. W. Lancaster, S. E. Staniforth and S. M. Cooper, *Chem. Commun.*, **1677** (1999).

[78] G. W. Turner, E. Tedesco, K. D. M. Harris, R. L. Johnston and B. M. Kariuki, *Chem. Phys. Lett.*, **321**, 183 (2000).

[79] S. Habershon, K. D. M. Harris and R. L. Johnston, *J. Comput. Chem.*, **24**, 1766 (2003).

[80] K. D. M. Harris, R. L. Johnston and S. Habershon, *Struct. Bond.*, **110**, 55 (2004).

[81] E. Dova, R. Peschar, M. Sakata, K. Kato, A. F. Stassen, H. Schenk and J. G. Haasnoot, *Acta Crystallogr. Sect. B*, **60**, 528 (2004).

[82] E. Dova, R. Peschar, M. Sakata, K. Kato and H. Schenk, *Chem. Eur. J.*, **12**, 5043 (2006).

[83] S. Hirano, S. Toyota, F. Toda, K. Fujii and H. Uekuasa, *Angew. Chem., Int. Ed.*, **45**, 6013 (2006).

[84] Z. J. Feng and C. Dong, *J. Appl. Crystallogr.*, **40**, 583 (2007).

[85] A. J. Hanson, E. Y. Cheung and K. D. M. Harris, *J. Phys. Chem. B*, **111**, 6349 (2007).

[86] Z. Zhou, V. Siegler, E. Y. Cheung, S. Habershon, K. D. M. Harris and R. L. Johnston, *ChemPhysChem*, **8**, 650 (2007).

[87] Z. Zhou and K. D. M. Harris, *Phys. Chem. Chem. Phys.*, **10**, 7262 (2008).

[88] K. Fujii, Y. Ashida, H. Uekusa, F. Guo and K. D. M. Harris, *Chem. Commun.*, **46**, 4264 (2010).

[89] K. Fujii, H. Uekusa, N. Itoda, G. Hasegawa, E. Yonemochi, K. Terada, Z. Pan and K. D. M. Harris, *J. Phys. Chem. C*, **114**, 580 (2010).

[90] G. K. Lim, Z. Zhou, K. Fujii, P. Calcagno, E. Tedesco, S. J. Kitchin, B. M. Kariuki, D. Philp and K. D. M. Harris, *Cryst. Growth Des.*, **10**, 3814 (2010).

[91] K. Fujii, M. T. Young and K. D. M. Harris, *J. Struct. Biol.*, **174**, 461 (2011).

[92] E. Courvoisier, P. A. Williams, G. K. Lim, C. E. Hughes and K. D. M. Harris, *Chem. Commun.*, **48**, 2761 (2012).

[93] P. A. Williams, C. E. Hughes, G. K. Lim, B. M. Kariuki and K. D. M. Harris, *Cryst. Growth Des.*, **12**, 3104 (2012).

[94] D. V. Dudenko, P. A. Williams, C. E. Hughes, O. N. Antzutkin, S. P. Velaga, S. P. Brown and K. D. M. Harris, *J. Phys. Chem. C*, **117**, 12258 (2013).

[95] G. Reck, R. G. Kretschmer, L. Kutschabsky and W. Pritzkow, *Acta Crystallogr. Sect. A*, **44**, 417 (1988).

[96] N. Masciocchi, R. Bianchi, P. Cairati, G. Mezza, T. Pilati and A. Sironi, *J. Appl. Crystallogr.*, **27**, 426 (1994).

[97] R. E. Dinnebier, P. W. Stephens, J. K. Carter, A. N. Lommen, P. A. Heiney, A. R. McGhie, L. Brard and A. B. Smith, *J. Appl. Crystallogr.*, **28**, 327 (1995).

[98] R. B. Hammond, K. J. Roberts, R. Docherty and M. Edmondson, *J. Phys. Chem. B*, **101**, 6532 (1997).

[99] V. V. Chernyshev and H. Schenk, *Z. Kristallogr.*, **213**, 1 (1998).

[100] C. C. Seaton and M. Tremayne, *Chem. Commun.*, **880** (2002).

[101] S. Y. Chong and M. Tremayne, *Chem. Commun.*, **4078** (2006).

[102] A. Altomare, N. Corriero, C. Cuocci, A. Moliterni and R. Rizzi, *J. Appl. Crystallogr.*, **46**, 779 (2013).

[103] H. M. Rietveld, *J. Appl. Crystallogr.*, **2**, 65 (1969).

[104] L. B. McCusker, R. B. Von Dreele, D. E. Cox, D. Louër and P. Scardi, *J. Appl. Crystallogr.*, **32**, 36 (1999).

[105] C. R. A. Catlow and G. N. Greaves eds., *Applications of Synchrotron Radiation*, Blackie, Glasgow, 1990.

[106] P. Willmott, *An Introduction to Synchrotron Radiation: Techniques and Applications*, John Wiley, New York, 2011.

[107] J. P. Attfield, *Mater. Sci. Forum*, **228**, 201 (1996).

[108] J. L. Hodeau, V. Favre-Nicolin, S. Bos, H. Renevier, E. Lorenzo and J. F. Berar, *Chem. Rev.*, **101**, 1843 (2001).

[109] H. Palancher, S. Bos, J. F. Bérar, I. Margiolaki and J. L. Hodeau, *Eur. Phys. J.–Spec. Top.*, **208**, 275 (2012).

[110] D. A. Perkins and J. P. Attfield, *J. Chem. Soc. Chem Comm.*, **229** (1991).

[111] J. K. Warner, A. K. Cheetham, D. E. Cox and R. B. Von Dreele, *J. Am. Chem. Soc.*, **114**, 6074 (1992).

[112] E. Y. Cheung, K. D. M. Harris and B. M. Foxman, *Cryst. Growth Des.*, **3**, 705 (2003).

[113] R. E. Dinnebier, F. Olbrich, S. van Smaalen and P. W. Stephens, *Acta Crystallogr. Sect. B*, **53**, 153 (1997).

[114] E. J. Mittemeijer and P. Scardi, *Diffraction Analysis of the Microstructure of Materials*, Springer, 2004.

[115] B. C. Giessen and G. E. Gordon, *Science*, **159**, 973 (1968).

[116] R. J. Nelmes and M. I. McMahon, *J. Synchrotron Radiat.*, **1**, 69 (1994).

[117] Y. W. Fei and Y. B. Wang, in *High-Temperature and High-Pressure Crystal Chemistry*, Vol. **41**, Mineralogical Society of America, Chantilly, VA, 2000, p. 521.

[118] M. I. McMahon, *Top. Curr. Chem.*, **315**, 69 (2012).

[119] J. S. O. Evans and I. R. Evans, *Chem. Soc. Rev.*, **33**, 539 (2004).

[120] R. I. Walton and D. O'Hare, *Chem. Commun.*, **2283** (2000).

[121] N. Pienack and W. Bensch, *Angew. Chem., Int. Ed.*, **50**, 2014 (2011).

[122] T. Friščić, I. Halasz, P. J. Beldon, A. M. Belenguer, F. Adams, S. A. J. Kimber, V. Honkimäki and R. E. Dinnebier, *Nature Chem.*, **5**, 66 (2013).

[123] J. W. Couves, J. M. Thomas, D. Waller, R. H. Jones, A. J. Dent, G. E. Derbyshire and G. N. Greaves, *Nature*, **354**, 465 (1991).

[124] A. J. Dent, M. P. Wells, R. C. Farrow, C. A. Ramsdale, G. E. Derbyshire, G. N. Greaves, J. W. Couves and J. M. Thomas, *Rev. Sci. Instrum.*, **63**, 903 (1992).

[125] G. Sankar, P. A. Wright, S. Natarajan, J. M. Thomas, G. N. Greaves, A. J. Dent, B. R. Dobson, C. A. Ramsdale and R. H. Jones, *J. Phys. Chem.*, **97**, 9550 (1993).

[126] J. M. Thomas, *Chem. Eur. J.*, **3**, 1557 (1997).

[127] G. Sankar, J. M. Thomas and C. R. A. Catlow, *Top. Catal.*, **10**, 255 (2000).

[128] D. Grandjean, A. M. Beale, A. V. Petukhov and B. M. Weckhuysen, *J. Am. Chem. Soc.*, **127**, 14454 (2005).

[129] E. Boccaleri, F. Carniato, G. Croce, D. Viterbo, W. van Beek, H. Emerich and M. Milanesio, *J. Appl. Crystallogr.*, **40**, 684 (2007).

[130] V. Prevot, V. Briois, J. Cellier, C. Forano and F. Leroux, *J. Phys. Chem. Solids*, **69**, 1091 (2008).

[131] T. Hashida, K. Tashiro, K. Ito, M. Takata, S. Sasaki and H. Masunaga, *Macromol.*, **43**, 402 (2010).

[132] Y. Fukuyama, N. Yasuda, H. Kamioka, J. Kim, T. Shibata, H. Osawa, T. Nakagawa, H. Murayama, K. Kato, Y. Tanaka, S. Kimura, T. Ohshima, H. Tanaka, M. Takata and Y. Moritomo, *Appl. Phys. Express*, **3**, 3 (2010).

[133] D. Giron, *J. Therm. Anal. Calorim.*, **68**, 335 (2002).

[134] A. Kishi and H. Toraya, *Powder Diffr.*, **19**, 31 (2004).

[135] Y. Nishimoto, Y. Kaneki and A. Kishi, *Anal. Sci.*, **20**, 1079 (2004).

[136] C. Weiyu, K. Tashiro, M. Hanesaka, S. Takeda, H. Masunaga, S. Sasaki and M. Takata, *J. Phys. Chem. B*, **113**, 2338 (2009).

[137] E. Y. Cheung, K. Fujii, F. Guo, K. D. M. Harris, S. Hasebe and R. Kuroda, *Cryst. Growth Des.*, **11**, 3313 (2011).

[138] J. Rodríguez-Carvajal, *Physica B*, **192**, 55 (1993).

[139] G. E. Bacon, *Neutron Diffraction*, Clarendon Press, Oxford, 1975.

[140] T. Chatterji, *Neutron Scattering from Magnetic Materials*, Elsevier Science, 2005.

[141] R. J. Harrison, in *Neutron Scattering in Earth Sciences*, Vol. **63**, edited by H. R. Wenk, The Mineralogical Society of America, 2006, p. 113.

[142] D. Albesa-Jové, B. M. Kariuki, S. J. Kitchin, L. Grice, E. Y. Cheung and K. D. M. Harris, *ChemPhysChem*, **5**, 414 (2004).

[143] Z. Pan, M. Xu, E. Y. Cheung, K. D. M. Harris, E. C. Constable and C. E. Housecroft, *J. Phys. Chem. B*, **110**, 11620 (2006).

[144] E. J. MacLean, M. Tremayne, B. M. Kariuki, K. D. M. Harris, A. F. M. Iqbal and Z. Hao, *J. Chem. Soc., Perkin Trans. 2*, 1513 (2000).

[145] Z. Pan, E. Y. Cheung, K. D. M. Harris, E. C. Constable and C. E. Housecroft, *Cryst. Growth Des.*, **5**, 2084 (2005).

[146] S. Meejoo, B. M. Kariuki, S. J. Kitchin, E. Y. Cheung, D. Albesa-Jove and K. D. M. Harris, *Helv. Chim. Acta*, **86**, 1467 (2003).

[147] K. Fujii, A. L. Garay, J. Hill, E. Sbircea, Z. Pan, M. Xu, D. C. Apperley, S. L. James and K. D. M. Harris, *Chem. Commun.*, **46**, 7572 (2010).

[148] J. L. Sun, C. Bonneau, Á. Cantin, A. Corma, M. J. Diaz-Cabañas, M. Moliner, D. L. Zhang, M. R. Li and X. D. Zou, *Nature*, **458**, 1154 (2009).

[149] F. Gramm, C. Baerlocher, L. B. McCusker, S. J. Warrender, P. A. Wright, B. Han, S. B. Hong, Z. Liu, T. Ohsuna and O. Terasaki, *Nature*, **444**, 79 (2006).

[150] C. Baerlocher, F. Gramm, L. Massuger, L. B. McCusker, Z. B. He, S. Hovmoller and X. D. Zou, *Science*, **315**, 1113 (2007).

[151] C. Baerlocher, D. Xie, L. B. McCusker, S. J. Hwang, I. Y. Chan, K. Ong, A. W. Burton and S. I. Zones, *Nature Materials*, **7**, 631 (2008).

[152] D. Xie, L. B. McCusker and C. Baerlocher, *J. Am. Chem. Soc.*, **133**, 20604 (2011).

[153] R. J. Francis, S. O'Brien, A. M. Fogg, P. S. Halasyamani, D. O'Hare, T. Loiseau and G. Férey, *J. Am. Chem. Soc.*, **121**, 1002 (1999).

[154] S. L. James, C. J. Adams, C. Bolm, D. Braga, P. Collier, T. Friščić, F. Grepioni, K. D. M. Harris, G. Hyett, W. Jones, A. Krebs, J. Mack, L. Maini, A. G. Orpen, I. P. Parkin, W. C. Shearouse, J. W. Steed and D. C. Waddell, *Chem. Soc. Rev.*, **41**, 413 (2012).

[155] E. Y. Cheung, S. J. Kitchin, K. D. M. Harris, Y. Imai, N. Tajima and R. Kuroda, *J. Am. Chem. Soc.*, **125**, 14658 (2003).

[156] B. Q. Ma, K. L. Mulfort and J. T. Hupp, *Inorg. Chem.*, **44**, 4912 (2005).

[157] T. Friščić, E. Meštrović, D. Š. Šamec, B. Kaitner and L. Fábián, *Chem. Eur. J.*, **15**, 12644 (2009).

[158] K. Fujii, Y. Ashida, H. Uekusa, S. Hirano, S. Toyota, F. Toda, Z. Pan and K. D. M. Harris, *Cryst. Growth Des.*, **9**, 1201 (2009).

[159] S. Karki, L. Fábián, T. Friščić and W. Jones, *Org. Lett.*, **9**, 3133 (2007).

[160] X. H. Ma, G. K. Lim, K. D. M. Harris, D. C. Apperley, P. N. Horton, M. B. Hursthouse and S. L. James, *Cryst. Growth Des.*, **12**, 5869 (2012).

[161] P. J. Beldon, L. Fábián, R. S. Stein, A. Thirumurugan, A. K. Cheetham and T. Friščić, *Angew. Chem. Int. Ed.*, **49**, 9640 (2010).

[162] J.-P. Zhang, Y.-B. Zhang, J.-B. Lin and X.-M. Chen, *Chem. Rev.*, **112**, 1001 (2012).

[163] J. Bernstein, *Polymorphism in Molecular Crystals*, Oxford Science Publications, 2002.

[164] E. Dudognon, F. Danede, M. Descamps and N. T. Correia, *Pharm. Res.*, **25**, 2853 (2008).

[165] P. Derollez, E. Dudognon, F. Affouard, F. Danede, N. T. Correia and M. Descamps, *Acta Crystallogr. Sect. B*, **66**, 76 (2010).

[166] P. A. Williams, C. E. Hughes and K. D. M. Harris, *Cryst. Growth Des.*, **12**, 5839 (2012).

[167] I. R. Evans, J. A. K. Howard, T. Sreckovic and M. M. Ristic, *Mater. Res. Bull.*, **38**, 1203 (2003).

[168] F. Wöhler, *Pogg. Ann.*, **87**, 253 (1828).

[169] J. D. Dunitz, K. D. M. Harris, R. L. Johnston, B. M. Kariuki, E. J. MacLean, K. Psallidas, W. B. Schweizer and R. R. Tykwinski, *J. Am. Chem. Soc.*, **120**, 13274 (1998).

[170] E. J. MacLean, K. D. M. Harris, B. M. Kariuki, S. J. Kitchin, R. R. Tykwinski, I. P. Swainson and J. D. Dunitz, *J. Am. Chem. Soc.*, **125**, 14449 (2003).

[171] F. Guo, M. Casadesus, E. Y. Cheung, M. P. Coogan and K. D. M. Harris, *Chem. Commun.*, 1854 (2006).

[172] G. Bähr, in *FIAT Review of WWII German Science, 1939–1946. Inorganic Chemistry Part II*, edited by W. Klemm, Dieterichsche Verlagsbuchhandlung, Wiesbaden, 1948, p. 155.

[173] R. Dinnebier and J. Müller, *Inorg. Chem.*, 42, 1204 (2003).

[174] L. V. Interrante, G. A. Sigel, M. Garbauskas, C. Hejna and G. A. Slack, *Inorg. Chem.*, 28, 252 (1989).

[175] M. Edgar, V. J. Carter, D. P. Tunstall, P. Grewal, V. Favre-Nicolin, P. A. Cox, P. Lightfoot and P. A. Wright, *Chem. Commun.*, 808 (2002).

[176] F. Guo and K. D. M. Harris, *J. Am. Chem. Soc.*, 127, 7314 (2005).

[177] D. Albesa-Jové, Z. Pan, K. D. M. Harris and H. Uekusa, *Cryst. Growth Des.*, 8, 3641 (2008).

[178] B. Stöger, M. Weil, E. J. Baran, A. C. González-Baró, S. Malo, J. M. Rueff, S. Petit, M. B. Lepetit, B. Raveau and N. Barrier, *Dalton Trans.*, 40, 5538 (2011).

[179] H. Nakanishi, N. Nakano and M. Hasegawa, *J. Polym. Sci. B: Polym. Lett.*, 8, 755 (1970).

[180] Y. Sasada, H. Shimanouchi, H. Nakanishi and M. Hasegawa, *Bull. Chem. Soc. Jpn.*, 44, 1262 (1971).

[181] F. Guo, J. Martí-Rujas, Z. Pan, C. E. Hughes and K. D. M. Harris, *J. Phys. Chem. C*, 112, 19793 (2008).

[182] G. M. J. Schmidt, *Pure Appl. Chem.*, 27, 647 (1971).

[183] M. D. Cohen and G. M. J. Schmidt, *J. Chem. Soc.*, 1996 (1964).

[184] J. M. Thomas, *Philos. Trans. R. Soc. A*, 277, 251 (1974).

[185] J. M. Thomas, *Nature*, 289, 633 (1981).

[186] M. Brunelli, J. P. Wright, G. R. M. Vaughan, A. J. Mora and A. N. Fitch, *Angew. Chem. Int. Ed.*, 42, 2029 (2003).

[187] W. H. Zachariasen and F. H. Ellinger, *Acta Crystallogr.*, 16, 369 (1963).

[188] K. Shankland, W. I. F. David and D. S. Sivia, *J. Mater. Chem.*, 7, 569 (1997).

[189] R. Hoffmann, *Liebigs Ann. Chem.*, 102, 1 (1857).

[190] A. Kekulé, *Liebigs Ann. Chem.*, 105, 288 (1858).

[191] H. Ehrenberg, B. Hasse, K. Schwarz and M. Epple, *Acta Crystallogr. Sect. B*, 55, 517 (1999).

[192] N. Masciocchi and A. Sironi, *J. Chem. Soc., Dalton Trans.*, 4643 (1997).

[193] T. Straasø, S. Kapishnikov, K. Kato, M. Takata, J. Als-Nielsen and L. Leiserowitz, *Cryst. Growth Des.*, 11, 3342 (2011).

[194] S. Hanna, P. D. Coulter and A. H. Windle, *J. Chem. Soc., Faraday Trans.*, 91, 2615 (1995).

[195] R. J. Pazur, P. J. Hocking, S. Raymond and R. H. Marchessault, *Macromol.*, 31, 6585 (1998).

[196] S. Bruckner, S. Destri and W. Porzio, *Macromol. Rapid Commun.*, 16, 297 (1995).

[197] N. Masciocchi, M. Moret, P. Cairati, A. Sironi, G. A. Ardizzoia and G. Lamonica, *J. Am. Chem. Soc.*, 116, 7668 (1994).

[198] N. Masciocchi, M. Moret, P. Cairati, A. Sironi, G. A. Ardizzoia and G. Lamonica, *J. Chem. Soc., Dalton Trans.*, 1671 (1995).

[199] N. Masciocchi, P. Cairati, L. Carlucci, G. Ciani, G. Mezza and A. Sironi, *J. Chem. Soc., Dalton Trans.*, 3009 (1994).

[200] N. Masciocchi, P. Cairati, L. Carlucci, G. Mezza, G. Ciani and A. Sironi, *J. Chem. Soc., Dalton Trans.*, 2739 (1996).

[201] P. Lightfoot, M. A. Mehta and P. G. Bruce, *J. Mater. Chem.*, **2**, 379 (1992).

[202] P. Lightfoot, M. A. Mehta and P. G. Bruce, *Science*, **262**, 883 (1993).

[203] P. Lightfoot, J. L. Nowinski and P. G. Bruce, *J. Am. Chem. Soc.*, **116**, 7469 (1994).

[204] Y. G. Andreev, G. S. MacGlashan and P. G. Bruce, *Phys. Rev. B*, **55**, 12011 (1997).

[205] G. S. MacGlashan, Y. G. Andreev and P. G. Bruce, *Nature*, **398**, 792 (1999).

[206] P. G. Fagan, R. B. Hammond, K. J. Roberts, R. Docherty, A. P. Chorlton, W. Jones and G. D. Potts, *Chem. Mater.*, **7**, 2322 (1995).

[207] A. V. Yatsenko, V. V. Chernyshev, L. A. Aslanov and H. Schenk, *Powder Diffr.*, **13**, 85 (1998).

[208] R. B. Hammond, K. J. Roberts, R. Docherty, M. Edmondson and R. Gairns, *J. Chem. Soc., Perkin Trans. 2*, 1527 (1996).

[209] M. U. Schmidt and R. E. Dinnebier, *J. Appl. Crystallogr.*, **32**, 178 (1999).

[210] R. I. Walton, T. Loiseau, R. J. Francis, D. O'Hare and G. Férey, *Mat. Res. Soc. Symp. Proc.*, **547**, 63 (1999).

2

X-Ray and Neutron Single-Crystal Diffraction

William Clegg

School of Chemistry, Newcastle University, Newcastle upon Tyne, UK

2.1 INTRODUCTION

When it can be achieved and the results are definitive and unambiguous – which is less often the case for inorganic materials than for discrete molecular compounds – a crystal structure obtained by diffraction methods from a single crystal usually provides a detailed, precise and well-defined model for the location of the constituent atoms and/or ions relative to one another and to the symmetry elements of a crystalline solid material. As well as atomic positions, displacement parameters describing vibrational motion are also obtained, leading to the possibility of generating extensive geometrical (distances, angles, conformations *etc.*) and limited dynamic information on the sample under study. This powerful family of techniques, based on and characterising long-range structural order, complements information available from spectroscopic, computational and other scattering methods and forms an essential component of the experimental and theoretical approaches to structure available to materials scientists and others in the physical and biological sciences.

It is almost exactly a century since the first demonstration, by von Laue, Friedrich and Knipping in 1912, that X-rays (themselves

Structure from Diffraction Methods, First Edition. Edited by Duncan W. Bruce, Dermot O'Hare and Richard I. Walton.
© 2014 John Wiley & Sons, Ltd. Published 2014 by John Wiley & Sons, Ltd.

discovered by Röntgen only 17 years earlier in 1895) could be diffracted by crystalline solids, followed rapidly by the first practical use of the phenomenon in determining a crystal structure – that of the archetypal inorganic material sodium chloride – by the Bragg father-and-son team in 1913. The centenaries of these discoveries and the consequent award of Physics Nobel Prizes in 1914 and 1915 are being celebrated by crystallographers and others, with 2014 being declared the International Year of Crystallography by the United Nations General Assembly. Although crystallography is actually a much broader spectrum of sciences than this, and the study of crystals long predates the discoveries of the early 20th century, the application of diffraction techniques now dominates the field.

Changes over the last 100 years have been remarkable, in theory, experiment, computation and application. Advances in fundamental understanding have provided a firm basis for the development of powerful methods for solving and refining crystal structures from diffraction data. Technological achievements have included more powerful and more reliable radiation sources and faster and more accurate devices for recording diffraction patterns (see Figure 2.1). Crystallographers have been at the forefront of the exploitation of massive computing improvements and the subject is ideal for the application of modern graphics and database systems. Each generation sees major steps forward. While arguably the biggest impact overall has been in the biomedical field, with massive investment in macromolecular crystallography and its contributions to genomics and drug design, materials physics and chemistry have by no means failed to reap their own benefits.

This chapter cannot hope to give a full account of single-crystal diffraction techniques or address in any detail their historical development, fundamental theory or practical implementation in general; such a task is for large-scale books and intensive courses. The intention is rather to outline the main basic principles and their applications, with specific reference to the structural characterisation of inorganic materials, largely but not exclusively non-molecule in nature. Such compounds present their own particular challenges, which differ from those of molecular organic and inorganic compounds and from those of biological macromolecules. They will be a major focus here. For a more general treatment, background material and further details, readers are referred to standard crystallography texts, of which the following (certainly not an exhaustive list) may be mentioned as particularly useful in various respects:

- J. P. Glusker and K. N. Trueblood, *Crystal Structure Analysis – A Primer*, 3rd edition, Oxford University Press, 2010. Aimed primarily

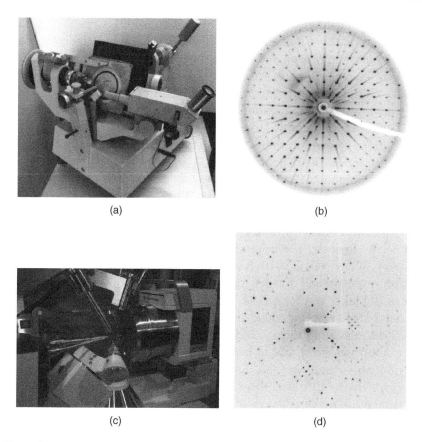

(a)

(b)

(c)

(d)

Figure 2.1 (a) X-ray precession camera from around 1970. (b) Typical precession photograph of an aligned crystal using filtered radiation, which gives rise to background streaks, noticeable particularly on the strongest reflections. This requires several hours of X-ray exposure. (c) Modern diffractometer with a charge-coupled device (CCD)-based area detector and dual-wavelength X-ray sources. (d) Typical small-range image (2D slice) of a resulting X-ray diffraction pattern taken with a few seconds of exposure.

at students of biology, with a main focus on molecular and macromolecular structures.

- J. P. Glusker, M. Lewis and M. Rossi, *Crystal Structure Analysis for Chemists and Biologists*, Wiley-VCH, 1994. Essentially an expanded version of an earlier edition of the above text.
- W. Clegg, *Crystal Structure Determination*, Oxford University Press, 1988. Written as an introductory text for chemistry undergraduates, minimising mathematical content. A new edition, probably somewhat expanded, is being written for publication in 2014.

- W. Massa (English translation by R. O. Gould), *Crystal Structure Determination*, 2nd edition, Springer, 2011. Originally in German. Similar in approach and level to the first and third books in the list.
- C. Giacovazzo, H. L. Monaco, G. Artioli, D. Viterbo, G. Ferraris, G. Gilli, G. Zanotti and M. Catti, edited by C. Giacovazzo, *Fundamentals of Crystallography*, 2nd edition, Oxford University Press, 2002. A considerably larger book, with somewhat disparate contributions by many authors. Covers a wide range of topics and includes some serious mathematics.
- C. Hammond, *The Basics of Crystallography and Diffraction*, 3rd edition, Oxford University Press, 2009. Focuses mainly on symmetry, crystallographic geometry and the theory and experimental procedures of X-ray and electron diffraction.
- D. W. Bennett, *Understanding Single-Crystal X-Ray Crystallography*, Wiley-VCH, 2010. A substantial and thorough treatment with an extensive and detailed mathematical foundation and derivations.
- A. J. Blake, W. Clegg, J. M. Cole, J. S. O. Evans, P. Main, S. Parsons and D. J. Watkin, edited by W. Clegg, *Crystal Structure Analysis, Principles and Practice*, 2nd edition, Oxford University Press, 2009. Comprises most of the lecture-presentation contents of the 2007 British Crystallographic Association (biennial) intensive course on crystal structure analysis, together with extensive tutorial exercises and a full set of answers.

2.2   SOLID-STATE FUNDAMENTALS

Crystalline solids are characterised by long-range internal order, described in terms of translation symmetry in all cases, as well as various possible combinations of rotation, reflection and inversion symmetry. Perfect order means zero entropy and is unachievable in practice; deviations from it are responsible for many of the real-life challenges in crystallography as an experimental science and analytical technique, as well as for a wide range of important physical and chemical properties of solid-state materials. In this section we set out the fundamental principles of crystallographic symmetry in the ideal case. We then outline briefly some of the chief deviations from ideal behaviour, leaving important details to later sections where they are of particular relevance.

2.2.1 Translation Symmetry

Solid inorganic materials with no rotation, reflection or inversion symmetry are rare. For illustration, we use here a 1 : 1 mixed-valence adduct of two rhenium-centred molecules, $[ReOCl_4] \cdot [ReO_3Cl]$.[1] In the absence of symmetry other than translation, the basic repeat unit of the crystal structure is one of each molecule. All $[ReOCl_4]$ molecules are identical and all $[ReO_3Cl]$ molecules are identical, not only in shape but also in orientation; they are arranged in a regular space-filling pattern repeating in three dimensions, which generates a complete single crystal. Even for a tiny crystal with micron-sized dimensions, there will be thousands of typical repeat units in each direction, so the pattern is effectively infinite on the repeat unit scale. Figure 2.2a shows a two-dimensional (2D) projection of this arrangement, in which a weak $Re \cdots O$ bonding interaction between the two chemically different

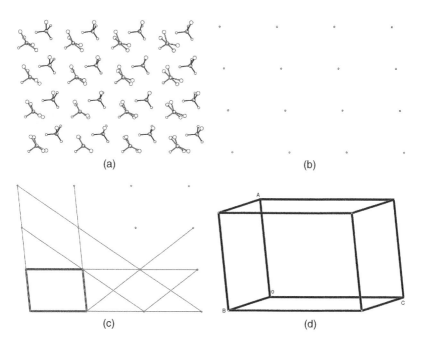

Figure 2.2 (a) 2D perspective projection of an inorganic crystal structure, showing 16 repeat units. (b) 2D lattice for this structure. (c) Pairs of parallel lattice lines, with the conventional 2D unit cell highlighted. (d) 3D unit cell for this structure, with the origin and unit cell axes labelled.

molecules is not included. Note that each molecule is surrounded by a number of close neighbours, and the selection of one particular pair of molecules to represent the repeat unit would be arbitrary in the absence of the weak interaction between them; such is the case for most ionic materials, for example.

In order to show the pure translation symmetry of this structure and provide a numerical representation of its repeat geometry, each pair of molecules (each complete repeat unit) is replaced by a single point, the points being chosen in an identical location in each unit, *e.g.* the Re atom of $[ReOCl_4]$. The points chosen are entirely arbitrary and the result is the same whichever are selected: a regularly spaced array of points that reveals the underlying repeat structure of the material but not the detailed contents of the repeat unit. Figure 2.2b is the 2D array of points corresponding to the pattern of Figure 2.2a; for the full crystal structure, of course, the array is 3D, with identical parallel layers of points occurring at regular intervals above and below the plane of Figure 2.2b. This array of identical points, all entirely equivalent due to the pure translation symmetry of the structure, is called the **lattice** of the structure. Note that this is the strictly correct, mathematical meaning of the word 'lattice'; it is often used, incorrectly, as a synonym for 'crystal structure' and, particularly unfortunately and inappropriately, in expressions such as 'lattice solvent', referring to solvent molecules located in spaces within a solid structure but not bonded covalently to the structure's main framework.

Sets of equally spaced, parallel planes can be drawn passing through lattice points. For the lattice of Figure 2.2b, one such plane is the page itself, and parallel planes contain each of the multitude of layers above and below this; other planes inclined to the page cut it in families of parallel straight lines, some of which are shown in Figure 2.2c (two lines from each family). If we choose three sets of planes, they divide the 3D lattice into identical parallelepipeds, shown in the 2D projection of Figure 2.2c as parallelograms. Provided that the planes are chosen so that each parallelepiped has a lattice point at every corner and no lattice points on its edges or faces or within its enclosed space, any one of these parallelepipeds can provide a possible description of the basic repeat unit of the structure, since the complete structure is formed by simple regular stacking of the parallelepipeds in three dimensions. Any one representative parallelepiped is called the **unit cell** of the structure. The chemical components of the structure usually lie across the faces and edges of unit cells rather than being contained neatly and precisely within them, so each unit cell (geometrical repeat unit) contains several fragments of different but identical components (chemical repeat units). The net contents of one unit cell amount to one chemical repeat unit – one $[ReOCl_4]$ and one [ReO$_3$Cl] molecule in this example – so the volume of the unit cell is the same for any appropriate and valid choice of three sets of defining planes.

Of course, sets of planes and the resulting unit cells may be chosen in many different ways, but there are standard conventions that depend on the presence of any non-translation symmetry in the structure (see later) and, subject to those symmetry conventions being satisfied, lead to a parallelepiped that is as near to orthogonal as possible, with the angles between its edges as close as possible to 90°. Once a unit cell has been chosen, its geometry, and hence the geometry of the lattice, is described completely by a set of six parameters: three lengths of non-parallel edges (unit cell edges) and the three angles formed by pairs of these edges meeting at one corner. Again, there are conventions regarding the choice of which corner of the unit cell to use to measure these angles (this corner is taken as the origin for the three axes and the lattice overall), the most widely used being that all three angles are either < 90° (a 'type 1 reduced cell') or ≥ 90° ('type 2'), with no mixture of acute and obtuse angles. In the absence of any non-translation symmetry, there are no restrictions on the values of the six unit cell parameters, but rotation and reflection symmetry in the structure do impose restrictions of equal and/or special values on axis lengths/angles. The conventional unit cell of the example structure is shown with thicker lines in the 2D projection of Figure 2.2c. The 3D unit cell of this structure has the following parameters: $a = 5.78\,\text{Å}$, $b = 6.02\,\text{Å}$, $c = 8.08\,\text{Å}$, $\alpha = 114.8°$, $\beta = 96.2°$, $\gamma = 95.0°$, with a unit cell volume $V = 251.0\,\text{Å}^3$; it is shown in Figure 2.2d.

Once a unit cell has been chosen to characterise the lattice geometry, every set of parallel lattice planes can be assigned a set of three **indices**: integers that specify the orientation of the planes relative to the unit cell axes. The definition of the indices for a set of planes may be expressed in different ways, one of which is to identify the plane passing through the unit cell origin and count how many interplanar spaces are crossed in going along each of the three unit cell axes in turn (a, b and c); this number may be zero (if the axis lies in the 'home' plane), positive or negative (look at the first plane out from the origin and note whether it intersects each axis in the + or − direction). The three sets of planes defining the unit cell faces are thus (100), (010) and (001). The three indices are conventionally given the symbols h, k and l. Other examples are (101), ($1\bar{1}1$), (312) *etc*; note the crystallographer's habit of writing minus signs above rather than in front of numbers and symbols (this is convenient and compact, and avoids ambiguities in some situations). By extension of this notation, with relevance later to the understanding of diffraction patterns, further planes can be interleaved between those defined in this way, to generate closer- (but still regularly) spaced families, some members of which pass only between and not through lattice points. Thus *e.g.* the planes in the (200) set have half the spacing of those in the (100) but the same orientation, and sets (300), (400), (500) *etc.* can be constructed in a similar way. Likewise (111), (222), (333) *etc.*

The spacing between planes in a parallel set can be calculated; it is a function only of the six unit cell parameters and the three indices h, k and l. In the general case, and using simple scalar algebra, the expression is cumbersome, involving sines and cosines of the cell angles. Vector notation is simpler and more compact. If the three unit cell axes are represented by vectors a, b, c then the unit cell volume is the scalar triple product:

$$V = a \cdot b \times c = b \cdot c \times a = c \cdot a \times b \qquad (2.1)$$

In order to generate a simple expression for interplanar spacings, and for important reasons that will be made clear when we look later at the equations that govern diffraction, we define a second lattice, called the **reciprocal lattice**, which is derived from the crystal lattice (or **direct lattice**) through the following vector relationships:

$$a^* = b \times c / V \qquad b^* = c \times a / V \qquad c^* = a \times b / V \qquad (2.2)$$

These relationships are equally valid in the reverse direction, from reciprocal to direct lattice, if all starred symbols are replaced by unstarred ones and *vice versa* (including the direct and reciprocal volumes, V and V^*). Note that the reciprocal lattice axes have units of Å^{-1} if the direct lattice is measured in Å; hence the name 'reciprocal lattice'. Using the reciprocal lattice, we now find, for a set of direct lattice planes with indices hkl:

$$d^*(hkl) = ha^* + kb^* + lc^* \qquad (2.3)$$

where d^* is a reciprocal lattice vector whose direction is the normal to the (hkl) planes, and the interplanar spacing $d(hkl)$ is the reciprocal of the length of the d^* vector:

$$d(hkl) = 1/|d^*(hkl)| \qquad (2.4)$$

From the definition, *directions* in the reciprocal lattice correspond to *positions* in the direct lattice and *vice versa*. It should also be noted that the following relationships follow directly from the definition of the reciprocal lattice:

$$a \cdot a^* = b \cdot b^* = c \cdot c^* = 1$$

$$a \cdot b^* = a \cdot c^* = b \cdot a^* = b \cdot c^* = c \cdot a^* = c \cdot b^* = 0 \qquad (2.5)$$

We will see later that lattice planes are an important concept in the theory and application of diffraction. They are also parallel to the planar faces observed in well-developed single crystals.

2.2.2   Other Symmetry

While pure translation symmetry is necessarily present in crystalline solids in the form of the crystal lattice, other symmetry may also occur, and in most solid materials it does so. We can distinguish types of symmetry that are familiar from discrete objects such as molecules, and further types of symmetry that combine familiar aspects with translation components in solids.

2.2.2.1   Rotation, Reflection and Inversion

Symmetry in individual objects such as molecules is described in many standard textbooks of chemistry and physics, and the application of mathematical group theory to molecular symmetry is an essential part of the understanding of spectroscopy, the physical properties of materials and some kinds of reactivity. Symmetry elements and operations are broadly divided into two types, referred to as **proper rotations** and **improper rotations**. Proper rotations are pure rotations about a particular axis, whereby an object presents two or more identical appearances in a complete 360° revolution; we thus have twofold, threefold *etc.* rotation axes, with conventional Schoenflies notation C_2, C_3, ... up to C_∞ for an object having cylindrical symmetry, such as a completely linear molecule. Improper rotations combine a rotation by a fraction of a complete revolution about an axis with *either* simultaneous reflection across a plane perpendicular to this axis *or* inversion through a point in the axis; there are two different definitions, using different symbols, but they describe exactly the same symmetry properties; one is preferred by spectroscopists and the other by crystallographers, for good reasons relating to specific features of these respective disciplines. A fourfold improper axis is written either S_4 or $\bar{4}$. $S_6 \equiv \bar{3}$ and is the symmetry about a line joining two opposite faces of a regular octahedron; $S_3 \equiv \bar{6}$ and is actually the same as having a C_3 (3) axis and a perpendicular mirror plane present at the same time. Two special cases of improper rotation are S_1 (also written σ) $\equiv \bar{2}$ (also written m), which is a pure reflection or mirror plane, and S_2 (also written $i) \equiv \overline{1}$, which is a pure inversion or 'centre of symmetry'. No other cases concern us here, because rotations, both proper and improper, with orders other than 1, 2, 3, 4 or 6, are incompatible with a crystal lattice and so are not found in true crystals (other symmetry, notably fivefold, does occur in quasicrystals, which do not have true long-range order).

2.2.2.2 *Rotation and Reflection Symmetry Elements with Translation Components*

It is a fundamental property of point-group symmetry (the symmetry of an individual object, in which all symmetry elements pass through at least one common point) that the repeated operation of any given symmetry element eventually restores the object, after a number of identical appearances, to precisely its original condition and orientation. While pure rotation (proper or improper), reflection and inversion work the same way in a crystalline solid, pure translation does not: it shifts the entire structure one or more unit cells along one or more cell axis directions (*i.e.* by a vector between any two lattice points). It is possible to combine either pure proper rotation or reflection with a translation component that is an appropriate *fraction* of a lattice vector in order to generate a type of symmetry element that is not possible with a discrete single object, but which in a crystalline solid produces the net effect of a pure lattice translation after a number of repeated operations. The combination of a rotation of order n (2; 3; 4; or 6) with translation along the rotation axis (by $1/2; \pm 1/3; \pm 1/4$ or $1/2$; or $\pm 1/6, \pm 1/3$ or $1/2$, respectively) produces what is known as a 'screw axis' because of its similarity to the operation of a screw thread; notation such as $2_1, 3_2, 6_2, 4_1$ is used, where the subscript indicates the translation (here $1/2, 2/3 \equiv -1/3, 2/6 = 1/3, 1/4$) as a fraction of the lattice vector in the direction of the rotation axis. Similarly, reflection may be combined with a translation (usually half a lattice vector, though there are some special cases of $1/4$) that occurs parallel to the reflection plane, either in a cell axis direction or along a diagonal, *i.e.* two axes at once. Conventional symbols here are the letters a, b, c, denoting the direction of translation or **glide**, with n for a diagonal direction and d for the special cases where the glide is $1/4$. Such a reflection with translation is called a **glide reflection** or **glide plane**. An example of a structure[2] in which normal mirror and glide reflections occur alternately is shown in Figure 2.3.

It can thus be seen that, while lattice fundamentals restrict rotation axis orders to just a few possible values, the potential to incorporate

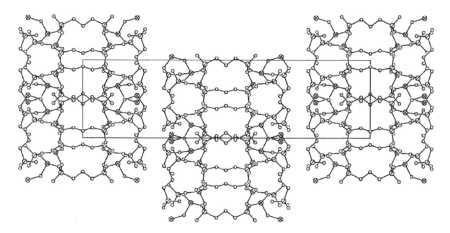

Figure 2.3 Part of a layer structure of a polymeric gallium phosphonate. H atoms are omitted. The unit cell is outlined. Mirror planes, seen edge-on, run vertically through the layers, while glide planes run vertically between the layers.

translation components means that crystal symmetry has a wealth of extra possibilities.

2.2.2.3 Consequences of Non-Translation Symmetry for Lattice Geometry

If symmetry is present in a crystal structure, it describes the geometrical relationship between components of the structure and applies to the arrangement of the components on the lattice. While inversion symmetry imposes no geometrical restrictions on a lattice (all lattices, ignoring the details of the structural motif repeated at each lattice point, have inversion symmetry anyway, since any lattice vector t is equivalent to its inverse $-t$), any rotation or reflection symmetry, including screw axes and glide planes, forces a lattice to adopt a particular form that displays that symmetry. Some examples, shown in 2D projections, are given in Figure 2.4; thus, for instance, mirror symmetry imposes orthogonality on some of the unit cell axes, so that two (or all three) of the cell angles must be exactly 90°, while fourfold rotation symmetry makes all three axes mutually perpendicular and two of them equal in length, the unit cell in this case being necessarily a square-based prism. As with point-group symmetry of individual molecules, symmetry elements in the solid state can be combined only in certain ways, not in arbitrary relative orientations. One result is that there is a small number of types of unit cell

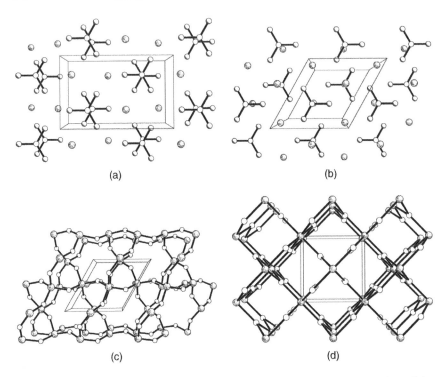

Figure 2.4 Crystal structures of the minerals (a) aragonite $CaCO_3$, (b) calcite $CaCO_3$, (c) quartz SiO_2 and (d) rutile TiO_2, showing the consequences of mirror, threefold, sixfold and fourfold rotation symmetry, respectively, on the unit cell shape.

shape that display different degrees of symmetry. These so-called **crystal systems**, of which there are seven, are listed in Table 2.1, together with their characteristic symmetries and the symmetry-imposed restrictions on their unit cell parameters. Note that a unique axis of highest symmetry, if present, is taken along c, except in the monoclinic system, where it is b according to the most widely used convention.

The importance of these characteristic unit cell shapes as indicators of the symmetry that is probably present (though accidental equalities, near orthogonalities and other apparently symmetrical shapes do sometimes occur in the absence of the corresponding expected underlying symmetry) is so great that the appropriate cell shape is usually chosen to describe a lattice even when a smaller cell would be possible and is suggested by the conventions so far discussed. As a 2D illustration (and the treatment is easily extended to 3D), consider the lattice for the structure of Figure 2.3, which is shown in Figure 2.5. The conventional unit cell would appear to be that shown at the top left, while an alternative

Table 2.1   The seven crystal systems. For the essential symmetry, each type of rotation axis (given as a number indicating the order of the axis) is generic; it can be a proper or improper rotation or a screw axis, and mirrors can also be glide planes. The centred cell types shown in parentheses can be converted into standard types (not in parentheses) by a different choice of axes. They are used in some cases in order to satisfy other conventions or conveniences regarding symmetry and geometry.

Crystal system	Essential symmetry	Unit cell restrictions	Cell types
Triclinic	None	None	P
Monoclinic	2 and/or m for one axis	$\alpha = \gamma = 90°$	P, C (I)
Orthorhombic	2 and/or m for three axes	$\alpha = \beta = \gamma = 90°$	P, C (A), I, F
Tetragonal	4 for one axis	$a = b$; $\alpha = \beta = \gamma = 90°$	P, I
Trigonal	3 for one axis	$a = b$; $\alpha = \beta = 90°$, $\gamma = 120°$	P (R)
Hexagonal	6 for one axis	$a = b$; $\alpha = \beta = 90°$, $\gamma = 120°$	P
Cubic	3 for four directions	$a = b = c$; $\alpha = \beta = \gamma = 90°$	P, I, F

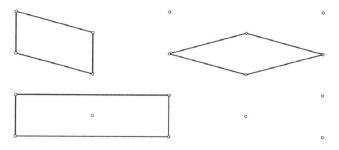

Figure 2.5   Lattice (in 2D projection) of the structure in Figure 2.3, showing two possible primitive unit cells and the conventional centred rectangular unit cell.

with angles further from 90° is to the right of this; the latter is a rhombus with two equal sides, $a = b$, enclosing an angle γ, which has no special value. In this case, the equality of the unit cell axes is due to the presence of mirror and glide symmetry in the structure. The characteristic marker of mirror symmetry for a 2D pattern is a rectangular unit cell ($a \neq b, \gamma = 90°$), not a rhombus. Such a unit cell can indeed be selected, with axes running horizontally and vertically in the diagram, as shown at the bottom, but this unit cell has lattice points not only at its corners but also exactly at its centre, and its area (the 2D equivalent of 3D volume) is twice as large as that of the original **primitive** cell; it thus contains not one but two fundamental repeat units of the structure.

In 3D, similar situations arise for some, but not all, of the crystal systems, generating a number of possibilities of **centred** unit cells as well as primitive ones. The type of unit cell centring (sometimes incorrectly referred to as 'lattice centring', but the lattice, as a fundamental property of the structure itself, remains constant no matter what arbitrary choice of unit cell is made!) is designated by a capital letter: P for primitive, one of A, B, C for lattice points at the centres of pairs of opposite cell faces (double volume), F for all of these at once (quadruple volume), I for a lattice point at the body centre of the cell (double volume) and R for a special case in which some kinds of trigonal structure (**rhombohedral**, for which the primitive unit cell is a symmetrical rhomboid with $a = b = c, \alpha = \beta = \gamma \neq 90°$, which may be thought of as a cube compressed or extended along one of the four body diagonals) are described on a unit cell of conventional trigonal/hexagonal shape (triple volume). Some types of centring are inconsistent with some of the crystal systems (*e.g.* cubic A) and some are never needed (*e.g.* triclinic centred of any kind). The full set of primitive and centred unit cells (which are known, illogically, as 'Bravais lattices') is included as a column in Table 2.1.

In summary, translation symmetry is always present in crystalline solids, and is described by the lattice and measured by unit cell geometry. Other symmetry may also be present, and leads to the classification of the seven crystal systems along with conventional unit cell shapes and the concept of centred as well as primitive unit cells. The total number of different ways in which symmetry can be combined in crystal structures is large but not infinite; it is exactly 230, and these different symmetry arrangements are called **space groups**, comparable to point groups for individual discrete objects, because the symmetry elements occur in regular parallel sets within the lattice, distributed throughout the space occupied by the crystal. Extensive sets of the properties of all 230 3D space groups, together with other important aspects of crystallographic symmetry, are available in Volume A of *International Tables for Crystallography*[3] (all volumes of these tables are also available online to subscribing institutions and individuals; see http://it.iucr.org/). It should be noted that the various space groups are by no means represented equally among known crystal structures; the distribution of observed space groups is quite different in materials of different kinds. Around a dozen space groups account for 90% of all known organic, organometallic and other organic-containing compounds in the Cambridge Structural Database (CSD) of about 650 000 structures.[4,5] Protein crystals are necessarily restricted to the 65 space groups that feature only rotation and translation symmetry, because they contain molecules of only one enantiomeric form. The incidence of space groups of the higher-symmetry crystal systems is far greater for inorganic materials, as can be seen from a survey of the contents of the Inorganic Crystal Structure Database (ICSD).[6,7]

In principle, the unit cell origin may be placed anywhere in a crystal structure. It is, however, usual (and it leads to considerable simplification of mathematical equations relating to diffraction) for the origin, and all other equivalent lattice points, to be located on symmetry elements, particularly on inversion centres where these are present. While inorganic materials of high symmetry often have atoms/ions lying on non-translation symmetry elements (these are called **special positions**), including the unit cell origin, this is not always the case; in some instances, as for most molecular crystal structures, most or all of the atoms lie in **general positions** of no point symmetry, and the symmetry elements of the structure are located between molecules, relating them to one another.

In the triclinic space group $P1$, which has no symmetry other than pure translation, the repeat unit of the structure is one complete unit cell. In all other space groups, only a fraction of the unit cell contents is unique and requires experimental determination; symmetry then generates the rest of the unit cell and lattice translation gives the entire crystal structure. The symmetry-unique fraction of the unit cell, which may be as little as $\frac{1}{192}$ of the cell volume for the highest-symmetry centred cubic space groups, is called the **asymmetric unit** of the structure. It may consist of one unit of the chemical formula (whether molecular or not), more than one unit (whereby these units are described as 'crystallographically independent' even though they may be chemically equivalent) or a fraction of one unit (in cases where atoms/ions lie in special positions such that the formula unit itself has crystallographic symmetry). The number of times the chemical formula appears in one complete unit cell is conventionally given the symbol Z, while the number in the asymmetric unit is often referred to as Z'.

The positions of atoms/ions in a crystal structure are usually expressed as dimensionless coordinates referred to the unit cell axes. Thus the origin is the position $0, 0, 0$ and the body centre of the unit cell is $1/2, 1/2, 1/2$. In principle, the contents of the asymmetric unit – all that is required, together with the space group and six unit cell parameters, to describe completely the geometrical arrangement of a fully ordered structure – can be expressed with all coordinates in the range 0–1; numbers outside this range refer to other unit cells. However, it is often more convenient to use coordinates that do fall outside this range, because groups of connected atoms/ions/molecules lie across unit cell boundaries and are better handled as single units than as fragments of separate units related by symmetry.

2.2.3   An Introduction to Non-Ideal Behaviour

As mentioned earlier, the ideal of perfect order is the most fundamental concept in understanding and describing crystalline solids and their diffraction patterns, but in reality various deviations from this ideal are observed. This is particularly the case for inorganic materials, even more so than for molecular solids. Particular deviations from perfect order will be treated in detail at appropriate stages later, but a brief summary is given here. Distinguishing these effects from each other is not always simple.

2.2.3.1   Atomic Displacements

Atoms in crystal structures are not stationary; they vibrate around a central position, the amount of vibration depending on the temperature and the strength of interactions within the solid, and generally not being equal in all directions (it is an **anisotropic** property, like many physical properties of most solids, except for those of cubic symmetry). Although there are cooperative atomic movements in solids, known as **phonons**, most atomic displacements are regarded as essentially independent. At any given instant, then, the perfect translation symmetry of the crystal lattice is broken by displacements of individual atoms from their central positions. This behaviour is modelled by a number of treatments of varying complexity, in which atomic displacement parameters are assigned to each atom. The simplest model is an **isotropic** one, in which each atom has just one displacement parameter, while the treatment most commonly applied for anisotropic behaviour has six parameters per atom, the displacement being modelled mathematically as a second-rank symmetric tensor and visually as an ellipsoid (3D equivalent of a 2D ellipse), as shown in Figure 2.6. The displacement parameters are determined along with atomic positions as part of a crystal structure analysis, derived from the diffraction pattern, in which intensities are influenced by atomic motion, since this motion spreads out the scattering power of the atoms. In many rigid inorganic framework structures, displacement parameters are quite small. The effect of large-scale displacement can be difficult to distinguish from static disorder.

2.2.3.2   Disorder

We have assumed that all unit cells within a given structure are identical, as is consistent with the lattice concept. This assumption underlies

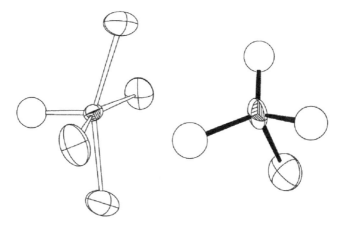

Figure 2.6 Graphical representation of the atomic displacements for the pair of molecules shown in the crystal structure in Figure 2.2. The oxygen atoms, with relatively low electron density and hence high uncertainties in parameters derived by X-ray diffraction, are modelled as isotropic and represented as spheres, while the heavier chlorine and metal atoms are modelled as anisotropic and represented as ellipsoids, with principal ellipses also shown; for the two metal atoms, one octant of the ellipsoid is cut out to show the principal radii. The spheres and ellipsoids are scaled to enclose 50% probability of the atomic electron density in each case, a commonly adopted convention in crystal structure illustrations.

the whole phenomenon of diffraction, which depends on periodicity of the scattering material. A lack of long-range order gives an amorphous rather than a crystalline material. A small degree of disorder within an otherwise ordered structure is a frequent occurrence and can take a number of forms: the possibility of more than one type of atom at a particular site, with a random distribution over the equivalent sites throughout the structure; incomplete occupancy of some positions, with vacancies; variation in the position and/or orientation of an atom or group of atoms; wholesale replacement of a chemical group by another of similar size in some parts of the structure; or a combination of these effects, as different atom types have different sizes and cause changes in their own positions and those of their neighbours. In each case it is important that the variation in the structure is a random one rather than a regular and systematic pattern; the latter would be correctly described by a larger repeat unit. Diffraction methods produce an image of the *average* asymmetric unit taken over the whole crystal, so the effect of disorder is as if all unit cells are indeed identical, but some atom sites are only partially occupied (not by a whole atom), or else chemical groups with reduced scattering power (apparent fractional occupancy) overlap and share the same space. The incidence of disorder complicates the description of the structure and can make the diffraction experiment itself more difficult, since it generally reduces diffraction intensities.

2.2.3.3 Twinning

In a single crystal, not only are all unit cells identical in form and content, they are also identical in orientation, giving a continuous single lattice throughout the crystal. In reality, there are actually small variations in the orientation of domains within the crystal – referred to as a **mosaic structure** by analogy to minor irregularities in a pattern of tiles on a wall – but these do not have any detrimental effect unless they are unusually large, when they cause a broadening in diffraction effects. It is however possible for domains within an apparently single crystal to have completely different orientations, giving what is actually a composite crystal. Each domain then produces its own diffraction pattern and the different patterns are superimposed. In cases where the two (or more) orientations are related to each other in a specific way connected with the crystal symmetry, some or all of the diffraction peaks of the different patterns coincide exactly or approximately, making it impossible to measure them as separate intensities. This phenomenon is called **twinning**. It can be treated satisfactorily if the relationship between the twin components is established, the true unit cell and space group are found and the diffraction patterns are handled appropriately in structure solution and refinement, allowing for the overlapped as well as the distinct reflections in the measured diffraction pattern.

2.2.3.4 Pseudosymmetry

It is by no means uncommon, particularly in inorganic framework materials, for parts of the structure to show, at least approximately, a higher space-group symmetry than the complete structure. Examples include high-symmetry inorganic frameworks with a less symmetrical, but ordered, set of guest molecules within their pores or channels. In many cases, the higher-symmetry substructure dominates the scattering and the true lower symmetry is evident only from careful examination of weaker features of the diffraction pattern. Problems encountered as a result of this include ambiguities of space group, potential errors in determining the unit cell geometry (if systematically weak diffracted beams are ignored), difficulties in locating the structural components of lower symmetry and a failure of the structure refinement to converge easily.

2.2.3.5   Modulated Incommensurate Structures

An ideal crystal structure has 3D periodicity, characterised by a lattice. Some structures contain a component, such as guest species within a host framework, a particular substituent or an atom displaced from an ideal position within a coordination polyhedron, that displays its own periodic positional behaviour with a repeat unit different from that of the main structure. If the secondary periodicity does not have a rational relationship to the main lattice repeat, the structure is said to be **incommensurate**. Whereas a commensurate secondary periodicity constitutes a genuine different true lattice (a substructure/superstructure situation) and a random variation is a form of disorder that affects the intensities but not the geometrical properties of the diffraction pattern, an incommensurate structure gives weaker secondary diffraction peaks around the main lattice-generated peaks. The spacing between these **satellite** peaks is related to the secondary periodicity and is not a simple fraction of the separations of the principal diffraction peaks. Measuring the complete pattern, including the satellites, requires the use of more than three dimensions and three indices, and characterising the incommensurate aspects of the structure involves the use of specialist software for refinement. If it is ignored, an incommensurate nature is effectively subsumed within some kind of dynamic and/or static disorder model that is not a true representation of the structure of the material in question.

2.3   SCATTERING AND DIFFRACTION

Radiation, whether electromagnetic or particulate (*e.g.* beams of electrons or neutrons), interacts in a number of different ways with physical matter, giving scattering, absorption, stimulated emission and photoelectric effects, and subsequent secondary processes in some cases. The cooperative scattering of radiation by a material with a periodic structure, involving constructive and destructive interference of the scattered radiation, is known as **diffraction**. Significant and useful diffraction occurs when the wavelength of the radiation is comparable to the size of the repeat unit of the structure. In the case of crystalline solids, this means selecting X-rays with wavelengths in the Å range, which is about the size of atoms and molecules, unit cell dimensions being typically $3 - 100\,\text{Å}$ for chemical systems and extending above this for biological macromolecules. Electron and neutron beams with appropriate velocities also have suitable associated wavelengths through the de Broglie relationship, $\lambda = h/\text{momentum}$, and can be used in crystallography. The use of electrons in diffraction by crystals is dealt with in Chapter 4, as it differs significantly in both theory and practical application. We consider here X-rays and neutrons, which have much in common in their use in crystal structure determination. The theoretical basis is first outlined with the assumption of **monochromatic** X-rays (one single wavelength). Specific differences with polychromatic X-rays and neutrons will then be described.

2.3.1   Fundamentals of Radiation and Scattering

Figure 2.7 shows some important properties of a radiation waveform. The wavelength, λ, measures the distance between adjacent maxima of the wave. It is what distinguishes different parts of the electromagnetic spectrum, from radio waves, with the largest wavelengths (measured in metres), to X-rays and γ-rays, with the smallest (down to fractions of an Å); X-ray wavelengths commonly used in crystallography are roughly in the $0.5 - 2.5$ Å range. The amplitude, here denoted $|F|$, is the 'size' of the wave; in physical terms it is related to the X-ray intensity, I, by $I \propto |F|^2$. In X-ray diffraction, the scattering is caused by interaction with electrons in the atoms of the crystal, so we measure scattered X-ray amplitudes relative to the scattering by a single electron, and we use electrons as units for $|F|$ when we wish to have them on a known rather than an arbitrary scale. The phase of a wave, ϕ, has no significance except in comparing two or more waves with the same wavelength; it

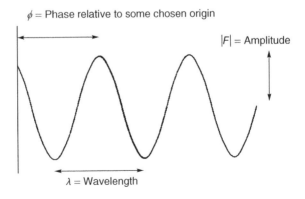

Figure 2.7   Properties of a wave.

is a relative rather than an absolute measurement. Two waves are completely in phase when their maxima coincide, so that they exactly match
crest-to-crest and trough-to-trough. They are completely out of phase
when the maxima of one match the minima of the other, crest-to-trough.
Relative phases are thus crucial to the description and understanding of
wave interference effects: constructive (in phase) and destructive (out of
phase). The phase of one wave relative to another, or relative to any
selected origin, may be expressed as a dimensionless fraction of a wavelength, but it is most commonly given as an angle, the range $0 - 360°$
(or $0 - 2\pi$ radians, the units in which phase angles appear in the mathematical equations for scattering and diffraction) corresponding to one
whole wavelength, so that $0°$ (or $360°$) is exactly in phase and $180°$ is
exactly out of phase. The relative phase for a wave, in general, can take
any value in this range.

Because a wave of a particular wavelength has two properties of interest, amplitude and phase, it can conveniently be represented as a complex
number for convenience of mathematical manipulation:

$$F = |F| \cos \varphi + i|F| \sin \varphi = |F| \exp(i\varphi) \tag{2.6}$$

Scattering itself varies in its detailed nature. The form of scattering of
X-rays exploited in most crystallographic work, and assumed in the
theory presented here, is elastic (no change in wavelength) and coherent
(a well-defined phase relationship between incident and scattered
X-rays). Inelastic and/or incoherent scattering also occurs (indeed,
inelastic scattering of neutrons forms the basis of an important form
of spectroscopy), and it contributes to measured levels of background
scattering, for which corrections need to be made to diffraction intensity
measurements.

2.3.2   Diffraction of Monochromatic X-Rays

We now develop a theoretical basis of X-ray diffraction, starting from a
single electron and building up to a crystalline solid material.

2.3.2.1   Scattering by One Electron

X-rays are scattered almost entirely by electrons; interactions with
atomic nuclei are negligible. The scattering of X-rays by electrons is

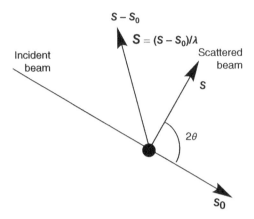

Figure 2.8 Definitions for use in scattering theory (terms are defined in the text below).

actually quite weak and hence inefficient; the vast majority of an X-ray beam incident on a sample passes straight through it without being scattered. The intensity of X-rays scattered is dependent on the angle of scattering, for which the standard symbol is 2θ (Figure 2.8):

$$I(2\theta) \propto I_0(1 + \cos^2 2\theta) \tag{2.7}$$

where I_0 is the incident intensity and $I(2\theta)$ is the intensity scattered in any direction inclined at 2θ to the incident direction; this gives a whole cone of scattered radiation. Figure 2.8 also defines incident and scattered beam directions as two vectors, s_0 and s respectively, which are taken to have unit length. For use later, we also define here the **scattering vector** $S = (s - s_0)/\lambda$. Reference to Figure 2.8 and some simple trigonometry shows that the length of vector S is $2(\sin\theta)/\lambda$. Note that the scattered intensity, even for a single electron, decreases as 2θ increases.

2.3.2.2 Scattering by Two or More Electrons

Next consider scattering by a pair of electrons (treated as localised objects), which is subject to interference effects. We take one electron to define the origin, while the other is at a position given by a vector r. Each electron scatters incident X-rays as above, giving two separate waves in any particular direction. The scattering effects need to be combined, allowing for a phase difference that is caused by the physical separation of the two electrons in space and which is different in different directions. In any particular direction s, the path difference between the X-rays scattered by the two electrons is $r \cdot (s - s_0)$, which gives a phase difference of $2\pi r \cdot (s - s_0)/\lambda = 2\pi r \cdot S$ in radians. The scattering from the two electrons in the direction described by the scattering vector S thus combines as:

$$F(S) = F_1 + F_2 = F_1 + F_1 \exp(2\pi i r \cdot S) \tag{2.8}$$

2.3.2.3 Scattering by One Atom

This treatment can be extended to any number of electrons. For a single atom, we need to regard the electron density distribution as a continuous density rather than a set of individual discrete points, so adding up scattering contributions turns from a summation to an integral, but it is still necessary to incorporate phase differences due to the finite size of the atom. If we write the electron density distribution of the atom as $\rho(r)$, where r is a position vector measured from any convenient origin (here the atomic nucleus is the obvious choice), then the scattering of incident X-rays by the atom in a particular direction is:

$$f(S) = \int \rho(r) \exp(2\pi i r \cdot S) \, dV \tag{2.9}$$

which can be seen as a generalisation of the specific two-centre scattering of Equation 2.8 to a continuous electron density distribution in an atom. This term $f(S)$ is called an **atomic scattering factor**.

Electron density distributions, $\rho(r)$, are known for all atoms (neutral and common ionised forms) from quantum mechanics, so scattering factors can be calculated; these are available in standard published tables[8] and are embedded in many crystallographic software packages. Note that the scattering factor of any given atom has units of electrons (it is measured relative to the scattering by a single electron) and that it varies with the scattering angle, 2θ. The value of $f(0)$ (i.e. the scattering factor of an atom at $2\theta = 0$, exactly in line with the incident beam) is equal to the total number of electrons in the atom, because the exponential term is unity when $S = 0$; all electrons of the atom scatter exactly in phase in this direction.

Atoms in a crystal structure are not stationary; they vibrate around their mean positions, and this vibration is described mathematically by **displacement parameters** (see Section 2.2.3.1). The vibration of atoms spreads out their electron density over a larger volume, and this has an impact on the scattering power of the atom for X-rays. The effect is to modify the scattering factor, $f(S)$, from its stationary-atom form and the usual treatment is to multiply this by a further exponential term in which the displacement parameters appear; for the simplest model of isotropic displacement, the extra term is:

$$\exp(-8\pi^2 U \sin^2\theta / \lambda^2) \tag{2.10}$$

where U is the mean-square amplitude of atomic vibration in $\mathring{A}^2$. It can be seen that atomic vibrations reduce scattering intensities, the effect increasing as the scattering angle increases, and this compounds the 2θ-dependent fall-off in intensity that already occurs as a consequence of destructive interference across the finite size of the atom and the polarisation-generated decay that applies even for a single electron. Typical atomic scattering factors are shown in Figure 2.9, including the effect of atomic displacements.

2.3.2.4   Scattering by a Group of Atoms

To a first approximation, the atoms in a crystal structure are treated as independent scatterers of X-rays. The total scattering for a molecule, an asymmetric unit, a unit cell or any other discrete group of atoms (leaving aside for the moment the arrangement of these entities on a crystal lattice) is obtained by an appropriate addition of all the separate atom scattering contributions, treated mathematically in vector notation to allow for phase differences produced by the relative positions of the atoms. In most cases, this approximation is a good one, since the bulk of electrons in

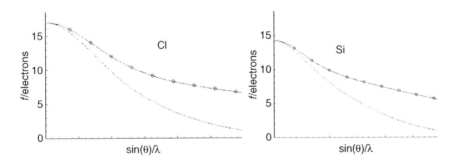

Figure 2.9   X-ray scattering factors for Cl and Si atoms: stationary atoms (upper curve) and vibrating atoms with arbitrary displacement parameters (lower curve).

atoms, especially heavier atoms, are core electrons belonging exclusively to one particular atom, and only a very small proportion of the total electron density of the structure is involved in bonding and other valence effects. We will consider later the impact of valence electron density and techniques for its investigation.

We choose an origin from which to measure the positions of all the individual atoms (the centres of their electron density distributions, generally assumed to coincide with their atomic nuclei, though this is not exactly the case for covalently bonded hydrogen atoms in particular); in the case of atoms in a crystal structure, the origin is that of the unit cell. Each atom, labelled j, lies at a position given by the vector r_j and has an atomic scattering factor f_j' (this is the stationary-atom scattering factor, f_j, modified as noted above to account for atomic displacements).

The total scattering from the group of atoms is then:

$$G(S) = \sum_j f_j' \exp(2\pi i r_j \cdot S) \tag{2.11}$$

This is a finite sum of discrete terms because we have chosen to model the structure as a collection of discrete atoms. We could alternatively treat the whole structure as a continuous electron density distribution, $\rho(r)$, in which case we would obtain a result similar in appearance to that of Equation 2.11, with an integral over the whole volume of the molecule, asymmetric unit or unit cell. In either case, the result we obtain is a continuous function of S and, unlike the scattering for a single atom with isotropic displacements, the intensity varies extensively and in a complicated way with direction and is not a monotonically decreasing function of the single variable 2θ.

2.3.2.5   Scattering (Diffraction) by a Crystal Structure

In the final step of this analysis, we repeat the contents of one unit cell at every point of the crystal lattice. Every unit cell is identical in contents and orientation in a single crystal. Every unit cell therefore makes exactly the same contribution to the overall observed scattering from the crystal, this contribution being $G(S)$, as given in Equation 2.11, and the remaining question is how these identical contributions combine with each other in terms of phase differences.

Scattering of waves by any periodic array with a sufficiently large number of points leads to strong constructive and destructive interference effects, which we know as diffraction, the observable result being zero intensity everywhere except in directions for which all the points give scattering that is exactly in phase. For a single row of points (a 1D lattice) with spacing a, the path length for rays scattered by two adjacent points is $a \cdot (s - s_0)$, and this must be equal to a whole number of wavelengths for constructive interference:

$$a \cdot (s - s_0) = n\lambda \qquad (2.12)$$

where n is an integer (positive, zero or negative).

For a 3D crystal lattice, any vector between two lattice points can be written as:

$$t = ua + vb + wc \qquad (2.13)$$

where u, v and w are integers. The condition for constructive interference is then:

$$t \cdot (s - s_0) = n\lambda$$

$$\text{or} \quad (ua + vb + wc) \cdot S = n \qquad (2.14)$$

This condition is satisfied when S is any vector between points of the reciprocal lattice and not otherwise. To demonstrate this, we express S as a reciprocal lattice vector in a form analogous to that of Equation 2.13 for a director lattice vector t:

$$S = ha^* + kb^* + lc^* \qquad (2.15)$$

where h, k and l are integers. Now we substitute S from Equation 2.15 into Equation 2.14 and refer back to the defined relationship between direct and reciprocal lattices in Equations 2.2 and 2.5:

$$(ua + vb + wc) \cdot (ha^* + kb^* + lc^*) = uha \cdot a^* + vkb \cdot b^* + wlc \cdot c^*$$

$$= uh + vk + wl = \text{integer} \qquad (2.16)$$

All the other six terms of the scalar product involve cross-terms such as $a \cdot b^*$ that are zero.

This means that scattered intensity is non-zero only in certain discrete directions, and these directions are specified by the integer indices h, k, l. The directions of diffraction by the direct lattice correspond to the points of the reciprocal lattice, so the reciprocal lattice itself is a geometrical representation of the diffraction pattern of a single crystal.

Equation 2.15 is effectively the well-known Bragg equation written in a vector form using the reciprocal lattice. To see this, remember that the length of vector S is $2(\sin\theta)/\lambda$ and note that the equation for S in terms of reciprocal lattice parameters is actually the same as Equation 2.3 for reciprocal lattice vectors $d^*(hkl)$; according to Equation 2.4, the length of d^* is $1/d(hkl)$, where the direct lattice planes with indices h, k, l have interplanar spacing $d(hkl)$. Thus:

$$2 \sin \theta / \lambda = |S| = |d^*(hkl)| = 1/d(hkl) \tag{2.17}$$

Rearrangement gives:

$$\lambda = 2d(hkl) \sin \theta \tag{2.18}$$

which is the familiar form of the Bragg equation relating diffraction angles to lattice plane spacings. Note that the Bragg equation is incomplete on its own as a condition for diffraction by a lattice. There is also the associated requirement that the normal to the lattice planes hkl bisects the angle between the incident and diffracted beams. Because of the geometrical form of the Bragg condition, which gives the appearance of reflection of X-rays from the lattice planes, diffracted beams are often referred to as **reflections** or **Bragg reflections**.

Combining all this information, we obtain an expression for the total scattering of X-rays in any direction by a single crystal:

$$F(S) = \sum_j f'_j \exp(2\pi i r_j \cdot S) \tag{2.19}$$

As we have just seen, lattice diffraction restrictions mean the intensity is zero except in directions corresponding to reciprocal lattice points for which $S = ha^* + kb^* + lc^*$. Each atom position r_j can be written as:

$$r_j = x_j a + y_j b + z_j c \tag{2.20}$$

in terms of its fractional coordinates x, y, z within the unit cell, and hence:

$$F(hkl) = \sum_j f'_j \exp \left[2\pi i \left(x_j a + y_j b + z_j c \right) \cdot \left(ha^* + kb^* + lc^* \right) \right] \tag{2.21}$$

which, because of the defined relationship between the direct and reciprocal lattices, simplifies to:

$$F(hkl) = \sum_j f'_j \exp \left[2\pi i \left(hx_j + ky_j + lz_j \right) \right] \tag{2.22}$$

where the summation needs be made over only one unit cell, the total intensity from the crystal being the intensity from one unit cell multiplied by the number of unit cells.

There are various ways of representing the geometrical requirements of diffraction in graphical form, including a construction known as the **Ewald sphere**, which demonstrates how and when individual diffracted beams (reflections) arise as a crystal (and its direct and reciprocal lattices) is rotated to different orientations in an incident X-ray beam with a fixed direction. Such treatments are useful for further visual understanding of the process and of the construction and operation of devices for recording diffraction patterns – cameras and diffractometers – but they are not necessary for our purpose here and there is insufficient space to cover them; reference can be made to standard crystallography texts such as those mentioned at the beginning of the chapter and also to the comprehensive compilation in Volume B of *International Tables for Crystallography*.[9]

A monochromatic X-ray diffraction pattern from a single crystal is thus a set of discrete beams of X-rays $F(hkl)$ in particular directions specified by the integer indices h, k, l, all other directions giving no diffracted intensity because of the lattice. The diffraction directions (the geometry of the diffraction pattern, ignoring the variation of individual intensities) are described by the reciprocal lattice, with each discrete reciprocal lattice point corresponding to one observed reflection. Indeed, some devices for recording diffraction patterns are constructed in such a way that the pattern appears as neat 2D slices through the reciprocal lattice, the reflections lying on equally spaced rows in regular 2D lattice arrays; the precession photograph in Figure 2.1b is an example. Modern diffractometers with area detectors give images on which rows of reflections can often be seen if the unit cell is not too small, though these rows are generally curved rather than straight because of the arbitrary orientation of the crystal; see Figure 2.1d for an example. The intensities of the individual reflections depend on the identity (f_j') and positions (r_j or x_j, y_j, z_j) of atoms within the unit cell of the structure, as given by Equation 2.22.

2.3.3   Diffraction of Polychromatic X-Rays

X-ray diffraction from single crystals is usually carried out with monochromatic X-rays. In this case, the geometrical restrictions imposed by the lattice, as encapsulated in the reciprocal lattice construction and the Bragg equation, are so severe that a stationary crystal gives very little diffraction, the conditions being satisfied for very few reciprocal lattice points or reflections. In order to record a complete diffraction pattern up to some particular maximum 2θ angle, or any significant proportion of the pattern such as a symmetry-unique part of it, the crystal must be rotated in the X-ray beam, the movement bringing different reciprocal lattice points successively at different times into a situation satisfying the diffraction conditions. It therefore takes some time to record a complete pattern, which is built up gradually during the rotation; depending on the sample, the incident X-ray intensity and the equipment used, the required time may range from seconds to days.

An alternative approach is to use a continuous range of X-ray wavelengths in what is known as **white radiation** by analogy with white light in the visible part of the electromagnetic spectrum. In this case a diffraction pattern can be recorded with a stationary crystal, or from a crystal in a number of discrete orientations for completeness, many reflections occurring simultaneously because the white radiation changes single scattering vectors, S, into ranges of vectors due to the $1/\lambda$ dependence of this defined vector. Different reflections in the pattern are caused by diffraction of X-rays of different wavelengths, and some coincidences occur; for example, the reflection h, k, l for wavelength λ_1 and the reflection $2h, 2k, 2l$ for wavelength $\lambda_1/2$ satisfy the diffraction conditions at the same time and produce reflections in exactly the same direction. The analysis of diffraction patterns of this kind is, therefore, more complicated, as are corrections to the measurements for simultaneous non-diffraction effects such as absorption. The polychromatic approach is generally known as **Laue diffraction**. Its major advantage is its speed.

2.3.4 Diffraction of Neutrons

The diffraction of beams of neutrons of appropriate wavelength, whether monochromatic or polychromatic, is based on the same principles as X-ray diffraction. All the geometrical aspects relating to diffraction by the lattice are exactly the same. The main difference is in the nature of the interaction of the two forms of radiation with physical matter, and this has major consequences for the diffraction intensities and for practical considerations.

While X-rays are scattered almost entirely by electrons, neutrons have their main interaction with atomic nuclei. Since nuclei are very small compared with the broad distribution of electron density in atoms, neutron scattering is much weaker than X-ray scattering and the technique generally requires larger crystals in order to give practically useful diffraction patterns. The small size of nuclei does, however, give one advantage to neutron diffraction: neutron scattering factors for stationary atoms have essentially no θ-dependence. There is no fall-off at higher angles, as there is for X-ray scattering factors. Although neutron diffraction intensities overall are lower than those for X-rays, the high-angle data are *relatively* stronger. There is still some decrease in intensity at higher angles because of atomic displacements, which introduce an exponential decay with $\sin^2\theta/\lambda^2$ in exactly the same way as with X-rays, but the overall effect is much smaller.

Scattering of X-rays is proportional to electron density; the scattering at $\theta = 0$ from an atom is given by its atomic number, and so X-ray scattering increases steadily and strongly across and down the Periodic Table. For samples containing atoms of widely different atomic numbers, such as oxides or, even more so, hydroxides of heavy metals, the heavy atoms strongly dominate the scattering and the lighter atoms are much more difficult to locate and refine with adequate precision (the structure illustrated in Figures 2.1 and 2.6 is a good example, the metal being Re, with an atomic number of 75; *cf.* 17 for Cl and 8 for O). For many inorganic materials, in contrast to organic compounds without heavy atoms, hydrogen atoms are almost or essentially invisible to X-rays. Neutron scattering by nuclei is quite different: there is no simple pattern of scattering power (the terms **scattering length** and **scattering cross-section** are generally used for neutrons, and the situation is complicated by the need to consider both coherent and incoherent scattering) in terms of the position of elements in the Periodic Table; neutron scattering factors vary substantially, and quite erratically, from element to element, but with a few exceptions the total range is considerably smaller than

Table 2.2 Relative X-ray scattering factors and neutron coherent scattering lengths of selected elements and isotopes. Each column has its own common scale, but the two sets of values are not on the same scale. Except where explicitly stated, the neutron values are for the naturally occurring mixture of isotopes.

Atom	X-ray (electrons)	Neutron (femtometres)
H (^{1}H)	1	−3.7
D (^{2}H)	1	6.7
O	8	5.8
Al	13	3.4
Si	14	4.1
V	23	−0.4
Cr	24	3.6
La	57	8.2
Ce	58	4.8

that for X-rays; and different isotopes of the same element have different scattering properties. Some selected values are given in Table 2.2 for illustration.

Among the practical consequences of this difference are the following:

- Neutron scattering by light atoms is not, in general, swamped by that from heavy atoms, so the lighter atoms are likely to be found more precisely and reliably; this can be seen from the ranges in the two columns of values in Table 2.2, and is particularly the case for H, which has a relatively large scattering length (albeit a *negative* one, which means H scatters out of phase rather than in phase with most other atoms, and it also has a high incoherent scattering effect), the isotope D being an even stronger, in-phase, neutron scatterer.

- Close neighbour elements in the Periodic Table in many cases have markedly different neutron scattering cross-sections and can thus be readily distinguished; with X-rays, the electron densities are rather similar, and assigning the correct atom types in some situations, especially when there is disorder, can be problematic, with other factors such as differences in structural chemical behaviour (atomic/ionic radii, preferred coordination geometry *etc.*) needing to be considered also. Three such pairs are given in Table 2.2.

- Neutrons, unlike X-rays, can distinguish different isotopes of the same element. Where these are disordered, as in the natural abundances of multi-isotope elements, a weighted-average neutron scattering factor applies.

- Neutrons locate atomic nuclei directly; this is what we usually mean by the positions of atoms in a structure. X-rays provide a map of electron density. While it is generally assumed that the electron density of an atom has the nucleus at its centre, this is not always the case, because of valence effects. The discrepancy is most marked for covalently bonded H atoms, for which the centre of electron density is displaced along the bond and towards the other atom by, typically, around 0.1 Å. X-ray diffraction thus gives H atom positions in most cases that are systematically in error. This is not usually a problem, especially if the effect is recognised and appropriate allowance is made in interpreting the results, but neutron diffraction does provide a solution in cases where it is needed (such as metal hydride complexes and unusual hydrogen bonding situations), provided suitable crystals are available for the technically more demanding technique.

- Although the main interaction of neutrons is with atomic nuclei, neutrons have a magnetic moment and this can interact with the magnetic moment of atoms with unpaired electrons. In the case of paramagnetic substances, the atomic moments are randomly orientated, so there is no periodic structure to the magnetism, and neutron diffraction gives essentially the same result as X-ray diffraction. For cooperative arrangements of atomic moments, however, such as antiferro- and ferrimagnetic behaviour, the repeat unit of the structure including magnetism may be different from the purely atomic lattice arrangement, and in this case the magnetism generates extra diffraction maxima in the neutron diffraction pattern, a larger unit cell giving a smaller reciprocal unit cell, because atoms with spins in different orientations are not truly equivalent by symmetry.

2.3.5 Some Competing and Complicating Effects

To complete this section, we consider briefly other effects that can occur simultaneously with coherent elastic scattering and which may have a significant impact on the diffraction experiment and its results.

2.3.5.1 Other Scattering

Inelastic and/or incoherent scattering generally contributes to background scattering that is superimposed on the principal diffraction pattern. Other contributions come from scattering of the X-ray beam by components of the experimental equipment (*e.g.* collimator, beam stop to trap the undiffracted direct beam, sample mount, air). The background level around reflections is measured and subtracted from reflection intensities. Another contribution to scattering known as **thermal diffuse scattering** comes from cooperative lattice vibrations; it gives effects that are nothing like as sharp as Bragg reflections but do peak at reflection positions. In most cases, this is treated as part of the background, but when it is more significant it needs to be measured and corrected more carefully. Some other effects that can add artificially to measured reflection intensities include the diffraction of harmonics ($\lambda/2, \lambda/3, ...$) present in a supposedly monochromatic X-ray beam (similar to the overlaps in Laue diffraction), and the possibility of multiple reflection, in which a diffracted beam acts as incident beam for a second reflection, giving a net effect that coincides with a third reflection, these three having related indices (this is known as the **Renninger effect**). These effects are most often seen as weak but significant observations of reflections that should have zero intensity as a consequence of symmetry (such **systematic absences** arise for screw axes and glide planes, as well as for centred unit cells, and contribute to procedures for establishing the correct space group, but the treatment is not given here). Appropriate countermeasures include avoidance through appropriate design of equipment and procedures, and correction following recognition and measurement of the effect.

2.3.5.2 Absorption

X-rays, like other electromagnetic radiation, are absorbed as they pass through matter. Absorption occurs at the same time as diffraction and reduces the intensity of the reflections. The main problem is that the amount of absorption depends exponentially on the path length through the sample, as well as on the absorption coefficient for the particular combination of material and wavelength, and different reflections have different path lengths, so absorption does not affect all reflection intensities equally, as illustrated in Figure 2.10. This is one of the most important sources of potential systematic errors in crystallography.

Absorption coefficients tend to increase with the atomic numbers of the elements present in the sample and with increasing X-ray wavelength, though neither of these relationships is by any means a smooth and monotonic one. The impact of absorption on intensity measurements can be assessed by calculating the absorption coefficient, μ, from the chemical formula and then $\exp(-\mu t_{max})$ and $\exp(-\mu t_{min})$ for the maximum and minimum crystal dimensions. If the ratio of these two numbers is close to unity then *relative* absorption effects due to crystal shape are

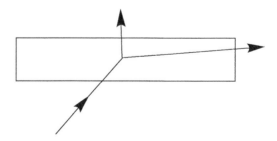

Figure 2.10   X-rays passing through a crystal in different directions may have very different path lengths, leading to significant variations in diffracted intensities, even for reflections that are equivalent by symmetry and so should have equal intensities.

not important, though the absorption overall may still be significant and will have an important θ dependence (even for a crystal ground to a spherical shape, absorption cannot be ignored if μ is high). It can be seen that the crystal shape is a particularly important factor here. In general, except for compounds containing only the lightest atoms and forming approximately equidimensional crystals, absorption should always be treated appropriately for inorganic materials. Appropriate treatment can include minimising absorption effects by selection of crystal size and shape and of X-ray wavelength, and corrections should be made to the intensities, using methods based explicitly on measurement of crystal shape and/or empirically on the observed variation of intensities of symmetry-equivalent reflections measured at different crystal orientations; these intensities would be equal in the absence of absorption and other systematic error effects, and an empirical absorption correction aims to find parameters describing the absorption such that the variations are minimised and then applies these parameters to the complete data set; a common approach is to model absorption effects using spherical harmonics. Inadequate absorption corrections have a particular effect on atomic displacement parameters, sometimes rendering them physically meaningless, and generally reduce the precision of any crystal structure determination. In extreme cases they can prevent structure determination with wildly incorrect intensities.

Absorption affects neutron diffraction considerably less than X-ray diffraction, since absorption coefficients are much lower in the former. Indeed, the high penetrating power of neutrons through solid matter is a factor that makes this technique particularly useful in investigating materials under conditions such as high pressures, special gas atmospheres *etc.*, for which X-ray diffraction relies on the use of very short wavelengths. Nevertheless, absorption effects should always be assessed with simple calculations, and appropriate steps should be taken if they prove significant.

2.3.5.3 Extinction

As mentioned earlier, X-ray scattering is generally weak and inefficient, in the sense that only a very small proportion of the incident beam is deflected from its original direction. In some cases, however, diffraction can be sufficiently strong that the incident beam is robbed of significant intensity. This means that parts of the crystal further along the beam receive lower incident intensity and therefore produce less diffracted

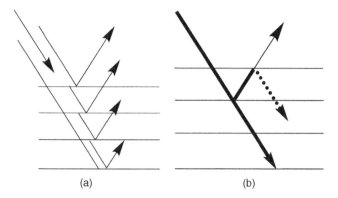

(a)   (b)

Figure 2.11 Two ways in which strong diffracted intensities can be lower than is predicted by straightforward theory: (a) attenuation of the incident beam by the diffraction process itself; and (b) reduction of the diffracted beam intensity by further diffraction.

intensity per unit volume. The net effect is a strong reflection whose intensity is lower than would be expected from the straightforward theory, the reflection intensity reduction being due to the process of diffraction itself. In a related phenomenon, the diffracted beam can serve as incident beam for subsequent reflection, yet again in the opposite direction, by the same set of lattice planes (the Bragg condition is still satisfied), and this also reduces the intensity. These effects, illustrated in Figure 2.11a and b respectively, are known as **extinction** and tend to occur particularly for strong, low-angle reflections from crystals with a high degree of perfection (large domains, low mosaic spread). They are usually recognised from the occurrence of reflections of this type with observed intensities lower than those calculated from the proposed structure. Depending on the severity of the problem, it may be treated by physical conditioning of the crystal, such as thermal shocks, by calculation of corrections as part of structure refinement with the inclusion of one or more extinction parameters or by omission of the worst-affected reflections from the refinement, if there are not many of them.

2.3.5.4 Resonant Scattering

X-ray diffraction depends for its success on the assumption that all electrons in atoms scatter X-rays with the same phase shift (or no shift at all). Detailed treatment shows that each scattering event actually gives a phase shift of 180°; since this is constant, it can be ignored.

This simple situation applies, at least in a sufficiently good approximation, provided the X-ray photon energy, $E = h\nu = hc/\lambda$, is well removed from the energy required to promote an electron in the atom to a higher available energy level or to remove an electron completely and produce ionisation and a photoelectron. When the X-ray photon energy is equal to such an electronic transition energy for the atom, this is said to be an **absorption edge** of the atom (there is a drastic increase in the absorption coefficient for this element at this wavelength). Under these circumstances, and for a range of energies above this, the atom introduces a significant phase shift for scattered X-rays. This effect can be treated mathematically by making scattering factors complex instead of purely real numbers, with an imaginary component (which represents a 90° phase shift) as well as a real part. The phenomenon, in addition to being appropriately called **resonant scattering**, is also widely known as **anomalous scattering** or **anomalous dispersion**. Each atomic scattering factor can be adjusted by the addition of two 'anomalous' terms, one real (which may be positive or negative) and one imaginary: $f + f' + if''$. The values of f' and f'' are strongly wavelength-dependent but do not vary with θ. In most cases they are much smaller than $f(0)$, but some combinations of elements and X-ray wavelengths give values equivalent to several electrons. Some example values are given in Table 2.3; see Section 2.4.1 for a discussion of the particular X-ray wavelengths chosen. Failure to apply these resonant scattering terms can significantly affect atomic displacement parameters and, in some structures having no inversion symmetry, they can cause atoms to be shifted significantly from their true positions.

Resonant scattering can have some important practical uses. Careful choice of wavelength at a synchrotron source (where a continuous spectrum of X-rays is available for monochromatic selection) can make the complex scattering factors of neighbouring elements significantly

Table 2.3   Real and imaginary terms (in electrons) for resonant scattering of X-rays for selected elements and two commonly used wavelengths.

Atom	f' and f'' (Mo $K\alpha$)		f' and f'' (Cu $K\alpha$)	
C	0.003	0.002	0.018	0.009
O	0.011	0.006	0.049	0.032
Cl	0.148	0.159	0.364	0.702
Co	0.349	0.972	−2.365	3.614
Ni	0.339	1.112	−3.003	0.509
Y	−2.796	3.567	−0.267	2.024
Bi	−4.108	10.257	−4.011	8.931

different from each other by setting it between their corresponding absorption edges, and it may even be possible in some cases to distinguish between different oxidation states of the same element, for which the electron density distributions and hence the complex scattering factors are different.

The most common application, however, is in the study of crystal structures lacking inversion symmetry. In the absence of resonant scattering effects, the intensities of two reflections, h, k, l and $\bar{h}, \bar{k}, \bar{l}$, are equal; this is known as **Friedel's law**, and the reflections are called **Friedel pairs** or **Friedel opposites**. This equality follows from Equation 2.22 and holds for all centrosymmetric structures, even if there is resonant scattering, but it is no longer true for non-centrosymmetric structures when there are significant f'' terms in any of the scattering factors, though the intensity differences are usually small. The important consequence is that, with sufficiently large imaginary contributions and careful intensity measurements, preferably including Friedel pairs, and with proper corrections for systematic effects such as absorption that could otherwise mask the small differences, it is possible to distinguish a structure from its non-identical inverse. This provides a method for determining the **absolute configuration** of chiral materials and for characterising other structures such as those with polar axes, though this is of practical value only when the so-called **absolute structure** determination can be related to an appropriate physical property (optical, magnetic *etc.*).

2.4   EXPERIMENTAL METHODS

In this section we consider the procedures used to obtain an X-ray or neutron diffraction pattern from a suitable single crystal and the corrections usually made to the raw measured intensities in order to give a set of data ready for solution and refinement of the structure.

2.4.1   Radiation Sources

2.4.1.1   X-ray Sources

There has been enormous development in laboratory X-ray sources over the last century. The basic principle remains the same, and many standard sealed X-ray tubes are still in use, as shown in schematic form in Figure 2.12. Electrons are emitted from a heated filament in an

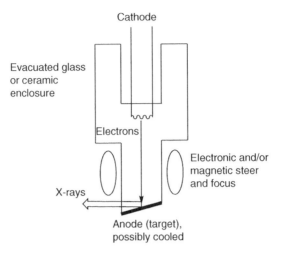

Figure 2.12 Schematic diagram of a typical X-ray tube. The components labelled to the right are additional features of microfocus tubes and are not found in conventional devices.

evacuated glass or ceramic enclosure and are accelerated rapidly through a potential of typically $40 - 60$ kV across a gap of the order of 1 cm, the filament serving as cathode and a cooled metal target (usually with a copper base) as anode. The anode is coated with a selected material, the most widely used being copper or molybdenum, and less often silver or chromium. Most of the kinetic energy of the electrons is lost as heat, but a small proportion is converted into X-rays by interaction with the target metal atoms. A broad, low-intensity continuous spectrum of **white radiation** is generated with a minimum wavelength (maximum photon energy) fixed by the accelerating voltage. Superimposed on this are some much more intense, sharp maxima caused by electronic transitions within the target atoms, generated when the incident electrons remove a core electron and the resulting core energy-level vacancy is filled by an electron from a higher orbital. Because of spin–orbit interactions, some of these maxima consist of finely spaced doublets. The sharp peaks are characteristic of the target element; they are labelled with a pair of letters (one Greek) indicating the two orbitals involved in their production: $K\alpha$, $K\beta$, $L\alpha$ etc. For monochromatic work, one of them, usually the most intense, is selected (see below) and used for diffraction. The widely used $K\alpha$ radiation is actually a doublet, but the $K\alpha_1$ and $K\alpha_2$ components have very close wavelengths with unequal intensities. They are not usually separated in single-crystal diffraction applications, but their presence does need to be recognised and the $\alpha_1 - \alpha_2$ separation must be incorporated in data-processing methods that analyse the profiles of X-ray reflections to improve intensity measurements. Laue diffraction can use the whole white radiation spectrum but is better performed with synchrotron radiation.

The main intensity-limiting factor for X-ray tubes is heat generation. Two approaches have been taken to get around this, which may be applied together. The first is to use a moving target in a **rotating anode** system, so that the electron impact spot is replaced continuously. The other is to focus the electron beam with electric and/or magnetic fields to produce a smaller impact spot from which the generated X-rays can be harvested more efficiently (coupled with X-ray optics as described below), so that similar or even higher usable intensities are obtained from a lower electron current. While rotating anode systems consume rather more power than conventional sealed tubes (typically around 15 – 20 kW, compared with around 3 kW for the latter), **microfocus tubes** use only tens of watts of power, giving a rather narrower but much brighter X-ray beam for use in diffraction.

Selection of a single X-ray wavelength was approximated in early equipment by the insertion of a metal foil as a **filter** in the beam; the filter metal was chosen to have an absorption edge between the desired $K\alpha$ wavelength and that of the main contaminant, $K\beta$, e.g. a Zr filter for Mo radiation. This reduced the $K\beta$ intensity drastically (and the white radiation component on this side of the absorption edge) and the $K\alpha$ intensity much less; the precession photograph in Figure 2.1b was obtained with Zr-filtered Mo$K\alpha$ radiation. A cleaner single wavelength is provided by a **crystal monochromator**. This is itself a single crystal, of a stable material that has a very strong Bragg reflection, such as the 002 reflection of graphite (for which all carbon atoms scatter in phase, because they lie in layers exactly in these lattice planes), and it is orientated in the X-ray beam so as to satisfy the Bragg equation for this reflection. The diffracted beam from the monochromator is used for the experiment, while the undiffracted beam (which has most of the intensity) is absorbed by the monochromator housing. A further development of this approach uses graded multilayers, with a composite material displaying a gradual change of chemical composition along its length, leading to a range of lattice spacings from one end to the other; this smooth variation is often correlated with a slight curve applied to the material. The overall effect is that the selected wavelength of radiation emerges from the multilayer monochromator with a small range of Bragg angles rather than one constant angle, and the resulting convergence compensates (or slightly overcompensates) for the natural divergence of an X-ray beam from an electron impact target to give a well-collimated parallel or slightly convergent incident beam at the sample crystal.

An alternative approach is to exploit the phenomenon of grazing-incidence reflection of X-rays, which occurs at very small angles of incidence to a surface (which may be curved). The reflection provides a focusing effect for the otherwise divergent beam and simultaneously selects one wavelength, in the output direction. X-ray optics of this kind have been in use at synchrotron sources since they were first employed in crystallography, but more recently they have been successfully developed for laboratory systems, particularly in combination with microfocus sources. Several orders of magnitude in intensity have thereby been gained in recent decades.

For even more intensity, to deal with the most challenging samples, we must turn to synchrotron radiation sources. These large-scale and expensive national and international facilities serve many different user communities, providing opportunities for diffraction and scattering, spectroscopy, imaging, lithography and other applications. The principles of operation of a synchrotron storage ring are not important here, except that it generates intense radiation right across the electromagnetic range from infrared to X-rays, the exact form of the output spectrum depending on the size, detailed construction and operating parameters of the machine, as well as on the particular magnetic components, known as **insertion devices**, with names such as 'wigglers' and 'undulators', that are used at different locations around the circumference to tune the output for a particular use. It should also be noted that the relativistic nature of the synchrotron particle beam (electrons or positrons) leads to beams of radiation that are highly collimated, usually very narrow (at least in one dimension) compared with laboratory X-ray beams and almost completely polarised in the horizontal plane, the main consequence of the polarisation being that diffraction equipment has to be used 'on its side' relative to standard laboratory practice, with diffraction by the sample occurring in a vertical plane.

The use of synchrotron radiation offers a number of potential advantages. The most obvious is the very high intensity, which not only allows much smaller and otherwise weakly diffracting crystals to be examined but also makes it easier to obtain results for samples in containers such as high-pressure cells and gas cells for *in situ* studies. The opportunity to select any wavelength from the range available rather than the restricted choice of characteristic radiation from X-ray tubes means that effects such as absorption, extinction and resonant scattering can be either minimised or enhanced for exploitation in distinguishing elements or determining absolute structure; a short wavelength compresses the diffraction pattern to lower Bragg angles and is helpful in experiments with special apparatuses that restrict access to the sample, such as high-pressure diamond anvil cells, while a long wavelength serves to resolve individual reflections from a large unit cell. Intense white radiation, uniquely available at a synchrotron, makes Laue diffraction feasible. Finally, the X-rays from a synchrotron source are not actually a continuous beam but are generated in short, rapid pulses, the time structure depending on the details of how bunches of electrons or positrons are produced and maintained in the ring; this pulse structure can be synchronised with pulses of laser radiation and possibly with mechanical and electronic gating of the incident beam or X-ray detector system, allowing excited states of crystalline materials to be structurally characterised in an approach generally known as **photocrystallography**.

Advantages do not usually come without costs or restrictions. In the case of synchrotron-based single-crystal diffraction, these include the need to compete for scarce resources in a peer-reviewed application process, the inevitable delays in allocation and scheduling that result from this and the inconvenience of using a non-local facility at times that may not be the most desirable.

2.4.1.2   *Neutron Sources*

There are two different kinds of neutron source, both of which are large-scale, central facilities like synchrotrons. One is a nuclear reactor, which may exist exclusively or primarily for use as a neutron source for research, or may be used for other purposes and have exploitation of neutrons as a secondary benefit. To obtain practically useful wavelengths, the neutrons are moderated (slowed down). They may be used polychromatically for Laue diffraction or conditioned with a monochromator, as for X-rays.

The other neutron source, which is becoming more widely available as new facilities are constructed, is a **spallation** facility. Here a small synchrotron (operated as a pulsed particle accelerator, not as a storage ring as for synchrotron radiation) produces short, rapid pulses of protons or other particles and fires them at a target, in which nuclear reactions produce a range of radiations and particles, including in particular neutrons and muons. One of the main differences from a nuclear reactor is that spallation neutrons are pulsed rather than continuous. Unlike X-rays and other electromagnetic radiation, for which velocity in a vacuum is constant, neutrons have an associated wavelength that is inversely proportional to their velocity, through the de Broglie relationship. This means that a pulse of neutrons with a range of wavelengths becomes spatially extended along its line of flight. The velocity, and hence the wavelength, of any particular neutron can be calculated from the time it takes to arrive at a detector following pulse generation, and this provides a way of analysing a polychromatic Laue neutron diffraction pattern that is not possible for X-rays. Laue diffraction at spallation neutron sources, using all available neutrons, is thus relatively more widely used than the corresponding technique at synchrotron sources. It makes the most of the relatively weak intensities seen in neutron diffraction.

The use of neutron facilities presents similar disadvantages to the use of synchrotrons in terms of restricted access to scarce resources. In addition, crystals generally have to be rather larger than for laboratory X-ray diffraction and much larger than for synchrotron study, experiments take longer and there are additional health-and-safety precautions associated with the use of neutrons and nuclear facilities.

2.4.2   Single Crystals

We have noted already that a single crystal is one in which all unit cells are identical in content and orientation, subject to the minor orientational variation of a small mosaic spread, and that common deviations from this ideal include various types of structural disorder and twinning, which can often be handled satisfactorily once recognised. Advice is given in standard crystallography texts on the desirable quality and size of a single crystal for diffraction studies, but it tends to be appropriate mainly for molecular – especially organic – compounds investigated using conventional laboratory X-ray equipment. Some caveats need to be given for inorganic materials and for other radiation sources.

First, examination with a polarising optical microscope can help to assess the quality of a crystal; a well-formed single crystal should transmit polarised light clearly and uniformly throughout a rotation of the sample stage, except for a sharp extinction of the light every 90°, in orientations that correspond to principal directions of the anisotropic refractive index of the crystal. To this general advice must be added the recognition that many inorganic materials are very dark or opaque, so that light transmission properties are of no practical use, while others belong to crystal systems of high symmetry: cubic crystals have completely isotropic optical properties, and isotropic behaviour is also observed when tetragonal, hexagonal or trigonal (**uniaxial**) crystals are viewed along their principal symmetry axes. In such cases there is no transmission of polarised light at all. Extinction by different parts of the crystal in different orientations indicates a composite crystal, which may or may not be a twin. It is not necessary for a crystal to have well-defined faces, although this does help in applying good absorption corrections if these can be measured and assigned indices once the unit cell is known. Ultimately, the best test of crystal quality is the diffraction pattern.

Regarding optimum crystal size, a number of contributing and competing factors apply, so 'one size fits all' is certainly not correct. Although it is not an exact formula, an estimate of the relative scattering power (the **scattering efficiency**)[10] of a crystal is given by the formula:

$$\frac{V_{crystal}}{V_{cell}^2} \sum_{cell} f^2 \tag{2.23}$$

A direct dependence of diffraction intensities on the crystal volume, $V_{crystal}$, is obvious. Two other factors are advantageous for many inorganic materials. The scattering is inversely proportional to the square of the unit cell volume; this is one reason why biological macromolecules give generally weak diffraction. Although some inorganic materials have rather complex structures with large repeat units, in many cases the cell volume is not particularly high, even for high-symmetry structures in which the asymmetric unit is a small fraction of the complete unit cell, because it is quite common for some atoms to lie in special positions. Furthermore, the scattering depends on the elements present in the structure *via* the $\sum f^2$ term. Inorganic materials often contain elements with relatively high atomic numbers and have a low hydrogen content compared with most organic compounds. The combination of these various effects leads to generally stronger diffraction by inorganic materials than by other chemical and biological systems, so that it is possible to obtain measurable diffraction patterns from relatively small crystals. Indeed, small crystals are preferable in many cases as they help reduce absorption effects, which depend exponentially on crystal linear dimensions, while diffraction intensities increase only in proportion to crystal volume. For the same reason, when significant proportions of heavier atoms are present, short wavelengths ($MoK\alpha$ close to $0.7\,Å$, $AgK\alpha$ close to $0.5\,Å$, or a similar short synchrotron wavelength) will be preferred to large ones ($CuK\alpha$ close to $1.5\,Å$, $CrK\alpha$ close to $2.3\,Å$). In avoiding serious absorption problems, the overall crystal size is less important than its shape, with extreme examples being thin plates or long needles.

As a general rule, a single crystal should be as large as possible in neutron diffraction, because of the weak interaction of neutrons with samples and the lack of significant absorption problems.

2.4.3   Measuring the Diffraction Pattern

Devices and techniques for recording X-ray and neutron diffraction patterns have undergone many developments and improvements. In the first half of the 20th century, most measurements were made using photographic film and subsequent 'offline' analysis to obtain diffraction geometry and intensities (one type of X-ray camera is shown in Figure 2.1a). The Weissenberg camera, based on a different geometry for the film and the crystal movement relative to the incident beam, was particularly widely used. For the next few decades the main equipment was small-sized electronic detectors (usually scintillators and photomultipliers) that recorded one reflection at a time, in so-called **serial diffractometers**; with these it was necessary to establish reliably the unit cell geometry and the orientation of the crystal mounted on the instrument, *via* analysis of a small number of reflections found in some kind of random or systematic search, before the full diffraction pattern could be measured. Experiment time depended linearly on the size of the asymmetric unit, and often only a symmetry-unique set of data was collected; data collection times of a few days were typical, somewhat reduced from the weeks often required for photographic methods.

The last 20 years have seen the widespread replacement of serial diffractometers by **area detector** systems, which record many reflections simultaneously; these are effectively electronic versions of photographic film. Colour-centre image plates were already in use, mainly by macromolecular crystallographers, whose samples needed long exposures and were therefore not greatly disadvantaged by the slow image read-out, and there were some other types of area detector, but from the mid-1990s **charge-coupled devices** (CCDs, the same technology used in digital cameras and camera phones) became dominant, which had the advantage of faster read-out. Various solutions to the problem of the small size of CCDs were developed, including tiled systems and fibre-optic tapers, which reduced the optical image on an X-ray phosphor plate to the CCD positioned behind it. A typical CCD-based area detector system is shown in Figure 2.1c, in which the relatively large detector face plate is coupled to the smaller CCD chip by a fibre-optic taper. Older serial point-detector systems were fairly similar in overall appearance, but had only a small detector in place of the large one, capable of measuring only one diffracted beam at a time and requiring more movement of the crystal in order to bring reflections individually into the horizontal (equatorial) plane in which the detector itself moved to the correct 2θ angle.

New technology that is now coming into use is the so-called **solid-state pixel detector**, in which the X-ray diffraction pattern is recorded directly on the detector surface and the measurements are made in real time rather than being accumulated and then read out subsequently. This development considerably increases the rate at which data can be collected, from the CCD's timescale of hours to minutes or even seconds for a complete diffraction pattern. This is a particular advantage for synchrotron facilities and at neutron sources, for which similar area detectors have also been developed.

With such rapid methods available, it is feasible to collect a complete diffraction pattern in a short time – including all symmetry-equivalent reflections – and not just a unique set of data appropriate to the space group. Indeed, the data set will usually include multiple measurements of the same (not just symmetry-equivalent) reflections. This high degree of **redundancy** (or **multiplicity of observations**) brings considerable advantages: it allows a thorough and careful examination of the symmetry of the diffraction pattern, which can help to recognise pseudosymmetry problems; it provides plenty of data for use in empirical absorption corrections; it allows significantly deviating intensities to be identified, so that they can be investigated, corrected or eliminated from the data; it improves the overall signal-to-noise ratio (SNR) of the data; and it can ensure that a full set of Friedel opposites is present for non-centrosymmetric structures.

Another major advantage of area detectors over serial diffractometers is that the whole explored region of reciprocal space is recorded, not just the reciprocal lattice points (Bragg reflections) with their immediate background neighbourhood. In cases that prove difficult, it is possible to examine the whole pattern carefully, and this may show that the true unit cell is larger than was originally thought (a subcell is found initially; further reflections lie between the points of its corresponding reciprocal lattice), or it may reveal a second twin component (or even more) or satellite reflections of an incommensurate structural component.

2.4.4 Correcting for Systematic Errors

As already noted, in many cases it is necessary or at least desirable to correct the raw measured data for absorption effects. Extinction corrections, by contrast, are usually made later as part of structure refinement. Other corrections that are generally applied before the structure is solved are for variations in the measured intensity due either to changes in the incident beam (these are particularly important for synchrotron data, where there may be small fluctuations and/or a slow decay, depending on the methods used to inject and maintain the synchrotron particle beam) or to changes in the diffracting power of the crystal caused by gradual decomposition or degradation of quality or by imperfect crystal centring in the X-ray beam, such that the volume of the crystal in the beam does not remain constant as it is rotated. All these corrections depend for their success on the high degree of data redundancy that characterises modern diffraction crystallography.

Corrections are also made at the same time for geometrical factors resulting from the particular type of equipment in use and for polarisation effects of the radiation source, optics and experimental sample itself. These are well established, routine and usually applied completely automatically by standard data-processing software.

2.5   STRUCTURE SOLUTION

The measured and corrected intensities $I(hkl)$, each together with an estimate of its precision $\sigma[I(hkl)]$, generated on the basis of expected statistical behaviour of the diffraction process and known as a **standard uncertainty**, s.u. (formerly called **estimated standard deviation**, e.s.d.), usually amount to several thousand data points. Each is labelled with its corresponding set of three indices h, k, l, which identify the reciprocal lattice point associated with this diffracted wave and its direction relative to the incident beam and the crystal. Remember that each wave has both amplitude, $|F(hkl)|$, and relative phase, $\phi(hkl)$, these two being represented together by the complex number, $F(hkl)$, which is called a **structure factor**. Unfortunately, all the phases are lost in the process of recording the diffraction pattern and only the amplitudes are available, on a relative scale, as experimental data:

$$I(hkl) \propto |F(hkl)|^2 \qquad (2.24)$$

We have already seen how the structure factors (diffracted waves) are related to the electron density of the crystal structure, expressed either as discrete atoms or as a continuous function:

$$F(hkl) = \sum_j f'_j \exp\left[2\pi i\left(hx_j + ky_j + lz_j\right)\right]$$

$$= \int_{cell} \rho(xyz) \exp\left[2\pi i\left(hx + ky + lz\right)\right] dV \qquad (2.25)$$

This equation represents what happens physically in the diffraction experiment. It can also be used to calculate the diffraction pattern expected for a structure for which we have a model in either atomic or electron density functional form. The calculation gives both amplitudes and phases of the structure factors, but the experiment provides only amplitudes, with no phases.

The structure factor in Equation 2.25 is an example of a **Fourier transform**, relating a function in direct space (the crystal structure) to a corresponding function in reciprocal space (the diffraction pattern). A fundamental property of Fourier transforms is that they are reversible, so we can write:

$$\rho(xyz) = \sum_{hkl} F(hkl) \exp\left[-2\pi i\left(hx + ky + lz\right)\right] \qquad (2.26)$$

This means that the electron density of the structure is obtained by adding together (superimposing) all the diffracted waves (structure factors) with their correct relative phases. The calculation needs to be performed only over the asymmetric unit of the unit cell, since values at all other points are then fixed by symmetry. This equation represents what would be achieved if we could take the diffraction pattern and physically recombine the waves with lenses to give an image, as occurs in an optical microscope; unfortunately, this is physically unachievable, as lenses for X-rays are almost non-existent (they are highly specialised devices that can be used in particular uncommon circumstances).

Not only can we not use a physical X-ray microscope but the simple use of Equation 2.26 (the **reverse Fourier transform**) to calculate an image directly from the measured diffraction pattern is also impossible, because we have the amplitudes but no phases; this is the notorious **phase problem** of crystallography. Experimental measurement of phases is also a very non-routine procedure, with limited practical feasibility and application.

All these principles and equations are equally valid for neutron diffraction, the main difference being that 'electron density' must be replaced by 'neutron scattering density', which is modelled in an atomic representation by individual neutron scattering lengths appropriately modified to account for atomic displacements; a 'neutron scattering density map' obtained by application of Equation 2.26 (if we had the correct phases) would look rather similar to an electron density map, with peaks in the 3D density distribution corresponding to atoms, but the peaks would be considerably sharper (the widths representing only the vibrational displacements of the atomic nuclei and possibly some structural disorder), would have heights corresponding to neutron scattering lengths rather than atomic numbers and in some cases (for atoms with negative scattering lengths) would actually be holes rather than peaks.

One major task in X-ray or neutron single-crystal diffraction analysis is thus to solve the problem of the missing phases by somehow obtaining estimates of at least some of the phases reliable enough to generate a recognisable image of a significant proportion of the structure. This may be done in a number of ways.

2.5.1   Direct Methods

A whole family of approaches is known collectively as **direct methods** and is used widely to solve molecular crystal structures, especially of organic compounds. These methods are based broadly on the recognition that the reverse Fourier transform (Equation 2.26) is simply a sum of waves, each of which contributes positive values in some regions of the structure and negative values in others, the positive and negative contributions giving an overall net value of zero. The only term in Equation 2.26 that makes a net positive contribution is $F(000)$, equal to the total in-phase scattering power of the unit cell (or asymmetric unit, depending on the volume over which the sum is being performed). Thus, each of the other terms in the sum depletes scattering density in some regions and enhances it in others. The largest impact clearly comes from the highest structure factor amplitudes, allowing for the general decay in amplitudes with increasing Bragg angle; direct methods usually *normalise* the amplitudes, using the observed θ-dependent decay to estimate relative amplitudes that would be expected for stationary point atoms (all electron density at the nucleus). Although *individual* phases of these relatively strong reflections are unknown, the requirement that the summation gives a result that is nowhere negative and that has density concentrated in discrete atoms does lead to *relationships among* the phases (sums and differences of small numbers of phases) of reflections with related indices. These relationships have associated probabilities of being correctly predicted, the probabilities themselves depending on the amplitudes: the stronger the reflections in the relationship set, the higher the probability of the relationship being correct. Usually, for structures of small to moderate size, there are rather more relationships with acceptably high probabilities than there are unknown phases of strong reflections, and a variety of approaches use the relationships to find more-or-less consistent sets of phases for a small subset of the reflections in the complete data set. Many of these take a 'brute force' approach based on initially assigned random phases, large numbers of such starting points being processed, often in parallel, by modern fast computers in search of a solution. Alternatively, the process may be started by randomly positioning atoms or electron density and using the forward Fourier transform (Equation 2.25) to obtain a calculated diffraction pattern, including initial phases, to use with the relationships; the starting point for the calculations may be either in direct space (random atoms) or in reciprocal space (random phases), interconversion being *via* Fourier transforms.

Such direct methods often succeed for inorganic materials, locating some if not all of the atoms, but it should be no great surprise when they do not, because the underlying theories of direct methods that lead to the phase relationships they exploit are based on the assumption of an essentially random distribution of atoms of similar scattering power in the structure. In cases of very disparate scattering powers ('unequal atom structures') and/or high symmetry or strong pseudosymmetry, this assumption is far from valid.

2.5.2   Patterson Synthesis

An alternative approach well suited to unequal atom structures is the use of the Patterson function, defined as:

$$P(uvw) = \sum_{hkl} |F(hkl)|^2 \cos\left[2\pi\left(hu + kv + lw\right)\right] \qquad (2.27)$$

This is similar in form to Equation 2.26, except that squared amplitudes are used and all relative phases are set to zero (effectively ignored; all waves are combined in phase). The Patterson function uses only information that is available experimentally (amplitudes and indices, no phases) and so can be calculated immediately from any measured set of data. It has a form similar to an electron density map but with broader peaks, many of which are not resolved from each other. Its practical value comes from another general property of Fourier transforms, which states that the Fourier transform of a product of two functions is a convolution of the separate Fourier transforms of the two functions. Calculating a Patterson function, as defined by Equation 2.27, is mathematically equivalent to forming the Fourier transform of the product of the diffraction pattern (the set of complex numbers $F(hkl)$) with its complex conjugate (in which the signs of all imaginary components are reversed). The complex conjugate of the diffraction pattern is actually the diffraction pattern that would be observed for the inverse of the crystal structure being investigated (*i.e.* invert the signs of all x, y, z coordinates of all atoms). The Patterson function is therefore the convolution of the crystal structure with its inverse (also called the autocorrelation function of the electron density), which may be expressed in this way:

$$P(uvw) = \int_{\text{cell}} \rho(xyz)\rho(u - x, v - y, w - z)dV \qquad (2.28)$$

What this means is that peaks occur in the Patterson map at positions corresponding not to atoms, but to vectors between pairs of atoms. For N atoms in the unit cell, there are N^2 such vectors, though N of them coincide at the position 0,0,0, giving a very large peak there. Many peaks overlap and are not resolved, but those corresponding to two atoms with large scattering powers are likely to be prominent, because Patterson peak sizes are proportional to the product of the two atomic scattering factors (X-ray or neutron).

While a Patterson map for an equal atom structure is usually rather featureless, unequal atom structures give maps with a small number of relatively large peaks standing out of the background. When these peaks are caused by pairs of symmetry-related atoms, they lie in special positions and can often be readily identified. The process of solving a structure by Patterson map analysis consists in proposing positions for some of the atoms, usually the heaviest, that are consistent with the main Patterson peaks.

2.5.3   Symmetry Arguments

In the structures of many inorganic materials, some atoms, often strongly scattering ones, lie in special positions on symmetry elements. In some cases the positions of these atoms can be deduced, or at least a small number of possibilities for them can, purely from consideration of the symmetry, the number of formula units in the unit cell (Z) and perhaps the preferred coordination geometry. As an example, consider the cubic form (there is also a known lower-symmetry polymorph) of the salt $[Cr(NH_3)_6][HgCl_5]$, which is isostructural with its analogues having Cu and Cd instead of Hg.[11] The space group is $Fd\bar{3}c$, for which the asymmetric unit is $\frac{1}{192}$ of the unit cell. Simple calculation from the measured unit cell parameter ($a = 22.653$ Å) and the density shows that there are 32 cations and 32 anions in the unit cell; both ions must therefore lie in special positions at which several symmetry elements intersect, and only in the case of N and H are there enough atoms present in the cell for these to lie in general positions. Reference to space group tables shows that the only possible sites for the Cr and Hg atoms have either $\bar{3}$ (S_6) or 32 (D_3) point-group symmetry. The former is consistent with the expected octahedral coordination for the $[Cr(NH_3)_6]^{3+}$ cation (S_6 is a subgroup of O_h) and the latter imposes exact regular trigonal bipyramidal geometry on the $[HgCl_5]^{3-}$ anion (D_3 is a subgroup of D_{3h}), with the Cl atoms also in special positions, three of them on twofold rotation axes and the other two on threefold rotation axes. The asymmetric unit contains only one atom each of Cr and Hg (both with fixed coordinates and with constraints on their anisotropic displacement parameters), two Cl atoms (with some positional and displacement parameter constraints) and one N and three H atoms in general positions. Thus the metal atoms can be located entirely from symmetry considerations, without the need for direct methods or a Patterson synthesis.

2.5.4 Charge Flipping

A structure solution method that has become very popular since its introduction a few years ago is **charge flipping**.[12] This is one of a number of methods (which include more recent enhanced direct methods, often known as **dual space methods**) that involve successive use of forward and reverse Fourier transforms (Equations 2.25 and 2.26, respectively) with some kind of adjustment of the generated electron density in direct space (*e.g.* based on expected interatomic distances or other geometry, suppression of negative electron density or other so-called 'map modification') to give an updated trial structural model, as well as the combination of observed amplitudes (the known experimental data in reciprocal space) with phases calculated from the current model in each cycle, these phases possibly being modified to satisfy expected phase relationships more closely. Charge flipping begins with randomly assigned electron density, obtains from it calculated phases to combine with the observed amplitudes, then generates a new electron-density map, in which the sign is inverted (flipped) for all electron density with less than a specified small positive threshold value, which gives the new electron density for the next cycle. A large number of cycles are performed; during each one the observed and calculated amplitudes are compared and an overall numerical measure of their disagreement is calculated. Successful structure solution is usually recognised by a sudden marked decrease in this value.

The method seems improbable as an idea, but it works surprisingly well for something so simple, even where other methods fail. Usually a complete set of data (a whole sphere of reciprocal space) is used in the calculation, the symmetry-unique reflections being expanded by application of the initially estimated symmetry, and calculations cover the entire unit cell rather than just the asymmetric unit; this has the advantage that no symmetry is actually assumed in the method itself, and the true symmetry is revealed in the solution that is found. This is particularly useful in cases of ambiguous space group choices and for pseudosymmetry problems.

2.5.5   Completing a Partial Structure Model

In many cases the initial structure solution, by whatever method, reveals the approximate positions (exact for some special positions) of some but not all of the atoms; in general, atoms with lower scattering power are likely to be missing, and H atoms will not yet be found by X-ray diffraction. Some refinement (see later) of the positions and displacement parameters of the located atoms can be carried out at this stage, in conjunction with the search for the remaining atoms. This search is usually done by some kind of Fourier recycling procedure, although details can vary. The approach is shown in Figure 2.13. The atoms so far located serve as a currently incomplete structural model, for which a corresponding diffraction pattern can be calculated, giving a full set of amplitudes, $|F_c|$, and phases, ϕ_c, for each reflection. Comparison of $|F_c|$ with the observed amplitudes, $|F_o|$, indicates how well the model fits the experimental data; a number of **residual indices** can be calculated, examples being the following:

$$R1 = \frac{\sum ||F_o| - |F_c||}{\sum |F_o|} \qquad R2 = \left[\frac{\sum (F_o^2 - F_c^2)^2}{\sum (F_o^2)^2} \right]^{1/2}$$

$$wR1 = \frac{\sum w^{1/2} ||F_o| - |F_c||}{\sum w^{1/2}|F_o|} \qquad wR2 = \left[\frac{\sum w(F_o^2 - F_c^2)^2}{\sum w(F_o^2)^2} \right]^{1/2} \qquad (2.29)$$

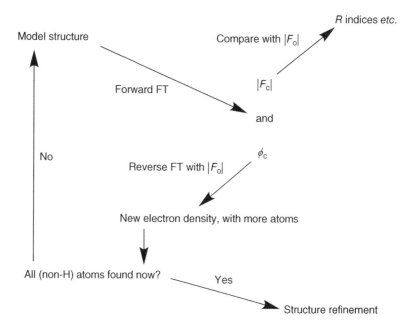

Figure 2.13 Schematic flowchart for the completion of a partial model structure.

notation being taken from the most widely used program for structure refinement (programs are described below). The $R1$ and $wR1$ residuals are similar to mean deviations, while $R2$ and $wR2$ are similar to root-mean-square deviations. In the weighted residuals $wR1$ and $wR2$, a weight is assigned to each observed reflection in view of its perceived relative reliability; usually these weights are derived from the standard uncertainties of the intensities, a simple case being $w = 1/\sigma^2(F_o^2)$. These residuals are defined as dimensionless so that they can be compared for structures of different sizes, and they may be expressed either as shown here or as percentages, by multiplication by 100%. Which are the most appropriate residual indices to use depends on the details of the structure refinement methods. A typical well-determined crystal structure of an inorganic material, after refinement is complete, should have $R1$ approximately in the range 0.01–0.05; $R1$ is widely used, having become established early in the development of crystal structure analysis. The $R2$ indices, based on F^2 rather than $|F|$ values, are usually somewhat higher, typically 0.04 – 0.15. An initial structural model will have considerably higher residuals, but their values will decrease as the model is improved.

So much for the amplitudes, $|F_c|$, of the calculated diffraction pattern. The most important result, however, is that we have a set of phases, ϕ_c.

These are not the correct phases for the structure, of course, as the model structure is not correct and complete, but they are the best estimate we currently have; they are (probably) better than nothing! Combining the *calculated* phases ϕ_c with the *observed* amplitudes $|F_o|$ allows us to perform a reverse Fourier transform and obtain an electron-density map. This will show features of the model structure, but there will be significant differences because $|F_o|$ rather than $|F_c|$ was used in the calculations. Correctly identified atoms of the model structure are retained; incorrect atoms are likely to appear with higher or lower scattering density and/or shifted in position. New peaks show where atoms that are missing from the model should be added. These features allow us to make an improved model, with which the cycle of forward and reverse transforms can begin again, the process being repeated until the model essentially reproduces itself with no further significant changes.

If the reverse transform is calculated using $|F_o|-|F_c|$ instead of $|F_o|$, in conjunction with the phases ϕ_c, a **difference Fourier map** is obtained. Here the appearance of the correct atoms in the model structure is suppressed (not always completely) and new peaks corresponding to missing atoms stand out more clearly. This is the approach most widely used, together with variants of it, in which other, more complicated combinations of $|F_o|$ and $|F_c|$ appear and the contributions of individual reflections to the sum may be weighted to enhance particular features or to reduce the impact of expected errors in the calculated phases. Some of the main difficulties encountered in this stage of a structure determination involve identifying and modelling the disorder components of a structure, and these have to be dealt with on a case-by-case basis.

Although this procedure for completing an initial partial structure model works smoothly for most molecular materials, there is a particular problem that can arise when the atoms of the model structure, taken on their own, display a higher symmetry in their arrangement than the true space group symmetry of the complete structure, and this occurs more often for inorganic materials, especially when the atoms first located lie in special positions. In such cases, the extra incorrect symmetry elements of the partial model are carried through the Fourier transform calculations and appear in the new difference or full electron density map, which therefore contains not only the features that are being sought, but also a false second image (or even more than one extra image) of these features related by the extra symmetry (the false symmetry is most often twofold, rather than higher; *i.e.* inversion, reflection or a twofold rotation axis). Selection of one correct, self-consistent set of atoms from the double image may be done manually on the basis of recognised sensible geometry or gradually by adding just one new atom from a related pair and performing another Fourier cycle, in which one image should appear more strongly than the other; alternatively, it can be done by programs that perform direct methods of some kind to locate the missing part of the structure using some atoms already in place.

The simplest example of this situation is a structure in space group $P1$, for which the model structure consists of a single **heavy atom**, by which we mean an atom with a significantly higher scattering factor than those of others (the origin in $P1$ is completely arbitrary as there is no symmetry other than pure translation; if there is one heavy atom in the structure, it can be placed at the origin, thereby defining the origin). This model structure, however, has inversion symmetry; the heavy atom on its own conforms to the higher-symmetry space group $P\bar{1}$. The electron density map, or difference map, calculated by Fourier recycling from the model, also has inversion symmetry and contains both the true structure and its inverse superimposed. Similar arguments apply for many other space groups, and inversion is not the only possible false symmetry.

A similar problem is found when the atoms of the model structure all lie in special positions such that they make no net contribution to certain subgroups of reflections; the model structure has a smaller effective unit cell than the complete structure (it forms a subcell). It therefore gives no calculated phases at all for these reflections, to which only the as-yet missing atoms contribute scattering. Note that one consequence of this is that such reflections are, on average, significantly lower in intensity than the reflections to which all atoms contribute. This can lead to an incorrect (too small) unit cell being found, a problem that affected serial diffractometers much more than it does area detectors, because unit cells were often determined from relatively few reflections found in an initial search, and these inevitably tended to be strong reflections on the whole. The situation can usually be recognised by careful visual inspection of the diffraction pattern and by examining the statistics on the intensities of reflections as a function of indices, index parity classes (different even/odd combinations) *etc.*, and appropriate care can be taken in approaches to solving the structure. This is one of many situations, encountered particularly in studying inorganic materials, where automated methods are likely to fail and human crystallographic experience and expertise are essential. The earlier example of $[Cr(NH_3)_6][HgCl_5]$ in the cubic space group $Fd\bar{3}c$ presents such a case: the metal and axial chlorine atoms contribute only to certain classes of reflections, which are generally strong, while other reflections, receiving scattering from only equatorial Cl, N and H atoms, are weak. A much more frequently encountered situation arises with structures in the common monoclinic space group $P2_1/c$ (which may also be given as $P2_1/a$ or $P2_1/n$ with different choices of axes) with $Z = 2$ and heavy atoms lying on inversion centres; in this case, these atoms contribute to only half of the reflections, which correspond to a pseudocentred rather than the correct primitive unit cell.

2.6   STRUCTURE REFINEMENT

Once all the atoms of the asymmetric unit have been located (though not usually H atoms at this stage, nor possibly some atoms subject to disorder), the structure is essentially solved; the model structure gives a calculated diffraction pattern with amplitudes in reasonable agreement with the observed ones, and Fourier recycling gives no significant changes. Some refinement of atom positions and displacement parameters may already have been undertaken between Fourier calculations, but now the refinement is taken further. Refinement means systematically adjusting the various numerical parameters that describe the model structure in order to give the best fit between observed and calculated amplitudes. What then is meant by 'best fit'? Although there are a number of possible criteria, the one most widely used is known as **non-linear least-squares refinement** and is a well-established mathematical and computational procedure for fitting a set of parameters to data that are related to them by non-linear equations such as Fourier transform equations. Linear least-squares fitting, or linear regression, is a simple version of this, familiar in fitting straight-line plots to data. Where equations are non-linear, the theory involves approximations that mean the process gives shifts in parameters rather than the values of parameters themselves (which is why crystal structures have to be solved before they can be refined) and has to be iterated in cycles until the shifts become negligible (**convergence** of refinement). Note that refinement can only adjust existing parameters and cannot of itself introduce new ones; during refinement, difference maps should be generated and examined to see if further parameters need to be added, *e.g.* for disorder modelling, atoms not previously identified (H atoms, low-occupancy disorder), extinction *etc.*

2.6.1   Minimisation and Weights

The function to be minimised in least-squares refinement is usually one of the following, commonly described as 'refinement on F'and 'refinement on F^2'.

$$\sum w(|F_o| - |F_c|)^2 \text{ or } \sum w(F_o^2 - F_c^2)^2 \tag{2.30}$$

Here each reflection is assigned a weight w. Weighting functions vary considerably in different refinement programs and procedures, but they are usually based on the standard uncertainties of intensities derived from the data processing carried out during and after data collection. From initial $\sigma(I)$ values, simple calculations give $\sigma(F_o^2)$ and $\sigma(|F|)$ for use in refinement. Weighting schemes recognise that these standard uncertainties come from analysis of random errors only, and that there are also systematic errors that need to be accommodated. Thus a typical weighting scheme for F^2 refinement could be:

$$w = 1/[\sigma^2(F_o^2) + \text{extra terms}] \tag{2.31}$$

where the extra terms may be functions of F_o, F_c, θ *etc*. Optimisation of these extra terms is usually itself part of the refinement, with the aim of having a minimisation function that does not vary significantly and systematically with intensity, with Bragg angle or with indices.

2.6.2   Parameters, Constraints and Restraints

The parameters that can be refined consist mainly of three coordinates (x, y, z) for each atom, one (isotropic), six (standard anisotropic model) or more displacement parameters for each atom and an overall scale factor that puts the calculated values $|F_c|$ or F_c^2 on the same scale as the observed values $|F_o|$ or F_o^2, with $F(000)$ fixed at the number of electrons per unit cell to define an absolute scale. In the case of disorder, the occupancy of each atom site may also be a refined parameter instead of being fixed at 1. There may be further parameters, *e.g.* for extinction. A fixed occupancy of 1 for an ordered atom fully occupying its site is an example of a **constraint**. Constraints may also be required for some other parameters or combinations of parameters, *e.g.* atoms on special positions are subject to constraints on one or more of their coordinates and, possibly depending on the actual site symmetry, on their anisotropic displacement parameters. One approach to positional disorder, for which unconstrained refinement often gives unsatisfactory results, is to constrain the geometry of a particular group of atoms such as a solvent molecule with a well-known and rigid shape, and to allow only its orientation to vary. In X-ray diffraction, H atoms are often constrained to maintain expected bond lengths and angles; they are made to **ride** on the atom to which they are bonded.

There are situations in which rigid constraints are inappropriate but free refinement is unsuccessful. Here **restraints** (also known as soft constraints) may be applied. Some feature of the structure, such as a bond length or angle, is assigned a desirable target value, and the difference between this value and the one calculated from the parameters of the current model is added, squared and with an appropriate weight, to the minimisation function. The target value thus acts effectively as an extra observation along with the diffraction data, and the refined parameters adjust to give the best fit to the total set of observations. Restraints may be applied to various geometrical features, *e.g.* the expectation that a group of atoms should be in a common plane or that non-bonded atoms should not lie too close together, and also to displacement parameters, *e.g.* similar displacement amplitudes of two atoms of similar size along the bond connecting them or similar displacements of two disorder components of an atom on sites close together. Restraints are very useful, particularly for disordered structures and when atoms of very different scattering powers are present, and their success depends critically on finding appropriate target values and on assigning sensible weights that represent a proper balance between diffraction data and restraints, which are supposed to complement each other, rather than compete. It is important that constraints and restraints are clearly described as part of the structural model that is eventually presented as the outcome of the experiment.

2.6.3   Refinement Results

Refinement produces a set of parameters, together with an s.u. for each refined parameter; constrained parameters have zero uncertainty. The s.u. values will reflect the use of restraints as well as constraints; strongly restrained parameters will have lower s.u. than if they were unrestrained. They inevitably also reflect the relative scattering powers of the atoms concerned, and are generally higher for atoms affected by disorder. The overall level of s.u. values indicates the precision and quality of the structure determination; values are calculated by formulae that include the minimisation function (itself related to the residual indices) and depend inversely on the excess of number of unique data over number of refined parameters.

As far as structural geometry is concerned, which is likely to be the main interest in a crystal structure determination, the primary result of refinement is a set of atomic positions relative to the unit cell origin and axes. From the atomic positions, together with the cell parameters and space group symmetry, interatomic distances (whether bonded or not) and angles can be calculated, each with an associated s.u., as well as aspects of conformation, coordination geometry, network topologies, the shapes and sizes of any unoccupied void spaces in the structure *etc*. Here there is plenty of scope for interpretation (and misinterpretation!), description and both numerical and graphical representation. For structures with any complexity, the convergence of refinement and the satisfactory lack of any significant features in a final difference map (an important indication that the structure is complete) by no means signal the end of the task of structure determination.

2.6.4   Computer Programs for Structure Solution and Refinement

While many programs are written specifically for one of other of these stages, the two are not entirely separate and some structure solution programs in particular do perform a limited initial refinement as part of their assessment of whether a structure is likely to have been solved successfully. New programs are written from time to time, while once-popular programs fall into relative disuse when they are not further developed and they are overtaken. Factors such as ease of use, the provision of graphical user interfaces and the availability of versions for particular computer hardware and operating systems are also important. More recently, integrated packages have been developed that have a user interface and framework within which other well-established older programs are embedded to do the main work. They provide utilities that link these together, possibly with some special individual features of their own for structure solution or refinement. Examples of these include WINGX[13,14] and OLEX2.[15]

Currently, the most widely used public-domain programs for direct methods and Patterson analysis are probably SHELXS[16,17] and members of the SIR family.[18,19] DIRDIF[20] applies direct methods to solve a 'difference structure' once parts of the structure have been found that do not provide an adequate phasing model for the rest – one of the problems discussed above. More advanced **dual space** direct methods are found in SHELXD[16,17] and in the earlier SnB (Shake-and-Bake) program.[21,22] Charge flipping is performed by the original SUPERFLIP,[23,24] and similar procedures are included in a number of packages, such as WINGX, OLEX2, and PLATON.[25,26] In addition, as for refinement programs, commercial software is supplied by manufacturers of X-ray diffraction equipment systems.

The most popular refinement programs are SHELXL[16,17] and CRYSTALS.[27,28] The program JANA[29] provides extensive facilities for handling incommensurate structures.

2.7 PROBLEM STRUCTURES, SPECIAL TOPICS, VALIDATION AND INTERPRETATION

A number of factors can render a crystal structure determination non-trivial, and it is by no means always obvious from the outset that a particular structure will present problems. Difficulties are encountered more often with inorganic materials than with molecular compounds. A few of the more common problems are described here, together with some additional topics. We also consider some issues of interpretation of results and ways in which the correctness of a structural result can be assessed.

2.7.1 Disorder

Truly random disorder leads to a model structure that describes an average unit cell in which some atom sites are not fully occupied; in the real structure, of course, every site either has an atom or does not, but the presence or absence varies in a non-systematic way from one cell to the next.

Partial occupancy of a single site can occur for uncharged atoms and for ions of an element displaying more than one possible charge, but overall charge balance must be maintained; here the site occupancy factor is treated as a refined parameter. Occupancy factors and displacement parameters are fairly strongly correlated, although the scattering power depends linearly on one and exponentially on the other, so it is usually possible to obtain satisfactory results if the atom concerned has a sufficiently high scattering factor and we have data to a sufficiently high maximum Bragg angle.

If two atoms/ions are disordered substitutionally over a single site, or if an atom/ion is disordered over two possible positions, the occupancy factors can be refined, but their sum must be constrained to unity so there is actually only one independent parameter, not two. It is often necessary to constrain the displacement parameters of two or more atoms sharing a site so they are equal. The same can be done for atoms disordered over positions that are so close together that their scattering densities overlap strongly, making it impossible to refine individual displacement parameters freely; an alternative approach is to use restraints so that the atoms have different but related parameters. Restraints can also be very effective in maintaining essentially the same geometry (bond lengths and angles, but still allowing some conformational flexibility) for a whole group of atoms disordered over two overlapping positions, such as an organic ligand or a solvent molecule. A good example is the cyclic ether tetrahydrofuran, which is often disordered but is flexible and not usually planar, so that rigid constraints are inappropriate.

In the most extreme cases of disorder, individual atoms cannot be resolved, at least not with any sensible geometry, and a difference map shows diffuse density rather than discrete peaks. Another approach here is to calculate a Fourier transform of the diffuse density region and subtract this contribution of the disordered component from the observed structure factors, then continue the refinement with the modified data and with an empty void in the model structure. The calculation, most often performed by the SQUEEZE option of PLATON, provides estimates of the total diffuse density and the volume occupied by it, from which it may be possible, with knowledge of the chemical synthesis conditions, to identify the chemical nature and amount of the disordered species, which is often a solvent.

2.7.2 Twinning

Disorder involves random alternative arrangements of atoms within the unit cell. Twinning, by contrast, arises from alternative orientations of the unit cell that are not random but related systematically. In a single crystal there are **domains** of ideal structure, with the components arranged regularly on the lattice, but defects occur between them, called **grain boundaries**, which lead to slight variations in the orientations of the domains, giving the mosaic structure of what is referred to as an **ideally imperfect single crystal**, the type encountered normally. Domains or grains may also come together in crystal growth in a random way, giving a specimen that is a conglomerate or multiple crystal, of no use for single-crystal diffraction studies.

There is, however, another possibility that is not a single crystal but not a random conglomerate either. One form of this can occur when a **supercell** (a cell with a greater volume, having lattice points both at its corners and lying between them) is found that has a shape appropriate to a higher-symmetry crystal system, either exactly or (more usually) approximately. Under such circumstances, and if interactions within the structure do not prevent it energetically, a block of unit cells (a domain) may be rotated (or reflected or inverted) to a different orientation. The two orientations are related by a symmetry operation that could exist in the supercell but does not actually occur in the true structure; the domains fit together reasonably well in the two related orientations. A simple example could be an orthorhombic structure in which $2a \approx b$. A pair of unit cells side by side forms approximately a square when viewed along the c-axis. If such a pair of cells is rotated by $90°$ about the c-axis (a fourfold rotation, consistent with the tetragonal system appropriate to but not required by a square shape) it can still fit approximately into the same space, so it is possible for domains of this structure to co-exist in two orientations related by $90°$ rotation – symmetry that does not belong to the orthorhombic system. Such a situation, in which two (or more) parts of a structure are related by a symmetry operation that is inconsistent with the true unit cell but consistent, or approximately consistent, with a supercell, is called **non-merohedral twinning**. Some reflections of the two diffraction patterns, one for each twin component, overlap with each other, though they have different indices, while other reflections do not: they belong to only one of the domains.

A simpler situation arises when a unit cell has a shape that could almost belong to a higher crystal system, such as a monoclinic structure with β close to $90°$ (almost the right shape for orthorhombic). Here again domains of the structure can occur with different orientations related by symmetry appropriate to an orthorhombic but not a monoclinic structure, and this gives **pseudomerohedral twinning**, for which all reflections of the domains overlap each other, though they may not coincide exactly.

Exact **merohedral twinning** occurs when the symmetry relationship between domains is one that is possible for the crystal system but does not actually occur in the space group of the structure itself; for this purpose, trigonal (excluding rhombohedral) and hexagonal systems are considered together, as they have the same basic unit cell shape. Here reflections belonging to different domains coincide exactly, and this form of twinning is impossible to detect from the geometrical form of the diffraction pattern alone. A particular special case of merohedral twinning is twinning by inversion, in which domains of a non-centrosymmetric structure and of its inverse occur together in a twinned crystal; this has a bearing on the question of absolute configuration of chiral molecules (and related phenomena such as crystal polarity), which is considered further in Section 2.7.4.

Twinning, if not recognised, results in poor structure refinement and in some cases may prevent the structure from being solved or may lead to an incorrect unit cell and/or space group. Details of how to handle twinned structures are too complicated and specialised to be described here. In general, a two-component twin is fully characterised by its **twin law** (a 3 × 3 matrix expressing the pseudosymmetry relationship between the twin component orientations, which is most often a twofold rotation about a direct or reciprocal lattice vector) and its **twin ratio** (the relative amounts of the two domains, which is usually a parameter in the structure refinement); the twin law and twin ratio need to be established as part of solving and refining the structure.

2.7.3   Pseudosymmetry, Superstructures and Incommensurate Structures

A problem not uncommon for inorganic materials is that the crystal structure is best described in one particular space group but approximates to another space group with more symmetry operations. In many cases, a significant proportion of the structure fits the higher symmetry, while this pseudosymmetry is broken by the rest of the structure; common examples include high-symmetry frameworks with lower-symmetry arrangements for guest species within them. The question sometimes arises whether the correct description of the structure is as ordered in a lower-symmetry space group or as disordered in a higher-symmetry space group. It is not always easy to answer, since the difference in practice between the two models may be quite small. Refinement of pseudosymmetric structures can be problematical, as there are high correlations between pairs of related parameters and reducing the symmetry means increasing the number of refined parameters; restraints are often needed, though sometimes they can be relaxed or removed once the model moves far enough away from a false symmetry. Of course, pseudosymmetric structures are also often difficult to solve in the first place.

An example in which the pseudosymmetry is inversion is the mixed-valence molecular complex $[Co_3(acetate)_4(salen)_2]$, where salen is a tetradentate ligand with a chiral component of known handedness.[30] The true triclinic space group is $P1$ but almost the entire structure fits the higher symmetry of $P\bar{1}$, the exception being the fact that both salen ligands have identical chirality, whereas inversion symmetry would require them to be enantiomers. Indeed, the structure can be refined in $P\bar{1}$ with twofold disorder of just a few atoms of the salen ligand, but this result is known to be incorrect from the chemical synthesis.

Where the extra symmetry that may not be genuine is pure translation, the situation is a superstructure problem. If, for example, a group of atoms can adopt two different positions or orientations, the random appearance of the two possibilities is a form of disorder, but regular alternation of the two along a particular direction gives a superstructure with the unit cell doubled in size in this direction. Layers of reflections in the reciprocal lattice are alternately strong and weak, and there is a risk that the weak layers will be overlooked, a half-size unit cell found and a regular superstructure refined incorrectly and described as randomly disordered. An example in which the superstructure arrangement involves an approximately tripled unit cell axis, combined with pseudo-inversion symmetry in the same structure, is a $CdBr_2$ complex of a sulfur-donor ligand.[31] The asymmetric unit of this non-centrosymmetric structure contains six independent ordered molecules, but it is just about possible to refine (rather unsatisfactorily) a model having one molecule with rather disordered ligands, in a centrosymmetric space group and a unit cell of one-third the correct volume.

If, instead of two (or more) possible arrangements alternating to give a superstructure, there is a more complex but systematic variation with a repeat unit different from that of the crystal lattice in this direction then we have a modulated incommensurate structure, giving a diffraction pattern with weak satellite reflections that do not lie at reciprocal lattice points. This requires special procedures for the data processing and for structure refinement.

It can be seen that the phenomena of disorder, twinning, pseudosymmetry and incommensurate structures have some features that are similar in their fundamental nature and in their impact on diffraction patterns. It is possible to confuse them and use the wrong model, and there are cases where it is difficult to distinguish between them, as this often depends on the measurement and interpretation of weak features in the diffraction patterns. Evidence from outside crystallography (spectroscopy, chemical synthesis details *etc.*) may be important in resolving the problem.

2.7.4   Absolute Structure

This expression is used to cover a number of more specific terms relating to the distinction between a non-centrosymmetric crystal structure and its inverse, generated by changing the signs of all coordinates of every atom, together, in some cases, with conversion of a space group to its enantiomorphic partner, *e.g.* $P3_1$ and $P3_2$ or $P4_12_12$ and $P4_32_12$, in which the hand is reversed for chiral screw axes; in a small number of space groups, inversion must be carried out through a point other than the conventional unit cell origin. Exactly what these two inversion-related structures represent depends on the particular space group.

For space groups having no improper rotation symmetry, the contents of the asymmetric unit are, in principle, chiral, and all copies of the asymmetric unit in the unit cell and in the entire structure have the same chirality. Indeed, chiral molecules can crystallise only in such space groups, of which there are 65 in total, and these, the so-called Sohnke space groups, are the only ones encountered for naturally occurring biological macromolecules. Distinguishing between the two chiral crystal structures thus provides an experimental method of establishing the **absolute configuration** of such compounds, and this is an important use of crystallography, especially in pharmaceutical and medical research. Chirality is not restricted to molecular compounds, but may also be displayed by inorganic materials; one of the simplest examples is quartz (Figure 2.3c), which may exist in the enantiomorphic space groups $P3_121$ and $P3_221$, depending on the handedness of the $Si - O$ helical chains found in its structure.

Not all non-centrosymmetric crystal structures are chiral, however; some contain inversion axes and mirror or glide planes, and so they contain both right- and left-handed structural motifs and are racemic. In most such cases the structure contains one or more **polar axes**, directions in the structure that are not symmetry-equivalent to the opposite directions. As an example, the non-linear optical material KTP (potassium titanyl phosphate) has space group $Pna2_1$, containing glide planes (designated by the letters n and a) perpendicular to the orthorhombic a and b axes; thus the directions $+a$ and $-a$ are equivalent by symmetry, as are $+b$ and $-b$. The c-axis, however, is polar: $+c$ and $-c$ are not equivalent, and features of the structure are seen to point in one of these directions but not the other, leading to the particular physical properties of this material that would not be exhibited by a centrosymmetric structure. Determining the absolute structure of such a compound means identifying the direction of the polar axis and thus distinguishing physically one side of the individual crystal from the other side in this direction.

Whatever the precise meaning of the term 'absolute structure' for a particular compound, its determination *via* X-ray diffraction depends on exploiting resonant scattering, and so there must be a significant resonant scattering contribution from at least one element present for the particular X-ray wavelength being used. The choice of wavelength can thus be important.

The most common computational approach is to treat the crystal as if it were an inversion twin containing domains of both the model structure and its inverse, and to refine the twin ratio as a parameter.[32] This provides not only a numerical indication of the correctness or otherwise of the model structure in this respect but also a measure of its reliability through the associated s.u.;[33] it also allows for crystals that are genuine inversion twins, by no means an uncommon result for non-centrosymmetric inorganic materials. The refined twin ratio is defined as the fraction of the inverted structure present in the crystal, so a value differing insignificantly from zero means the model structure is correct and a value of essentially 1 means it should be inverted. Any intermediate value with a small s.u. indicates an inversion twin, with 0.5 representing equal contributions by the two forms. An s.u. greater than around 0.1 shows that the resonant scattering effects are too small to give a definitive answer. This is a topic in which there is considerable current development, seeking to enhance the power of discrimination between opposite absolute structures in cases where the resonant scattering contributions are relatively weak.[34]

2.7.5 Distinguishing Element Types, Oxidation States and Spin States

One reason for choosing neutrons rather than X-rays for a single-crystal diffraction experiment, despite the cost and technical difficulties, may be the very different relative scattering powers of the atoms present in the crystal. The issue could be locating light atoms precisely in the presence of heavier atoms (*e.g.* for oxides of heavy metals, or for locating H atoms – even better if the material is deuterated) or distinguishing between atoms of similar electron density (*e.g.* C, N and O when much heavier atoms are also present, or two metals adjacent in the Periodic Table, especially if they may be disordered). It is also possible to enhance the X-ray contrast between atoms of similar electron density by selecting an X-ray wavelength, probably at a synchrotron, that lies between the absorption edges of the elements concerned, so that one of them gives much more resonant scattering. In some cases it is even possible to use this technique to distinguish between different oxidation states of the same element, as their absorption edges do not exactly coincide.

For materials that have an ordered arrangement of magnetic moments due to unpaired electron spins, antiparallel spins for atoms that are otherwise symmetry-equivalent make them inequivalent for neutron diffraction, because neutron scattering is affected by atomic magnetic moments. In such cases, the unit cell or the space group (or both) is different for X-rays and neutrons. Neutron diffraction thus allows a description of the spin ordering. Although there is an effect with X-rays also, it is extremely weak and requires specialised synchrotron facilities for its study.

2.7.6   Valence Effects

Neutron diffraction locates atomic nuclei accurately and is unaffected by electron density, except for the magnetic interactions described above. In X-ray diffraction, the assumption is made that the nucleus lies at the centre of the observed electron density for each atom. Except for the small but well-recognised systematic error this produces for the location of covalently bonded H atoms, the approximation is a good one for most purposes of structure determination. The combination of the two techniques provides an experimental method for investigating valence effects such as covalent bonds and lone pairs of electrons. If the same material is studied by both techniques at the same temperature then the neutron diffraction results can be used to calculate what the electron density would be for this arrangement of atoms in their correct locations without any valence interactions. Subtraction of this from the electron density actually obtained by X-ray diffraction gives a deformation density map showing the valence contributions.

It is actually possible to obtain similar results without using neutrons at all. For this, X-ray diffraction data need to be collected to a much higher Bragg angle than for standard structure determination, and at a temperature as low as possible in order to minimise the effects of atomic displacements. Valence effects contribute mainly to scattering at low angles, while high-angle data are principally the consequence of scattering by core electron density. Refinement with low-angle data down-weighted thus produces results very similar to those from neutron diffraction (except for the different atomic scattering powers). The normal approach to these so-called **charge density** studies is to include more parameters in the refinement, each atom being assigned parameters to describe its electron density distribution in terms of a collection of **multipoles** rather than a single spherical distribution. Because of the resulting large number of parameters and high correlations among them, data must be extensive and significant to high angles, and well corrected for systematic errors such as absorption. The availability of very short X-ray wavelengths (typically $0.3 - 0.5$ Å) and high intensities at synchrotron sources makes this work feasible. Structures with significant disorder are not suitable for such studies.

2.7.7   Diffraction Experiments under Non-Ambient Conditions

Through most of the 20th century, most non-biological X-ray diffraction experiments were carried out at room temperature. Crystals were generally exposed to the atmosphere, attached to the end of a fine glass fibre, unless they were known to be air-sensitive, in which case they were sealed in thin-walled glass capillary tubes. The development of relatively inexpensive and reliable low-temperature devices using a cryogenic gas flow, usually cold nitrogen obtained from liquid nitrogen, brought major advantages, and data collection in the $80 - 150$ K range is now considered routine, both in the laboratory and at synchrotron sources. Lower temperatures reduce atomic displacements, enhance high-angle intensities, make disorder modelling easier, improve crystal stability, and enable simpler handling of air-sensitive materials; the only significant disadvantage these days is the possibility of a major phase transition on cooling, with loss of the single crystal.

Lower temperatures can be achieved with helium as coolant, either in open-flow systems or in closed cryostats. Helium cryostats are commonplace at neutron facilities and are often available at synchrotrons, where the greater penetration of the former and higher intensity of the latter reduce the problem of absorption by the apparatus. Temperatures above ambient can be attained with a variety of heating devices; the main challenge is finding a suitable and reliable way of mounting crystals for study under these conditions.

Environmental cells and other facilities are also available for the investigation of samples under high pressure (small diamond anvil cells are the most popular and can be used in a standard laboratory as well as at central synchrotron and neutron facilities), in electric or magnetic fields, under laser or other irradiation (for studying excited states) and in controlled atmospheres or gas flows. In most cases, these devices restrict access to the sample, so short wavelengths are used to compress the diffraction pattern to lower angles and, in the case of X-rays, to reduce absorption by the apparatus; neutrons have a particular advantage here, with their high penetration.

2.7.8 Issues of Interpretation and Validation

With the increasing use of automated systems and software by non-experts, the ability of modern systems to give results for structures with serious problems such as disorder and twinning, the growing application to ever more complex and demanding materials and the unfortunate appearance of cases of deliberate scientific fraud and falsification of data in recent years, the question inevitably arises: how can we be sure a crystal structure is correct? It must be recognised that the result presented in every case is not a unique and direct outcome of the experiment but rather a model structure proposed as a fit to the data and refined using tools and features chosen by the particular experimenter. The accuracy and precision of the model as a true representation of the unknown structure will depend on the quality of the crystal, the equipment and the diffraction data, the validity of data processing and corrections, the appropriateness of refinement tools and procedures, the skill of the crystallographer and the interpretation of the output of the refinement. Inorganic materials are generally open to a wider range of interpretations and presentations than molecular compounds, for which the types and geometries of interactions between atoms are more restricted and rather better understood.

One widely used resource for structure validation, which also features a rich palette of other tools for visualising, manipulating and interpreting crystal structure, is the program PLATON,[25,26] which is constantly being updated and developed to incorporate fresh ideas, requirements and suggestions from the crystallographic community. A version of it is also promoted by the International Union of Crystallography,[35] and its use is mandatory for structures submitted for publication in the Union's journals. Validation tools are also used by the curators of structural databases (of which the most relevant are the Inorganic Crystal Structure Database[6,7] and the Cambridge Structural Database[4,5]) and by other journal publishers.

It should be recognised that some of PLATON's validation tests are less relevant to inorganic materials than to molecular compounds, because they are based on the assumption of directed covalent bonding and because interatomic distances in inorganic materials are much more variable and less predictable. Some warning alerts issued by the program may thus be of limited applicability, but nevertheless it is extremely valuable as an independent check that a structure determination has been completed successfully and appropriately.

One concept that is particularly useful in validating and interpreting inorganic crystal structures is that of **bond valence sums**,[36−38] in which a numerical measure of bonding strength is calculated for each near-neighbour of a particular central atom on the basis of atom types and distances, using parameters that have been developed to give a self-consistent treatment. Adding up these **bond valences** for the central atom gives a bond valence sum for that atom, for comparison with its assigned formal charge (oxidation state). There should not be any major discrepancies; if there are, possible explanations include a wrongly assigned atom type, an incorrect oxidation state or missing H atoms. Such arguments may be used *e.g.* to decide whether a supposed O atom is actually oxide, O^{2-}, hydroxide, $[OH]^-$, or water, OH_2, in a structure for which H atoms cannot be reliably located from the diffraction data; an illustration of this application is the use of bond valence sums to decide which oxygen atoms are protonated in polyoxometalates.[39] Another example is the correct assignment of oxidation states (including the possibility of disorder) in a mixed-valence compound.

Finally, one question that always needs to be asked is whether the particular individual single crystal selected for study, which may be only microns in size in some cases, is representative of the bulk material or whether it is one structural phase among two or more present in the sample. It is certainly not unusual for the best-looking crystals to be a minor product. It is a simple matter to calculate a powder diffraction pattern from a single-crystal result and to compare this with an experimental pattern measured from a randomly selected portion of the bulk sample. This can show that the sample is essentially pure and homogeneous (monophasic) or that it contains other materials, or it might show that the obtained structure is of a material so low in concentration that it is not seen in the powder pattern. The most frustrating outcome is sometimes the unusually interesting structure of a minor product that subsequently proves difficult or impossible to obtain either pure or in significant quantities. In contrast, it is highly satisfying to find a novel structure for a pure compound that can be made readily and reproducibly, providing excellent material for a high-impact publication!

SOFTWARE ACKNOWLEDGEMENTS

The structures shown in the figures in this chapter were generated with the program XP, which is part of the commercial Bruker SHELXTL package;[40] this incorporates versions of SHELXS, SHELXD and SHELXL (a new powerful structure solution program SHELXT is under development and will be released during 2014), together with proprietary software for data processing and graphics.

The X-ray scattering factor curves of Figure 2.7 were produced with the help of an online tutorial provided by Bernhard Rupp.[41]

Other figures were drawn with ChemDraw Ultra 12.0.[42]

REFERENCES

[1] A. J. Edwards, *J. Chem. Soc. Dalton Trans.*, 2419 (1976).
[2] M. P. Attfield, Z. Yuan, H. G. Harvey and W. Clegg, *Inorg. Chem.*, **49**, 2656 (2010).
[3] *International Tables for Crystallography, Volume A*, 5th edition with corrections, International Union of Crystallography, Chester and John Wiley, Bognor Regis, 2005.
[4] http://www.ccdc.cam.ac.uk (last accessed 14 January 2014).
[5] F. H. Allen, *Acta Crystallogr. Sect. B*, **58**, 380 (2002).
[6] http://www.fiz-karlsruhe.de/icsd.html (last accessed 14 January 2014).
[7] A. Belsky, M. Hellenbrandt, V. L. Karen and P. Luksch, *Acta Crystallogr. Sect. B*, **58**, 364 (2002).
[8] *International Tables for Crystallography, Volume C*, 3rd edition, International Union of Crystallography, Chester and John Wiley, Bognor Regis, 2004, Table 6.1.1.1.
[9] *International Tables for Crystallography, Volume B*, 3rd edition, International Union of Crystallography, Chester and John Wiley, Bognor Regis, 2008.
[10] M. M. Harding, *J. Synchrotron Rad.*, **3**, 250 (1996).

[11] W. Clegg, D. A. Greenhalgh and B. P. Straughan, *J. Chem. Soc. Dalton Trans.*, 2591 (1975).

[12] G. Oszlányi and A. Suto, *Acta Crystallogr. Sect. A*, **60**, 134 (2004); L. Palatinus, *Acta Crystallogr. Sect. B*, **69**, 1 (2013).

[13] L. J. Farrugia, *J. Appl. Crystallogr.*, **32**, 837 (1999).

[14] http://www.chem.gla.ac.uk/~louis/software/wingx/ (last accessed 14 January 2014).

[15] http://www.olex2.org/ (last accessed 14 January 2014).

[16] G. M. Sheldrick, *Acta Crystallogr. Sect. A*, **64**, 112 (2008).

[17] http://shelx.uni-ac.gwdg.de/SHELX/ (last accessed 14 January 2014).

[18] M. C. Burla, R. Caliandro, M. Camalli, B. Carrozzini, G. L. Cascarano, C. Giacovazzo, M. Mallamo, A. Mazzone, G. Polidori and R. Spagna, *J. Appl. Crystallogr.*, **45**, 357 (2012).

[19] http://wwwba.ic.cnr.it/content/sir2011-v10 (last accessed 14 January 2014).

[20] http://www.xtal.science.ru.nl/dirdif/software/dirdif.html (last accessed 14 January 2014).

[21] J. Rappleye, M. Innus, C. M. Weeks and R. Miller, *J. Appl. Crystallogr.*, **35**, 374 (2002).

[22] http://www.hwi.buffalo.edu/SnB/ (last accessed 14 January 2014).

[23] L. Palatinus and G. Chapuis, *J. Appl. Crystallogr.*, **40**, 786 (2007).

[24] http://superflip.fzu.cz/ (last accessed 14 January 2014).

[25] A. L. Spek, *Acta Crystallogr. Sect. D*, **65**, 148 (2009).

[26] http://www.cryst.chem.uu.nl/platon/ (last accessed 14 January 2014).

[27] P. W. Betteridge, J. R. Carruthers, R. I. Cooper, K. Prout and D. J. Watkin, *J. Appl. Crystallogr.*, **36**, 1487 (2003).

[28] http://www.xtl.ox.ac.uk/crystals.html (last accessed 14 January 2014).

[29] http://jana.fzu.cz/ (last accessed 14 January 2014).

[30] T. R. J. Achard, W. Clegg, R. W. Harrington and M. North, *Tetrahedron*, **68**, 133 (2012).

[31] N. A. Bell, W. Clegg, S. J. Coles, C. P. Constable, R. W. Harrington, M. B. Hursthouse, M. E. Light, E. S. Raper, C. Sammon and M. R. Walker, *Inorg. Chim. Acta*, **357**, 2091 (2004).

[32] H. D. Flack, *Acta Crystallogr. Sect. A*, **39**, 876 (1983).

[33] H. D. Flack and G. Bernardinelli, *J. Appl. Crystallogr.*, **33**, 1143 (2000).

[34] S. Parsons, H. D. Flack and T. Wagner, *Acta Crystallogr. Sect. B*, **69**, 249 (2013).

[35] http://checkcif.iucr.org/ (last accessed 14 January 2014).

[36] D. Altermatt and I. D. Brown, *Acta Crystallogr. Sect. B*, **41**, 244 (1985).

[37] I. D. Brown, *The Chemical Bond in Inorganic Chemistry*, Oxford University Press, Oxford, 2002.

[38] I. D. Brown, *Chem. Rev.*, **109**, 6858 (2009).

[39] R. J. Errington, S. S. Petkar, P. S. Middleton, W. McFarlane, W. Clegg, R. A. Coxall and R. W. Harrington, *J. Am. Chem. Soc.*, **129**, 12181 (2007).

[40] SHELXTL software, Bruker AXS Inc., Madison, WI, USA (2008 and subsequent versions).

[41] http://www.ruppweb.org/new_comp/scattering_factors.htm (last accessed 14 January 2014).

[42] http://www.cambridgesoft.com/Ensemble_for_Chemistry/ChemDraw/Default.aspx (last accessed 14 January 2014).

3

PDF Analysis of Nanoparticles

Reinhard B. Neder

Crystallography, Department of Physics, University of Erlangen, Erlangen, Germany

3.1 INTRODUCTION

In the last decade, the literature on nanoscaled structures has increased tremendously. These structures, such as energy-related materials and metal–organic frameworks (MOFs), very often show different ordering concepts at the very local scale of a few nanometres than at the meso- and macroscale, and many of their properties depend on these different ordering schemes. The topic of this chapter, the **pair distribution function** (PDF), is rapidly becoming a very popular tool in the study of the local order of increasingly complex matter.

In the strictest sense of its definition, a crystal structure consists of a motif of atoms that is repeated infinitely by translational symmetry. Within the motif, all atoms are at rest, and the content of each unit cell is identical to that of all the others. It is obvious that such a crystal does not exist, simply from the fact that no crystal can be of infinite size. A real crystal always contains a number of defects, *i.e.* deviations from the ideal perfect crystal structure. The simplest defect is its surface, as the translational symmetry is broken there. Even under thermodynamic equilibrium conditions, many defects are present in the form of interstitial atoms or atoms replaced by another chemical species. Local excitations such as phonons should also be considered defects of the perfect crystal structure.

Structure from Diffraction Methods, First Edition. Edited by Duncan W. Bruce, Dermot O'Hare and Richard I. Walton.
© 2014 John Wiley & Sons, Ltd. Published 2014 by John Wiley & Sons, Ltd.

Beyond these very simple defects, a large range of other types exist, such as small clusters within a host structure, guest molecules within a host, stacking faults and modulated structures, to name just a few.[1−3] Many of the physical and chemical properties of crystals depend on the presence of defects and a thorough understanding of these defects is necessary to understanding these physical properties. Examples of such properties include colour, which can arise if impurities are present, diffusion, which is greatly enhanced if vacancies are present, and the conductivity of doped semiconductor materials. For nanomaterials, a deviation from the bulk structure manifests itself not only through the limited size but also through significant defects such as surface relaxations, an increased number of stacking faults or complex twinning that is not observed in the corresponding bulk structures.

In order to describe defects, one has to consider two factors: the type of local deviation from the average structure and the distribution of the defects throughout the crystal structure. The local deviation will usually consist of two different aspects. The first is that part of the average structure is replaced by a different structure, which can be a simple point defect, a cluster of atoms or an extended defect. The second concerns a shift of the atom position with respect to the average structure. If an atom species is replaced by another with different radius, the surrounding atoms will experience a shift in position and this local strain may extend over considerable distance. The distribution of the defects may be a simple random distribution. This will be the case, for example, for impurities at small concentrations within an otherwise almost perfect crystal, such as a high-quality silicon crystal. At higher concentrations, however, defects will tend to influence each other. This influence may result from a small charge imbalance or a slight distortion of the local environment. As a consequence, the defects will be distributed more regularly, in what is called **short-range order**.[1−3]

The left part of Figure 3.1 shows a schematic sketch of a hypothetical AB_2 structure in a high-symmetry phase, as might be observed at elevated temperatures. As the temperature is lowered below the point of the phase transition, the A atom shifts from its local high-symmetry site along the horizontal axis, as illustrated by an atom in the lower part of the structure. Such a shift might be induced by a change in the bond angles in the neighbouring B atoms. In the high-temperature phase, such a localised shift would be a defect of the otherwise perfect crystal structure. As a consequence of the shift, we now have a locally different structure in which the A atom has three different bond distances to its nearest B

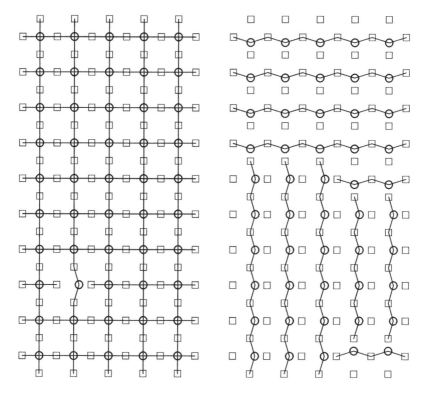

Figure 3.1 Example of local defects. The left side shows a single local defect in the high-symmetry structure. As one atom has slightly shifted to the right, bond distances to first neighbours have changed. The right side shows domain-type extended defects. The individual defect from the left side has induced the growth of a coherent domain. At the top, a second independent domain has nucleated.

neighbours. This shift will likely induce a local strain and cause the surrounding A atoms to also shift in the same direction. As a consequence, we will observe a different phase in which all A atoms are shifted slightly to the right, as in the right panel of Figure 3.1. Since the strain induced by the original shift is small, atoms that are far from the original atom may shift along any of the four axes. Owing to this independent nucleation and growth of the low-temperature phase, four different intricately inter-grown nanometre-sized domains will form, in which the A atom is shifted along either of the $\pm x$ or $\pm y$ directions. These domains may have irregular boundaries, as illustrated in Figure 3.1, but that is not important here. Due to the slight shift assumed in this example, adjacent domains will still diffract radiation coherently.

If one refines the crystal structure of such a crystal based on Bragg reflection intensities by either single-crystal or powder data, one will obtain the best result, with a high-symmetry model in which atom A is on a split $x, 0, 0$ position with one-quarter occupancy. If one averages the four domains, *i.e.* projects all atoms into a single unit cell, this is the correct description. As the right panel of Figure 3.1 illustrates, however, this is only a quarter of the truth. A much better description would depict the local structure as a low-symmetry structure and include a measure of the domain size and shape. In a more complex structure, the local structure might even include more than one defect type.

Such a description cannot be obtained from Bragg reflection data but must use the full diffraction pattern, including additional weak intensities between the Bragg reflections, which will occur as a consequence of the presence of defects. One approach is to analyse these additional weak intensities and derive a corresponding structural model. It is not usually easy to determine what kinds of change have occurred in the average crystal structure to have caused them. In this chapter, the focus shall be on analysis of the PDF, which is calculated by a Fourier transform of the measured intensities, which includes both the Bragg reflections and any further weak intensities.

The PDF is essentially a measure of all interatomic distances within a material. No model is required to determine these. The PDF of our example structure would show just one A–B bond length in the high-temperature phase and three different ones in the low-temperature phase. Under favourable circumstances, one might even observe that one of these latter distances occurs twice as often as the other two. One would also observe that the distances between first-neighbour A atoms are the same in both phases. This is a very strong indication of a locally ordered low-symmetry domain, rather than the high-symmetry model with random split positions.

Remember that the diffraction pattern of a crystal is proportional to the square of the amplitude of the scattered wave, described here as the structure factor of the whole crystal:

$$I(hkl) \propto |F_C(hkl)|^2 \tag{3.1}$$

with:

$$F_C(hkl) = \sum f_k e^{2\pi i h r_k} = \sum f_k e^{2\pi i (hx_k + ky_k + lz_k)} \tag{3.2}$$

hkl are the Miller indices of the plane from which the initial beam is diffracted, respectively the coordinates of the diffraction vector $h = ha^* + kb^* + lc^*$ in reciprocal space. f_k is the atomic form factor if X-ray or electron scattering is being considered or the scattering length in the case of neutron scattering. r_k is the vector describing the position of atom number k. The sum must be taken over all M atoms in the crystal. If the crystal is strictly periodic, the coordinates of an atom can be separated into a vector R_l to the origin of the given unit cell and a vector r_j from the origin of the unit cell.

$$r_k = R_l + r_J = i_1 a + i_2 a_2 + i_3 a_3 + r_J \qquad (3.3)$$

As the unit cells form the lattice, the indices i_1, i_2 and i_3 are restricted to integer numbers. With this replacement, the structure factor can be written as:

$$F(hkl) = \sum\sum f_j e^{2\pi i h(R_l + r_J)} = \sum\sum f_j e^{2\pi i h R_l} e^{2\pi i h r_J}$$
$$= \left[\sum e^{2\pi i h R_l}\right] \left[\sum f_j e^{2\pi i h r_J}\right] \qquad (3.4)$$

The index l runs over all unit cells of the crystal, and j over all N atoms in a given unit cell. As the sums over i and j are completely independent, these two sums can completely be separated.

For a perfect, infinite lattice, the first sum runs over an infinite number of contributions and its value is non-zero only for vectors h with integer indices hkl. This sum is the Fourier transform of the direct lattice and is thus identical to the reciprocal lattice. The second sum is the familiar structure factor, which now runs over all N atoms within one unit cell. As the first sum is non-zero only for integer hkl, it follows that the diffraction pattern of a perfect crystal consists of infinitely sharp Bragg reflections, whose individual intensity is represented by the square of the structure factor $F(hkl)$. In the last equation, the structure factor $F(hkl)$ is calculated from the contents of just one unit cell, the usual definition of the structure factor. In between the Bragg reflections, the intensity is zero for the perfect crystal. It is helpful for the following discussion to keep in mind that the second sum is generally a smoothly varying function that is non-zero throughout all of reciprocal space, not just at the Bragg positions.

If the crystal is of finite size, the first sum will run over a finite set of contributions. The result is still a reciprocal lattice, but this time each reciprocal lattice point is broadened, due to the finite size of the direct lattice.[4] This widening of the reciprocal lattice points can be observed for nanostructured objects, whose Bragg reflections are widened compared to a bulk-sized crystal.

For a disordered but still infinitely sized crystal, the two sums can no longer be separated, as the contents of the different unit cells differ one from another. As a consequence, the diffraction pattern contains non-zero contributions between the Bragg reflections as well. This additional scattered contribution is usually not centred in sharp reflections like the Bragg reflections but broadly distributed throughout reciprocal space, and is therefore termed **diffuse scattering**. It is this diffuse scattering that contains the information on the type and distribution of the defects. If suitable large single crystals exist, one can measure the distribution of the diffuse scattering throughout reciprocal space and analyse its intensity distribution in order to determine the underlying disorder.[1−3]

Often, however, single crystals will not be available and one must resort to powder diffraction in order to analyse the diffuse scattering. As a powder diffraction pattern is the spherical average of reciprocal space, all points in reciprocal space that have the same distance to the origin are projected onto the same point in the powder pattern. As a consequence, it is much more difficult to unravel the information contained in the powder pattern. This process would be much easier if one could analyse the changes in the disordered structure compared to the perfect structure in direct space. In reciprocal space, each defect contributes to the intensity at each diffraction angle, a straightforward consequence of the Fourier transform that relates the structure in direct space to the diffraction pattern. In direct space, disorder causes a localised change of interatomic distances and often changes the average occupancy of different sites *etc*. These localised changes are usually easier to identify.

This process is made much less difficult by the use of the PDF, which is at the heart of this chapter.[5−8] As detailed below, the PDF can be obtained from diffraction data on crystalline powders, but it can equally well be produced from the diffraction patterns of amorphous materials such as glasses, liquids and even gases. These amorphous samples were the sole scope of the PDF from its very beginning in the 1930s until the 1990s, when its modern form was developed.

3.2 PAIR DISTRIBUTION FUNCTION

In order to obtain the full information on a crystal structure in direct space, a structure determination must be carried out. This is well established for the average crystal structure and can be carried out for both single-crystal and powder diffraction data. An analogous technique to direct methods (see Chapter 2) has not been established for disordered structures. A few special algorithms exist for limited types of defect, such as short-range order in binary alloys or stacking faults in closed packed structures. As the number of possible deviations from an average structure and of the possible types of distribution of such local defects are very large, a general theory does not seem to be available. Thus the actual structure of a disordered material cannot be obtained directly. Aside from this lack of a general theory, one can, however, directly determine information about the distribution of all interatomic distances from a diffraction pattern.

Recall that the structure factor is essentially the Fourier transform of the scattering density:

$$F(h) = \text{Fourier}(\rho(r)) \qquad (3.5)$$

where $\rho(r)$ is the electron density in the case of X-ray scattering and the nuclear scattering density in the case of neutron scattering. Thus the diffracted intensity is given by the product of the structure factor with its conjugate complex value:

$$I(h) = F(h) \cdot F^*(h) \qquad (3.6)$$

If we now calculate the inverse Fourier transform not of the structure factor but of the intensity, we obtain:

$$\text{Fourier}[I(h)] = \text{Fourier}[F(h) \cdot F^*(h)] \qquad (3.7)$$

The Fourier transform of a product of two functions is equal to the convolution of the individual Fourier transforms and we obtain:

$$\text{Fourier}[I(h)] = \text{Fourier}[F(h)] \cdot \text{Fourier}[F^*(h)] \qquad (3.8)$$

The Fourier transform of the structure factor is the original scattering density and the Fourier transform of the conjugate complex structure factor is the conjugate complex of the scattering density at $-r$, which yields:

$$\text{Fourier}[I(h)] = \rho(r) \cdot \rho^*(-r) = \rho(r) \cdot \rho(-r) \qquad (3.9)$$

The last equality stems from the fact that the scattering function is a real valued function and thus its conjugate complex is identical to the function itself. The final result deduced from this relationship is that the Fourier transform of the intensity is the convolution of the scattering density with the scattering density inverted at the origin. This convolution product is the autocorrelation function of the scattering density. An autocorrelation function has a high value at position r if the original function has a high value at positions s and $s + r$. Thus the autocorrelation function of the scattering density will have high values at positions r that correspond to the length of interatomic vectors. If the Fourier transform of the intensity is calculated, we can expect to obtain immediate information about the distribution of interatomic distances, while information about the absolute position of the atoms will not be immediately available. This situation is similar to the Patterson function used in single-crystal structure determination. The Patterson function is the Fourier transform of the Bragg reflection intensity. It contains information on interatomic vectors, not on the absolute atom positions. By restricting the intensity to integer multiples of the reciprocal space vectors, the Patterson function becomes periodic in direct space. The topic of this chapter, the PDF, is essentially the Fourier transform of the intensity measured throughout reciprocal space and thus yields information in direct space on the distribution of interatomic distances. In contrast to the Patterson function, the PDF is no longer restricted to the periodicity of the unit cell.

The vast majority of present applications of the PDF technique use powder diffraction data. The technique is not limited to the use of powder data, however, and a first application to single-crystal diffraction data has been published.[9]

A powder pattern of any sample can be calculated by use of the Debye equation:

$$I(Q) = \sum \sum f_i f_j \frac{\sin(Qr_{ij})}{Qr_{ij}} \qquad (3.10)$$

where both sums run over all N atoms in the material. Q is 2π times the length of the scattering vector h, defined as $Q = 2\pi \frac{2\sin(\Theta)}{\lambda}$, and r_{ij} is the distance between atoms i and j. The Debye equation is derived by taking the square of the structure factor and then calculating the spherical average.[10-14] Note that the powder pattern calculated *via* the Debye equation depends on the interatomic distances, not on the absolute positions of the atoms. Another important point is that the structure for which the powder pattern is calculated *via* the Debye equation does not need to be a periodic crystal structure. As the sum runs over all interatomic distances r_{ij}, the structure can be a finite nanocrystal, an individual molecule or an amorphous structure.

The so-called **reduced PDF** is determined from the experimental data by the inverse process, the sine Fourier transform of the normalised scattering intensity, $S(Q)$:[5]

$$G(r) = \frac{2}{\pi} \int Q(S(Q) - 1) \sin(Qr)dQ \qquad (3.11)$$

The normalised scattering intensity, $S(Q)$, is obtained from the measured intensity by subtracting the background, correcting for absorption and scaling such that $<S(Q)> = 1$. The sine Fourier transform reflects the similarity to the Debye equation. In an ideal world, the integration would be carried out over data from $Q_{min} = 0$ to $Q_{max} = \infty$; in a real experiment it is limited to Equation 3.12:

$$Q_{max} = 2\pi \frac{2 \sin(\Theta_{max})}{\lambda} \qquad (3.12)$$

Figure 3.2 shows the experimental PDF obtained for crystalline CeO_2, which crystallises in the cubic fluorite structure with a lattice parameter

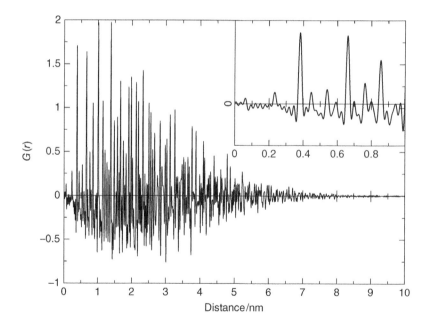

Figure 3.2 Experimental $G(r)$ for CeO_2. This reduced PDF, $G(r)$, was determined from X-ray diffraction data measured at beam line 11-ID-B at a wavelength of 0.02 128 nm. Significant and sharp peaks are visible up to some 8 nm. The insert shows the peaks within the first nanometre.

of 0.5407 nm. As this material is of good crystal quality, the peaks in the PDF extend to long distances. The insert in Figure 3.2 shows the first nanometre of the experimental PDF and will be used to describe a PDF's information content.

As a summary to guide the following discussion, one can determine the following properties directly by visual inspection of the PDF:

- the interatomic distances;
- the average number of neighbours around each atom;
- the width of the distance distribution between two atoms;
- for finite objects, the diameter.

Three main experimental factors also influence the PDF, and their combined effect is visible in it:

- the value of Q_{max};
- the instrumental resolution;
- the type of radiation.

As the PDF is the autocorrelation function of the scattering density, the peak positions correspond to all possible interatomic distances. One can think of the PDF as sort of a histogram of all interatomic distances. Figure 3.3 illustrates this schematically for a two-dimensional (2D) case. In the left part, all distances from the central atom to all other atoms are illustrated by rings of constant distance. These distances are then projected onto a 1D diagram: the PDF. The process is finally repeated for each of the atoms chosen as central atom.

CeO_2 crystallizes in the fluorite structure type, which is a face centred structure with four Ce atoms at (0, 0, 0) and eight O atoms at ($1/4$, $1/4$, $1/4$). The first maximum at 0.2314 nm in the PDF in Figure 3.2 corresponds to the first neighbour Ce–O distance i.e. to an atom pair separated by a vector [$1/4$, $1/4$, $1/4$]. The second maximum at 0.3825 nm corresponds to the Ce–Ce distance, a pair separated by [$1/2$, $1/2$, 0], the third at 0.44 nm to the Ce–O distance between atoms separated by [$1/4$, $1/4$, $3/4$] and the fourth at 0.5407 nm to the Ce–Ce distance between atoms separated by [1, 0, 0], i.e. the lattice parameter. Note that the peak heights vary considerably. Several factors affect the peak height in a PDF. As the autocorrelation function measures all distances from each atom, the peak heights are proportional to the average coordination number for each central atom. In the fluorite structure type, each Ce atom is surrounded by eight symmetrically equivalent O

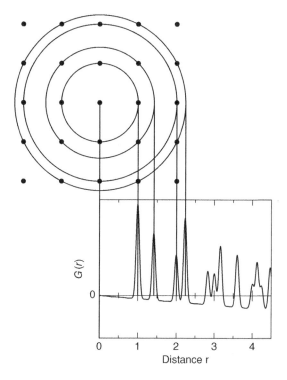

Figure 3.3   Schematic sketch to illustrate the relationship between the structure and the PDF. The circles represent the first four interatomic distances in a simple 2D lattice. The projection of these distances into a 1D graph corresponds to the positions of the peaks in the PDF. Maxima in the PDF are in addition weighted by the average coordination number and the product of the scattering lengths of the respective atom pairs.

atoms at distance 0.2314 nm and by twelve symmetrically equivalent Ce atoms at 0.3825 nm. This difference between eight and twelve neighbours is not enough to explain the large difference in peak height between the two peaks. Furthermore, if one familiarises oneself with the fluorite structure, it can be seen that each oxygen atom has six oxygen neighbours at $< 1/2, 0, 0 >$ at 0.2704 nm. At this distance, no peak can be discerned in the experimental PDF. As the PDF is obtained *via* sine Fourier transform of an experimental diffraction pattern, the individual atomic form factors (scattering lengths) determine the intensities in reciprocal space, and therefore the peak heights of the maxima in the PDF are proportional to the product of the two atomic form factors (scattering lengths). As this PDF is based on X-ray powder diffraction data, Ce ($Z = 58$) contributes much more to the PDF than O ($Z = 8$).

The strongest peaks in the PDF correspond to Ce–Ce distances, followed by Ce–O distances. O–O distances are too weak to be observed in the experimental PDF.

The peak height of the PDF is also influenced by the instrumental resolution function. In Figure 3.2 one can see a decrease in peak height with increasing r, and peaks essentially become negligibly small around 10 nm. Ceria has coherent domain sizes larger than 10 nm and this decrease of intensity is a direct effect of the angular resolution of the powder diffractometer. For finite objects, such as nanoparticles, the peak heights are also affected by the shape and diameter. If no structural correlation exists between neighbouring nanoparticles in the actual sample, distances between any given atom in one nanoparticle and any atom in any surrounding nanoparticle are distributed randomly. As a consequence, no peak is observed in the PDF at a distance beyond the nanoparticle diameter.

The third immediate piece of information in the PDF is the width of the maxima. This width corresponds to the width of the distribution of interatomic distances between a given pair of atoms, which will depend on the thermal motion of the atoms and the presence of local distortions that shift atoms within an individual unit cell. If the maxima have differing widths then the structure will contain different defects that influence the respective atom-pair distance distributions.

A fourth immediate observation is the presence of the small ripples around the first peaks in the experimental PDF. These are side effects of the finite value of Q_{max}, which has a decisive influence on the PDF. For an ideal Fourier transform, the upper limit in the integral in Equation 3.10 would be infinity. This will produce an ideal PDF. As the experimental data are limited at a finite value of Q_{max}, one can describe the effect of the finite Q_{max} by multiplying the ideal infinite powder pattern by a box function $BOX(Q)$ that is 1 for Q less than Q_{max} and 0 for Q larger than Q_{max}.

$$Powder_{finite} = Powder_{infinite} \cdot BOX(Q) \tag{3.13}$$

The Fourier transform of this product is identical to the convolution of the individual Fourier transforms:

$$PDF_{finite} = Fourier(Powder_{finite}) = Fourier[Powder_{infinite} \cdot BOX(Q)] \tag{3.14}$$

$$PDF_{finite} = Fourier[Powder_{infinite}] \cdot Fourier[BOX(Q)]$$

$$= PDF_{infinite} \cdot Fourier[BOX(Q)] \tag{3.15}$$

This shows that the effect of a finite value for Q_{max} is a convolution of the ideal PDF with the Fourier transform of the box function, which is:

$$\text{Fourier[BOX}(Q)] = \frac{\sin(Q_{max}\,r)}{r} \tag{3.16}$$

As the full width at half maximum (FWHM) of the function in Equation 3.16 is proportional to $1/Q_{max}$, the PDF maxima are widened as a smaller Q_{max} is used. A good experimental PDF is achieved for a value of Q_{max} larger than some $200\,nm^{-1}$.

In a complementary fashion, the PDF is influenced by the instrumental resolution function. For simplicity we will assume that the instrumental resolution is constant over the whole 2θ range. We can then describe the effect of the instrumental resolution on the powder pattern as a convolution of an ideal powder pattern with the instrumental resolution function:

$$Powder_{res} = Powder_{ideal} \cdot Resolution(Q)$$

The Fourier transform of this convolution is the product of the individual Fourier transforms:

$$PDF_{res} = \text{Fourier}(Powder_{res}) = \text{Fourier}[Powder_{ideal} \cdot Resolution(Q)] \tag{3.17}$$

$$PDF_{res} = \text{Fourier}[Powder_{ideal}] \cdot \text{Fourier}[Resolution(Q)] \tag{3.18}$$

$$PDF_{res} = PDF_{ideal} \cdot \text{Fourier}[Resolution(Q)] \tag{3.19}$$

If we assume for simplicity that the instrumental resolution function is a Gaussian function of width σ, we obtain the result that the PDF in direct space is multiplied by the Fourier transform of this Gaussian function, which is a Gaussian function of width $1/\sigma$. This multiplication in direct space means that peaks at larger interatomic distance r are dampened by the Gaussian function that decreases with increasing r. Thus, if the instrumental resolution function is set to a larger width, the PDF in direct space is already dampened at shorter interatomic distances r. The general considerations hold for other resolution functions and for a resolution function that varies with 2θ. As the measurement for the PDF in Figure 3.2 was carried out only up to $170\,nm^{-1}$, these Q_{max} ripples are spaced at $0.039\,nm^{-1} = 2\pi/Q_{max}$. The instrumental resolution also causes a distance-dependent broadening of the PDF maxima. The two parameters, the damping and the broadening term, can be determined for a particular instrument by refinement of the PDF of a standard with excellent crystal quality, such as CeO_2 or LaB_6. Thereafter, these parameters should be kept fixed during the refinement of an actual sample.

3.3   DATA COLLECTION STRATEGIES

In the description of the PDF, the following three experimental aspects were pointed out as being of most importance:

- The value of Q_{max} determines the width of the peaks in the PDF.
- The instrumental resolution determines the range of distances that is accessible in the PDF.
- The choice of radiation strongly influences the relative contributions of light *versus* heavy elements on the PDF.

These three aspects must be considered in the choice of instrument. The first question to be answered before choosing an instrument is, however, what aspects of the sample does one want to study? A few examples will illustrate the reasoning.

As the PDF is the Fourier transform of the measured intensity, the usual differences between X-ray and neutron diffraction need to be considered. The different dependencies of the form factor and of the scattering length on the atomic number will influence the experimental PDF. The search for local disorder induced by light elements in the presence of heavy elements is much easier if neutron diffraction data are used; in fact, it may only be possible for neutron diffraction data. Similarly, the distinction between elements close to one another in the Periodic Table will be possible if their neutron scattering lengths differ. In order to distinguish between such elements, one can also make use of anomalous X-ray scattering. If several data sets are collected close to one of the X-ray absorption edges of one of the elements in the sample, the atomic form factor of this particular element will change, while those of all other elements will remain at the same value. The difference between these experimental PDFs measured close to the absorption edge will consist of peaks that correspond to interatomic distances involving this element.[15]

As the quality of a PDF depends on Q_{max}, one has to keep in mind that the intensity in every X-ray experiment will decrease with increasing Q due to the atomic form factor, while the intensity in a neutron experiment is not affected by this decay. As a PDF experiment usually requires data measured up to a much higher Q_{max} than a standard powder diffraction experiment, this difference is much more important for the former than the latter. Fortunately, modern PDF beam lines at synchrotron sources offer sufficient flux at short wavelengths in combination with large area detectors to provide data in the high-Q part of the diffraction pattern, with an excellent statistical noise level. Data collection can be achieved within seconds, which even enables *in situ* studies.[16,17]

If the main scientific question concerns the local order/disorder, one usually needs to distinguish fairly similar interatomic distances in order to differentiate between different disorder models, then the value of Q_{max} should be chosen as high as possible. As the intensity falls off with increasing Q, the overall intensity available at the beam line in the high-Q region needs to be taken into account as well. Most PDF beam lines at synchrotron facilities use a primary radiation of 60–100 keV and a large area detector to allow fast data collection. The Q range at such an instrument will be some 200–250 nm^{-1}. Neutron sources at spallation sources allow much higher Q_{max}, in the range 400–1000 nm^{-1}. As the intensity of the neutron beam is lower and the interaction of neutrons with matter is much weaker compared to X-rays, counting times are a few hours. For this scientific goal, one will often be less interested in long interatomic distances beyond the first few nanometres. Therefore, one can relax the constraints on the instrumental resolution as this will usually increase the overall intensity at the sample. A higher flux means better counting statistics or less counting time.

The other extreme situation would be the study of domains within the sample. Here the main emphasis will be long distances in real space, and to achieve these the PDF must be obtained from high-resolution data. Domain sizes can be determined by the analysis of r-dependent fits to the data set. For these fits to be significant, the individual maxima have to be as narrow as possible. Therefore, a very large value of Q_{max} is required as well.

Somewhat in between these situations falls the study of nanoparticles. As sizes can be estimated from home laboratory diffraction data, one can choose the resolution of the beam line accordingly. In order to obtain reliable particle sizes, the instrumental resolution should be such that for a standard with very good quality the peaks drop off at about three times the particle diameter. In order to determine defects, one should pay attention to a good Q_{max} above some 200 nm^{-1}.

At most X-ray PDF beam lines, the primary X-ray optics do not provide a wide choice of wavelengths or instrumental resolution. For example 11-ID-B at APS and ID15 at the European Synchrotron Radiation Facility operate at energies close to 60 and 90 keV, respectively, and P02 at PETRA III, DESY, Hamburg uses a fixed energy at 60 keV. The wavelength distribution is mostly determined by the fixed monochromator crystals and is usually optimised with respect to a reasonable low $\Delta\lambda/\lambda$ of approximately 1×10^{-3}. As large flat-panel detector systems are almost commonplace, one can tune Q_{max} and the effective instrumental resolution by changing the sample-to-detector distance. As the detector is commonly placed with its centre on the direct beam direction, a decreasing sample-to-detector distance increases the maximum 2θ and thus Q_{max}. As a trade off, the effective instrumental resolution decreases.

As a last example, if the main scientific goal is a time-resolved measurement, all other aspects may have to be relaxed in order to obtain the required time resolution. Aspects like flux and especially the readout time of the detector become of utmost importance. Image plate-based systems that offer excellent intrinsic background have a readout time of 1 minute or more. Modern flat-panel detectors on the other hand can be read at up to 16 Hz. If the overall features of the sample can be deduced from high-quality PDF measured *ex situ*, one can relax requirements on Q_{max} and resolution in favour of short counting times.

3.4   DATA TREATMENT

The experimental PDF is obtained by calculation of the sine Fourier transform according to Equation 3.11. The argument of the integral in Equation 3.11 consists of the normalised structure factor, $S(Q)$, pointwise multiplied by Q, which is referred to as a **reduced structure factor**:

$$F(Q) = Q \cdot (S(Q) - 1) \qquad\qquad (3.20)$$

Note that this reduced structure factor, $F(Q)$, has to be carefully distinguished from the structure factor terminology used in single-crystal refinement. There the structure factor refers to the last term in Equation 3.4. As PDF has several roots in different scientific communities, the terminology varies across the literature. A very good summary has been written by Keen.[18]

In order to calculate the normalised structure factor, $S(Q)$, several corrections are needed.[5,19,20] The initial data are intensity data consisting of several contributions. In addition to the intensity scattered by the sample, the sample container and the sample environment will also contribute to the detected intensity. Thus the first correction is usually the subtraction of the background intensity. As the most common experiments currently are carried out with high-energy X-rays[19] or with neutrons,[20] the absorption correction often does not change the data very much.

Besides the pure elastic scattering process, inelastic scattering processes contribute to the overall intensity as well. For data collected with X-rays, the main background contribution is Compton scattering, which needs to be corrected accordingly. One should be aware of the exact experimental conditions. If an energy-discriminating detector was used, care must be taken if the energy over which the data were integrated includes the Compton signal over the full 2θ range. Most commonly these days, experiments are carried out with the use of an area detector. This detector type does not discriminate among the incoming energy, and thus both the Compton scattering and the elastic signal are integrated to yield a common experimental intensity at each 2θ point. Especially for compounds with light elements, the Compton signal increases with increasing Q and is often much more intense than the elastic scattering. While the Compton scattering process is well understood and the data can be corrected accordingly, this issue has to be kept in mind during data collection. As the normalised $S(Q)$ data are multiplied by Q, the final PDF will very much depend on the high-Q part of the diffraction pattern. After subtraction of the theoretical smooth Compton scattering function, the remaining signal will be fairly small, while all the experimental noise is still present in this signal. Thus care has to be taken to ensure that the experimental noise in the high-Q part of the diffraction pattern will still be tolerable even after some 80% of the signal has been subtracted.

For neutron scattering, a major background issue is presented if the sample contains hydrogen. As the incoherent cross-section of hydrogen is very large, an intense background is produced. This issue is especially important for spallation sources, as the incoherent background shows an even stronger Q-dependence than that observed for monochromatic instruments such as D4 at the Institute Laue Langevin. Empirical corrections of the incoherent background due to hydrogen have been successfully carried out, however.[21]

To evaluate the success of the corrections, one should inspect the PDF and the reduced structure factor, $F(Q)$. An indication of a good correction is given if $F(Q)$ oscillates around zero over the wide Q range without low-frequency oscillations. The PDF, $G(r)$, should not contain any strong peaks at distances below 0.1 nm. Peaks in this range are an indication that $F(Q)$ contains low-frequency oscillations. $G(r)$ should also be free of regular high-frequency oscillations that extend over the full distance range. Figure 3.4 shows the raw intensity for the CeO_2 sample. Data collection at 11-ID-B, Advanced Photon Source, Argonne, USA was carried out at room temperature at an energy of 58 keV. The sample was in a polyimide capillary of 0.8 mm diameter. The small scattering signal of the capillary can be seen as the small peak at $Q = 15\,nm^{-1}$. After background and Compton subtraction, the main next step is a division of the intensity by $<f^2>$. After suitable normalisation, $S(Q)$ oscillates around 1 and shows peaks all the way to $230\,nm^{-1}$ (Figure 3.5).

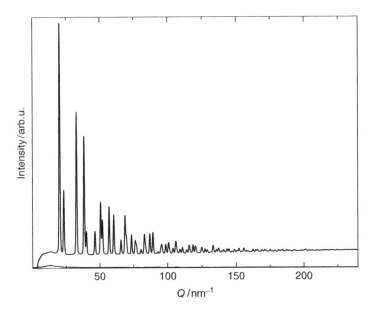

Figure 3.4 Raw intensity data from CeO_2 following integration of area detector data. Data below $5\,nm^{-1}$ are behind the beam stop. The small curve visible up to $25\,nm^{-1}$ is the background measured with an empty capillary.

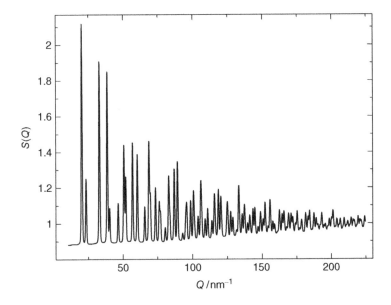

Figure 3.5 Normalised scattering intensity $S(Q)$ for the CeO_2 sample.

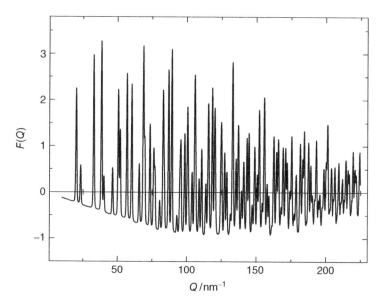

Figure 3.6 Reduced $S(Q)$ function, $F(Q) = Q \cdot [S(Q) - 1]$, for the CeO$_2$ sample. This properly corrected data set oscillates around $F(Q) = 0$.

Figure 3.6 shows the argument in the integral of Equation 3.20, $F(Q) = Q \cdot (S(Q) - 1)$. Owing to the stepwise multiplication by Q, the values in the high-Q part of the diffraction pattern have been increased considerably compared to $S(Q)$. Any experimental noise will likewise be multiplied by Q, and achievement of excellent counting statistics is an important requirement that cannot be overstressed. The PDF obtained after the sine Fourier transform has already been presented in Figure 3.2.

As an example of incorrect data, the CeO$_2$ $F(Q)$ has been modified in Figure 3.7 and the corresponding $G(r)$ is shown in Figure 3.8. The difference between the bad and the good $F(Q)$ has been chosen as a long-range oscillation. Similar, less obvious errors may be the result of incorrect background and absorption correction. Figure 3.8 shows the first nanometre of the resulting PDF, together with the good PDF from Figure 3.2. The main differences are the high-frequency oscillations that are dominant below 0.35 nm. The correct peak at 0.23 nm can no longer be distinguished from these oscillations. Even if $F(Q)$ and $G(r)$ look good right away, one should change the way $G(r)$ is obtained from the raw intensity. Features that change upon a change in data-correction procedures are likely artefacts of the Fourier transform and need to be interpreted with the utmost caution.

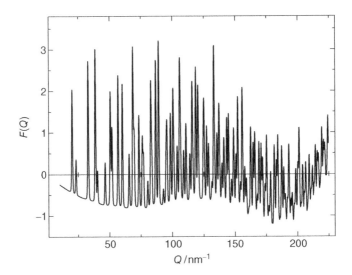

Figure 3.7 Reduced $S(Q)$ function, $F(Q) = Q \cdot [S(Q) - 1]$, for the CeO_2 sample, with an intentionally wrong data treatment. While this data set oscillates around $F(Q) = 0$, low-frequency oscillations are present that are not part of the sample scattering.

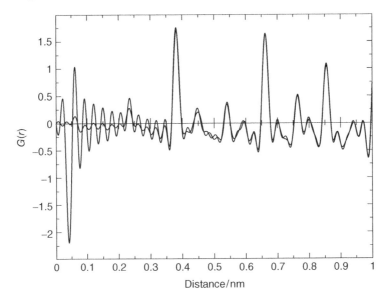

Figure 3.8 Reduced PDF calculated from the erroneous reduced scattering intensity, $F(Q)$, in Figure 3.7. The high-frequency oscillations at low distances are a consequence of the erroneous data correction. The properly corrected $G(r)$ does not show these strong oscillations. Note that the strong oscillations have the same frequency as the Q_{max} termination ripples in the good $G(R)$. This is caused by the cut-off for $F(Q)$ at $Q = 224.7\,\text{Å}$, where $F(Q_{max})$ is not equal to zero.

Besides the long-range oscillations present in $F(Q)$, another erroneous data treatment is illustrated in Figures 3.7 and 3.8. The value of $Q_{max} = 224.7\,nm^{-1}$ has been chosen at a point where $F(Q)$ is not close to zero. This type of cut-off introduces ripples into the PDF. For a proper data treatment, Q_{max} should be taken at a point where $F(Q)$ is equal to zero.

The differences between the two PDFs diminish rapidly with increasing distance, r. Fortunately, this holds for most errors in the data treatment. See reference [5] for the mathematical details of the error propagation. Generally, the absolute value of uncertainties in the PDF decreases with increasing distance. Thus, although the peak heights in the PDF decrease with increasing distance as well, the ratio $G(r)$:uncertainty remains fairly constant and even weak features at large distances can reliably be interpreted.

3.4.1   Calculation of $G(r)$ from a Structural Model

In order to calculate $G(r)$ from a structure model, the first step is to assemble the histogram of all interatomic distances between the N atoms in the model:

$$R(r) = \frac{1}{N}\sum\sum \delta(r - r_{ij}) \tag{3.21}$$

where r_{ij} is the modulus of the interatomic distance between atoms number i and j. Both sums run over all N atoms in the sample. This function, $R(r)$, is the radial distribution function of the sample. For a completely random structure, the number of interatomic pairs at a distance r is $4\pi\rho_0 r^2$, with ρ_0 the average number density of atoms per unit volume. It follows that on average, $R(r)$ will increase with r^2. The radial distribution function $R(r)$ is related to the pair density function $\rho(r)$ by:

$$R(r) = 4\pi r^2 \rho(r) \tag{3.22}$$

At large distances r, the pair density function oscillates around the average number density of the structure, while for very short distances its value is zero.

This pair density function is finally related to the PDF $g(r)$ by division through the average number density:

$$g(r) = \rho(r)/\rho_0 \tag{3.23}$$

This normalised PDF $g(r)$ will obviously oscillate around a value of 1 at larger distances r.

Finally, the relationship between $R(r)$ and the reduced PDF $G(r)$ is given by:

$$G(r) = \frac{R(r)}{r} - 4\pi\rho_0 r \qquad (3.24)$$

As the experimental function $G(r)$ is the Fourier transform of the normalised scattering function $S(Q)$, elements that contribute much to the scattering amplitude will likewise contribute much to the experimental $G(r)$. In the case of neutron diffraction, this is taken into account by multiplying the histogram by the scattering length of the atoms:

$$G(r) = \frac{1}{Nr}\sum\sum\frac{b_i b_j}{\langle b\rangle^2}\delta(r - r_{ij}) - 4\pi\rho_0 r \qquad (3.25)$$

where b_i and b_j are the individual scattering lengths of atoms i and j and $<b>$ is the average scattering length of all atoms. In the case of X-ray diffraction, one has to take into account that the atomic form factors vary as a function of Q. Fortunately, this Q-dependence does essentially cancel out in the quotient and one can replace the scattering lengths by the ordinal numbers, Z.

Equation 3.25 corresponds to a snapshot in time. In order to obtain a realistic $G(r)$ that can be compared to the experimental $G(r)$, the thermal vibrations in the sample must be modelled properly. If a large crystal has been modelled, each individual atom can be moved at random according to a distribution function with a width defined by its atomic displacement parameter (ADP). As a result, the interatomic distances will be distributed over a wider range. For a model crystal that consists of several thousands of atoms, the calculated $G(r)$ will be a reasonably smooth function. For small crystals, however, the distances sampled by the random shifts of the atoms are a subset of the large number of possible shifts and the calculated $G(r)$ will be subject to nonphysical noise.

As an alternative, one can convolute the δ distributions that are obtained from the model with the atoms at rest by a distribution function. If we assume an isotropic thermal vibration model, the distance r_{ij} in this last equation must be convoluted by:

$$D(r) = \frac{1}{\sigma\sqrt{2\pi}}e^{-\frac{1}{2}\frac{r^2}{\sigma^2}} \qquad (3.26)$$

If we further assume that both atoms vibrate independently, the sigma of the distribution is the sum of the two isotropic ADPs, u_i and u_j. This model is a very good approximation at longer distances r. At shorter distances, however, atoms tend to vibrate in a more correlated fashion, as the interatomic potential increases steeply with decreasing interatomic distance. Thus a movement of the two atoms towards each other is less likely than a movement in the same direction or in two normal directions. As a consequence, the distance distribution between two adjacent atoms will be narrower that the distance distribution between atoms that are far apart. This effect can be accounted for by correcting the sigma with a distance-dependent value, which can be modelled as a function proportional to $1/r$ or $1/r^2$:

$$\sigma(r) = \sigma - \frac{\gamma}{r}; \quad \sigma(r) = \sigma - \frac{\delta}{r^2} \qquad (3.27)$$

where δ and γ are sample-dependent parameters.[5]

Even with modern computers, a structure model will be very small compared to the actual experimental crystals. Finite size effects can therefore be expected and must be dealt with. These occur when one compiles the histogram of all interatomic distances between all atoms. Atoms at (or close to) the edge of the model crystal have no neighbours outside it, unlike those in the much larger actual sample crystal. For the sample crystal, only a negligibly small fraction of atoms is close to the surface. Periodic boundary conditions are routinely used to handle the finite model size. All vectors from one atom to another that point outside the model crystal are wrapped around it to point at an equivalent atom within it. Effectively, the model crystal becomes one unit cell of an infinite periodic superstructure. Keeping in mind that the PDF is centrosymmetric, this restricts the interpretation of correlations between the atoms of the model crystal to distances shorter than half the diameter. All longer distances are simply those within the periodic superstructure.

The actual sample crystal is of course finite as well, and thus atoms on the surface have fewer neighbours than those within the crystal. This deviation is, however, negligibly small for standard powder-sized crystals of approximately $1-5\,\mu m$ diameter.

A very different situation is encountered when the diameter of the sample crystal is just a few nanometres. Under these circumstances, the fraction of atoms close to the surface is significantly large and systematic deviations can be expected in the PDF. These deviations are a function of diameter and shape. Consider for simplicity a spherical particle that has a cubic primitive structure with just one atom at $(0, 0, 0)$. For short interatomic vectors, almost all atoms have six neighbours; only those within a monoatomic surface layer have fewer than six. If we increase the interatomic distance, the number of atoms that have fewer neighbours than those in the infinite bulk structure increases. For distances beyond the diameter, no atom has any neighbours. As a consequence, the radial distribution function, $R(r)$, is zero beyond the particle diameter. In order to derive an analytical function for the number of distances in a finite object compared to an infinite object, one has to integrate over all distances that end within the finite object, starting at every point within the finite object. One obtains an envelope function that gives the fraction of distances within the finite object compared to those within the infinite object. (These integrations have been carried out for a few special cases.[22−26]) The effect of the finite size can then be taken into account by multiplying the PDF of the infinite structure by the envelope function. As an example, for a sphere of diameter D one obtains the envelope function:

$$E(r, D) = 1 - \frac{3}{2}\frac{r}{D} + \frac{1}{2}\left(\frac{r}{D}\right)^3 \qquad (3.28)$$

For distances beyond the particle diameter, D, the envelope function is set to zero. Thus, if the finite object can be described as a limited volume within an infinite object of homogeneous density, one can readily calculate its PDF by calculating the PDF of the infinite object, which may be modelled as a much smaller model object, using appropriate finite boundary conditions. The PDF calculated for the infinite object is then multiplied by the envelope function to yield the final calculated PDF. Figure 3.9 illustrates this for a spherical CeO_2 particle of 3 nm diameter.

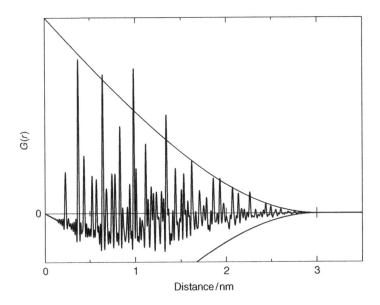

Figure 3.9 Calculated $G(r)$ for a spherical CeO_2 nanoparticle with diameter 3 nm. $G(r)$ is calculated from the content of a single unit cell, assuming periodic boundary conditions and applying the envelope function for a sphere. In order to illustrate the damping of the $G(r)$ peaks as a function of distance, the envelope function is plotted as well, though not to scale.

Compared to the experimental PDF of bulk CeO_2, the peaks in this calculated PDF quickly decrease in height, and they vanish beyond 3 nm.

This approach fails if the finite object cannot be modelled as a limited volume within a bulk structure. This is the case *e.g.* for a nanoparticle that is surrounded by a shell of stabilising organic molecules, for a general core–shell nanoparticle that consists of the overgrowth of two different structures or for any irregularly shaped finite object. The special case of spherical particles with a spherical shell could be treated as a bulk material, with the particles placed on the sites of a cubic or hexagonal closed packing. To avoid correlation between atoms in different particles, each particle would need to be rotated by a random angle. For irregularly shaped particles, one would have to derive an algorithm that achieved a reasonably close packing in the computer model. This can be done, but at great computational expense. Apart from the special spherical case, particles are best simulated as individual finite-sized model objects, and periodic boundary conditions must be disregarded. The radial distribution function of such an object will automatically consist of the distances within the object only. Thus, if $G(r)$ is calculated according to Equation 3.25, the result in Figure 3.10 is obtained. The calculated function oscillates around the $-4\pi\rho_0 r$ line, instead of around

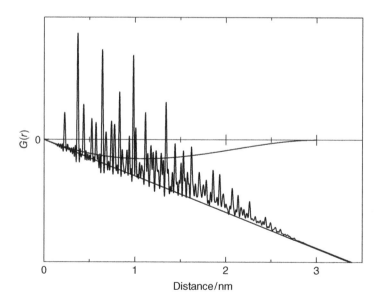

Figure 3.10   Calculated $G(r)$ for a single finite-sized CeO_2 nanoparticle with 3 nm diameter. No correction has been applied to the slope of the $-4\pi\rho_0 r$ line. The difference between this and the $G(r)$ obtained in Figure 3.9 is the smooth line, which is the $-4\pi\rho_0 r$ line multiplied by the envelope function for a sphere.

a value of $G(r) = 0$. The difference between Figure 3.10 and Figure 3.9 is the smooth function in the former, which is exactly the $-4\pi\rho_0 r$ line multiplied by the envelope function. Thus, one can model $G(r)$ for a single finite object by multiplying the $-4\pi\rho_0 r$ line by the envelope function. If the exact analytical function of the envelope function is not known, one can use an empirical function to minimise the difference between the experimental and calculated $G(r)$.[27−30]

In Equation 3.11, $G(r)$ was calculated from $Q_{min} = 0$ up to Q_{max}. So far, PDF experiments have ignored the small-angle scattering signal and have replaced the low-Q part of $F(Q)$ by an extrapolation from Q_{min} to $Q = 0$ by a straight line to $F(0) = 0$. If one calculates the small-angle scattering for the finite-sized model, one can derive a correction of the $-4\pi\rho_0 r$ line.[30]

Here it is instructive to compare this model calculation to the experimental situation. In an experiment, the sample consists of a powder of nanoparticles. To a good approximation, each nanoparticle is in a random orientation and no correlation exists from one particle to the next. Well-defined distance distributions exist only between the atoms within a single given nanoparticle. Owing to the random orientation, and most likely a slight random distance variation, the distances between atoms in one nanoparticle and those in its neighbouring nanoparticles are not correlated, resulting in a broad and smooth contribution to the experimental diffraction pattern and thus to the experimental $G(r)$. In the model that is based on a single particle, all these distances are neglected; they must be taken into account by multiplying the $-4\pi\rho_0 r$ line by the envelope function of this individual particle.

Other aspects that need to be taken into consideration are defects within the model. Usually, defects such as chemical short-range order, local distortions, stacking faults *etc.* are modelled using a random number generator. If a large model is simulated, the distribution of the defects will be a reasonably large subset of all possible conformations. As a consequence, the calculated diffraction pattern or calculated PDF is a good representation of the disorder model. If the model is calculated again, a different conformation will result, yet the calculated diffraction pattern or PDF will be very close to the previously calculated PDF. If, however, the model is an individual small finite object, the situation is quite different. Let's consider a Cobalt nanoparticle of 2 nm diameter. As the defect model, we will assume stacking faults. Cobalt crystallises in a hexagonal closed packed structure. Local cubic closed packed sequences are common defects in cobalt crystals. Figure 3.11 shows cross-sections through two model crystals that were modelled with identical parameters for size

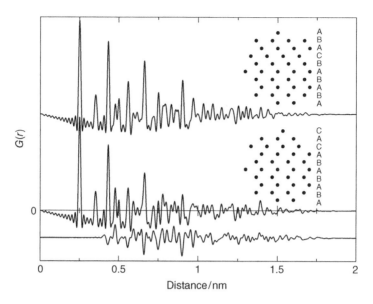

Figure 3.11 Cross-sections through cobalt nanoparticles and corresponding $G(r)$. The lower particle is almost defect-free, while the upper particle is switched from a hexagonal sequence to a cubic sequence at the fifth layer. Both particles were simulated with identical stacking fault parameters. They represent two of the many possible conformations. The difference curve between the two calculated PDFs shows large deviations. These are not observed for the first peaks, as the first and second coordination shells are identical across this particular fault.

and stacking fault probability. Even though identical probabilities were used to simulate the two crystals, the actual stacking sequences differ considerably. Figure 3.11 also shows the corresponding calculated PDFs, which are also very different from each other. As the c-lattice constant of Cobalt is 0.40686 nm, a 2 nm diameter sphere consists of 10 layers. While building up the model crystal, the simulation program has to decide nine times whether to use the hexagonal or the cubic stacking sequence. This is a very small number and one cannot expect that identical stacking fault sequences will result if the simulation is repeated just twice. A much larger crystal, consisting of several thousand layers, would be required to achieve a true representation of the stacking fault probability within a single simulated crystal. As the purpose of this simulation is the simulation of the actual nanoparticle at its small size, we are obliged to limit the crystal to 12 layers.

The solution to this contradiction between the need for a small size and the insufficient statistical representation of a defect parameter is to model the nanoparticle several times and to average the individual diffraction patterns or PDFs. This ensemble average has been success-fully tested for a number of nanoparticles.[31] The number of individual simulations must be large enough to ensure that two repeated runs of the ensemble average will yield results that do not differ significantly. The required number of individual simulations will depend on the number of defects present in the finite model and on the influence that different conformations will have on the PDF. One can determine a good estimate for this number through the following algorithm.[14] First, one simulates a very large number, N, of individual model crystals, as well as their PDF/diffraction pattern. This number should be close to the possible number of conformations, and be at least several thousand. The aver-age of all N calculated PDFs can be considered a good representation of the defect probability. Next, one draws a random subset of M individ-ual PDFs, averages these and calculates an agreement factor between the average of all N simulations and the average of the M simulations. For a given size of M, this procedure is repeated often, ideally such that M times the number of repetitions is close to N. Finally, this is repeated systemat-ically for increasing subset size M. For a small subset size, the agreement factors will scatter widely, while for larger subset sizes, the agreement factors will be more like each other. If M is equal to N, the agreement is of course perfect. As the subset size M is increased from 1 to larger sizes, the scatter of the agreement factors initially decreases rapidly and then levels off much more slowly. This saturation needs to be reached for a value of M that is much smaller than N. A reasonable compromise between the accuracy by which the defect parameters can be determined and the computational time required for the refinement is reached for a population size M that is slightly larger than the minimum population size at saturation level. Even at this population size, the agreement factors will be different for two separate populations. For an actual refinement, the upper limit of the R-values obtained for this subset size M represents an optimum agreement factor. An agreement factor that is lower simply means that by chance a special population was simulated that happens to better represent the data. It does not mean that the specific defect parameter is a better value for the description of the sample.

Through the ensemble average, one calculates the average of individual PDFs or diffraction patterns, each of which is obtained by use of ran-dom number generators. This corresponds to the experiment in which the diffraction patterns of individual (nanosized) crystals are averaged incoherently to the overall diffraction pattern. The ensemble average of individual PDFs or diffraction patterns leads to a loss of the algebraic derivative of structure simulation parameters with respect to the PDFs. As a consequence, standard least-squares refinement techniques cannot be used and population-based algorithms must be employed. These algorithms are computationally much more expensive. Ideally they will work in multiple processor facilities, such as parallel processors and large computer facilities. An emerging alternative is programs designed for the graphical processors of standard PCs.

3.4.2   Data Modelling

In a strict sense, the information contents of the diffraction pattern and of the PDF are identical, as the Fourier transform from $F(Q)$ to $G(r)$ does not add information. As the PDF contains direct information about interatomic distances, it is often more easy to interpret than the diffraction pattern itself. Since the types of deviation from the average structure are numerous, no general guide can be provided for the interpretation of a PDF. A few typical cases can, however, be used to develop guidelines. In all these cases it is presumed that the instrumental effects, such as Q_{max}, the damping coefficient and the broadening coefficient are known.

Currently three slightly different approaches are commonly used to evaluate PDF data. They differ in the way the structure is simulated and the way the defects are introduced. All three approaches are complementary to each other and the reader is encouraged to use all of them in the appropriate situations.

The first approach, which commonly is implemented using the program PDFgui,[32] could be called a 'small-box refinement strategy'. The user initially expands the asymmetric unit of a starting structure to a single unit cell or a supercell. Atom coordinates, anisotropic ADPs and instrumental parameters are then refined using a standard least-squares algorithm. The algorithm is a very fast one, and the program is intuitive to use. A particular strength is its ability to refine the structure to a limited distance range instead of the full PDF. This allows *e.g.* differentiation between a truly local structure and the average structure.

The second approach, which is usually implemented using either the program RMC_POT + +[33] or the program RMCprofile[34] could, in contrast, be called a 'large-box refinement'. The initial structure is expanded to a larger supercell, or alternatively a large starting structure can be imported. The main difference from the PDFGui approach lies in the choice of refinement strategy, however. Instead of a least-squares algorithm, the so-called **reverse Monte Carlo (RMC) algorithm** is used.[35,36] This minimises the difference between observed and calculated data by performing random modifications of the structure.

Such modifications might be to shift an atom from its current position or to switch two atoms. If such a move reduces the differences between observed and calculated data, it is accepted. Otherwise, it is accepted with a probability that depends on a Boltzmann distribution controlled by the amount by which the agreement has worsened. The advantage of such an approach is that the user does not have to know the type of defect structure present in the sample. The algorithm will automatically modify the structure to match the observed data. In contrast to the PDFGui or the DISCUS approach (see below), the user interprets these defects afterwards. A particular strength of the RMCprofile program is the analysis of highly disordered samples such as glasses. In contrast to PDFGui, RMCprofile can simultaneously refine to PDF and powder diffraction data. The latter help the reverse Monte Carlo algorithm to restrain the average crystal structure close to that which would be determined from a Rietveld approach.

The third approach, which uses the DISCUS program,[14] allows the user to expand an initial structure to either a small or a large box and to refine the local structure with respect to PDF, powder diffraction or single-crystal data. The particular strength of this approach comes from toolboxes that allow a large variety of small and extended defects to be introduced into the structure. Furthermore, in contrast to the two other approaches, one can simulate structures that cannot be described by the small- or large-box models, which both require periodic boundary conditions. The DISCUS approach does however require prior knowledge (or at least an educated guess) of the types of defect present. The large flexibility comes at the price of a less than intuitive program control. As the refinement algorithm is based on a population-based evolutionary algorithm, it typically requires quite a bit more computing time than PDFGui.

3.5 EXAMPLES

Rather than show successful PDF refinements, the remainder of this chapter will guide the reader through the steps required to develop an initial model based on the PDF observations and add further details to improve the refinement. In many cases, the average structure is known. From this average structure, the corresponding PDF can be readily calculated. The interpretation of the experimental PDF will now benefit from an inspection of the difference between experimental and calculated $G(r)$.

3.5.1   Local Disorder *versus* Long-Range Average Order

We shall first expand the example in the introduction to a full 3D problem. This illustrates the abilities of the PDF technique to distinguish local order from intermediate or long-range order.

The example uses a (cubic) perovskite-type structure, ABO_3 (Figure 3.12). In the high-temperature perovskite structure, the B atom occupies the position $(0, 0, 0)$ and is octahedrally coordinated by six oxygen atoms at $(1/2, 0, 0)$, while the A atom occupies the position at $(1/2, 1/2, 1/2)$ and is coordinated by twelve oxygen atoms. One of the distortions that is observed upon lowering the temperature is a shift of the B atom along one of the three symmetrically equivalent [100] directions. This reduces the symmetry from the cubic space group *Pm3m* to the tetragonal space group *P4mm* and is accompanied by a slight change of the lattice parameters. As a shift along any of the three axes is equivalent in the high-temperature phase, independent domains will nucleate and grow. In each of the domains, the tetragonal *c*-axis is parallel to the shift of the B atoms and may be orientated along any of the three axes that were symmetrically equivalent in the high-temperature phase.

The data set for this example was calculated from a simulated crystal of $50 \times 50 \times 50$ unit cells. In order to simplify the example a little, the tetragonal *c*-axis was limited to the positive and negative *a*-axis of the cubic phase. Thus the crystal consists of two domain types, chosen to be some 5–10 unit cells in diameter, with irregular boundaries between neighbouring domains and a broad size distribution. In each domain, the B atom was shifted along the positive, respectively negative cubic *a*-axis. All other atoms were kept at their original positions, with A at $(1/2, 1/2, 1/2)$ and oxygen at $(1/2, 0, 0)$. All lattice parameters were kept fixed at the 0.38 nm cubic lattice constant. While this presents a simplification compared to a real perovskite, it allows us to determine the PDF analysis and refinement strategies. For this crystal, the calculated PDF is given in Figure 3.13 and the first 0.6 nm in Figure 3.14.

Before we look at refined models, let us inspect the *experimental* PDF for clues to the local disorder. The maxima within the first 0.6 nm in Figure 3.14 do show hints that allow us to reveal the local order. The maxima at 0.27, 0.38 and 0.46 nm are all quite narrow, with the width that would be expected for a room-temperature data set. In contrast, the maxima at 0.19 and 0.33 nm are much wider and have an unusual peak shape. In the cubic high-temperature phase we would expect only a very few distances in this range: the 0.19 nm B–O distance of atom pairs separated by $[1/2, 0, 0]$, the 0.27 nm A–O and O–O distances of atom pairs separated by $[1/2, 1/2, 0]$, the 0.33 nm A–B distance of atom pairs separated by $[1/2, 1/2, 1/2]$ and the 0.38 nm distance between any atom and an identical atom separated by the unit cell parameter, *i.e.* the [1,0,0] vector. As the maximum at 0.27 nm is narrow, we can conclude that the A–O distance is not significantly changed compared to that found in the high-temperature phase. The narrow 0.38 nm maximum further indicates that locally the structure is periodic, with identical [100] distances between all atom pairs. The broad 0.19 and 0.33 nm maxima indicate that the local disorder is caused by a displacement of the B atom with respect to the A and oxygen atoms. This is already a lot of detailed information about our local structure.

Two different refinement strategies were used to determine the average and local structures of this model crystal. In the first model, all three atom types were fixed to their cubic positions. Only the thermal parameter for the B-type atoms was refined. As the calculated PDF was very similar to the experimental PDF, it is not reproduced in Figure 3.13; only the difference curve is presented, shifted down to $G(r) = -11$ for clarity. The difference curve shows a very good fit above *ca* 0.13 nm and an excellent fit above *ca* 3 nm. Thus the long-range structure can be described very well by a model in which the B atom is randomly displaced according to an isotropic ADP, although this parameter has a very high value of

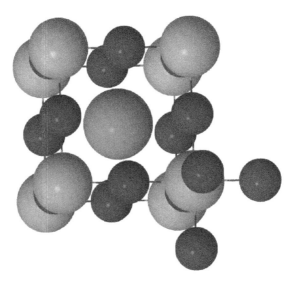

Figure 3.12 Perovskite structure, ABO_3. Atom type B (large light spheres) is octahedrally coordinated by six oxygen atoms (small dark spheres). Atom A (large lighter sphere) is located at the centre of the cube and is coordinated by twelve oxygen atoms.

$u = 0.125\,\text{Å}^2$, too high for real thermal motion, especially when compared to the value of $0.024\,\text{Å}^2$ for A and oxygen. At shorter distances, serious deviations exist between the experimental and calculated PDF, indicating that locally such a random displacement is not a good description. If such a situation is encountered, is it a good idea to restrict the refinement to a section of the experimental PDF. Here, the actual refinement was carried out only for the distance range from 1.5 to 5.5 nm. The full PDF was calculated afterwards and used to determine the difference curve in Figure 3.13. The very large ADP parameter and the lack of fit for distances below 0.15 nm call for a further model to be tested.

For the second model, the refinement was restricted to the very short distances up to 0.44 nm. In this model the B atom was allowed to shift along the a-axis and this displacement and the ADP parameter were refined. The difference curve shows that the local structure is matched perfectly in the range up to about 0.48 nm (see the left panel of Figure 3.14). Both the refined shift of $x = 0.05$ and the APD at $0.04\,\text{Å}^2$ perfectly match the parameters that were initially used to simulate the *sample* crystal: $x = 0.050$ and $u = 0.040\,\text{Å}^2$. In contrast, the first model cannot describe the local structure. Although the very large isotropic displacement parameter for B atoms in the first model manages to

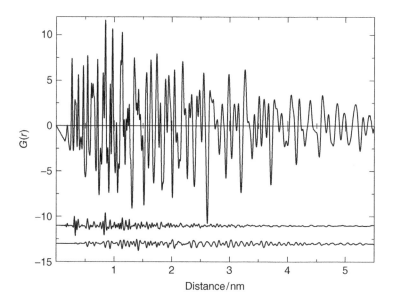

Figure 3.13 Experimental PDF for the disordered perovskite crystal. The difference curve at $G(r) = -11$ represents the difference between the experimental PDF and the PDF calculated for model 1. The lower difference curve corresponds to model 2. See text for details.

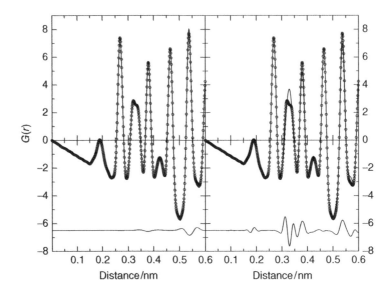

Figure 3.14 Experimental PDF given by open circles. Calculated PDFs are the continuous lines. Difference curves shifted down for clarity.

describe the shape of the 0.19 nm maximum reasonably well, the model fails to describe the peculiar peak shape at 0.33 nm. As the B-atom is shifted from $(0, 0, 0)$ to $(x, 0, 0)$ it is no longer located at the centre of the octahedron. As a consequence, the first maximum at 0.19 nm actually consists of three slightly different distance distributions: the short B–O distance to O at $(1/2, 0, 0)$, which is the length of $[1/2 - x, 0, 0]$, the long B–O distance to $O_{(-1/2, 0, 0)}$, which is the length of $[1/2 + x, 0, 0]$ and the four B–O distances to $O_{(0, 1/2, 0)}$, $O_{(0, -1/2, 0)}$, $O_{(0, 0, 1/2)}$, $O_{(0, 0, -1/2)}$, which all correspond to the length of $[x, 1/2, 1/2]$. Likewise, the maximum at 0.33 nm now corresponds to two slightly different B–A distances, for the four neighbouring A atoms with positive x-coordinates these distances are the length of $[1/2 - x, 1/2, 1/2]$, for the four A atoms with negative x-coordinate the distances are the length of $[1/2 + x, 1/2, 1/2]$.

The lower difference curve in Figure 3.13 shows that the second model does not describe the long-range structure well at all. Significant differences are observed for all distances above approximately 1 nm. In this distance range we face the situation that as the distance, r, is increased, more and more atom pairs become located within two different domains, with the B atom at $+x$ in the first domain and at $-x$ in the second. In our second model, all B atoms are at $+x$, and thus only one set of distances is calculated, although two different sets are required.

Both model refinements were carried out as small-box refinements. Just one unit cell was simulated, and periodic boundary conditions were used to calculate all interatomic distances. The combination of these two refinements allows us to derive a final model that describes the crystal with a domain model. Each domain is locally perfectly ordered, with a shift of B atoms along the positive or negative a-axis. The domain diameters are some 1.5–3.0 nm, as the upper difference curve in Figure 3.13 is negligibly small for distances above 3 nm. A full refinement with a big-box model would confirm this result.

3.5.2 ZnSe Nanoparticle

As a second example, consider the PDF of a ZnSe nanoparticle in Figures 3.15 and 3.16. The first obvious piece of information is that the peaks in the PDF that correspond to the nanoparticle become insignificantly small at around 3.5 nm, while those for the bulk ZnSe extend well beyond 6 nm. Thus one can deduce that this is a nanoparticle of approximately 3.5 nm diameter, in agreement with the particle size deduced from the powder pattern *via* the Scherrer equation or from profile analysis.[37,38] As the PDF is obtained from the diffraction

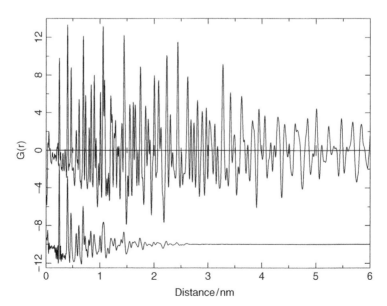

Figure 3.15 Experimental $G(r)$ for bulk (top) and nanocrystalline (offset to −10) ZnSe. Data were measured at BW5, DESY at 15 K at a wavelength of 0.01 nm.

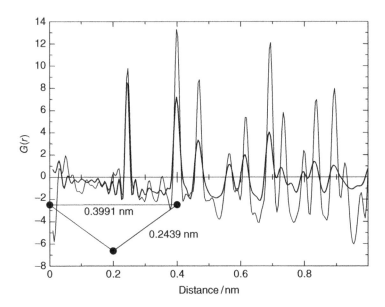

Figure 3.16 Detail of the ZnSe $G(r)$ (for key see Figure 3.15). The peak widths of the nanoparticle and the bulk structure are identical at the first peak. For further peaks, the relative widths increase for the nanoparticle relative to the bulk structure. The triangle illustrates the relationship between the first-neighbour distance at 0.2439 nm and the second-neighbour distance at 0.3991 nm. These distances indicate a 109.8° bond angle at the lower (central) atom.

pattern, it must be pointed out that the diameter calculated *via* the Scherrer equation or deduced from the longest observable distances in the PDF is strictly speaking the diameter of a structurally coherent domain within the sample; it is not necessarily the actual diameter of the nanoparticle, as observed *e.g.* through small-angle scattering or transmission electron microscopy (TEM). If several structurally coherent domains are irregularly intergrown, the size observed in the diffraction pattern or the PDF will not change, while the diameter calculated from the small-angle scattering signal will indicate a larger size. In the present case, TEM data support the 3.5 nm size.

The bulk structure of ZnSe is the zincblende structure, and an initial model calculation based on a spherical ZnSe nanoparticle of 3.5 nm size confirms that this is the correct structure for the ZnSe nanoparticles as well (Figure 3.17). This is also confirmed by the observation that the peak positions of the nanocrystalline sample and the bulk sample agree well. Independent of this prior knowledge, the local structure of the nanoparticles can be determined from the PDF data. Figure 3.16

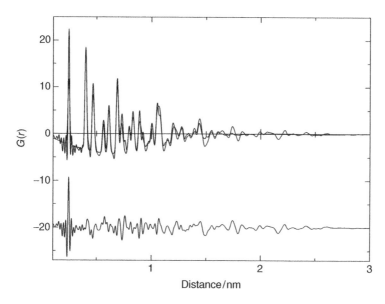

Figure 3.17 Comparison of the observed and calculated ZnSe nanoparticle $G(r)$. A perfect zincblende structure model with spherical shape was refined. The lower curve is the difference between experimental and calculated $G(r)$, offset for clarity.

shows that the first two peaks are at 0.2439 and 0.3991 nm. As this particular PDF was determined from high-energy X-ray data, far from any absorption edge, it does not contain straightforward evidence as to which particular atom pair forms a given distance. For a compound ZnSe, chemical experience indicates though that the shortest distance will be a Zn–Se bond and that the second shortest distance will correspond to a distance within the coordination shell of the central atom. Under these assumptions, the local structure in Figure 3.16 can be sketched out. As all bond distances in the triangle are known, the bond angle at the central atom is readily calculated as 109.80°, in very good agreement with an ideal tetrahedral angle of 109.47°. This leads to the conclusion that the local structure is tetrahedrally coordinated. The next two distances are at 0.466 and 0.564 nm, which can be expressed as 0.4660 nm = 2.439 nm $* \sqrt{\frac{11}{3}}$, 0.5640 nm = 0.2439 nm $* \sqrt{\frac{16}{3}} =$ 0.3991 nm $* \sqrt{2}$. A little structural experience indicates that these distances correspond to typical distances in a cubic crystal structure with lattice parameter 0.564 nm. Longer distances can be interpreted likewise. As this is a very simple crystal structure, the exercise is fairly straightforward. The general principle can however be applied to more complex crystal structures as well; see references [39,40].

The refinement whose result is shown in Figure 3.17 used a simple spherical model based on the zinc blende structure. Two prominent deviations between the observed and calculated PDF are obvious. First, the first maximum in the calculated PDF is much wider than the maximum in the experimental PDF. As detailed below, this indicates the presence of correlated displacements of first-neighbour atoms. Second, the heights of the maxima in the calculated PDF decrease at a much lower distance, r, than those in the experimental PDF. The spherical model apparently is unable to model the longer distances with sufficient intensity, as this would lead to much higher calculated peak heights at shorter distances. As a consequence, the current model underestimates the particle size.

Figure 3.18 shows a section of the nanoparticle PDF, as well as the PDF of bulk ZnSe measured under identical conditions. The first observation is that all peaks in the nanoparticle PDF are shifted to slightly shorter distances compared to those of the bulk structure. Thus we have clear evidence that the interatomic distances of nanocrystalline ZnSe are shorter by a factor of 0.9955. The main point here is the different peak widths, however. Note that the peaks at 2.0 and 2.2 nm have almost identical widths for the nanoparticle and the bulk structure, while the peaks

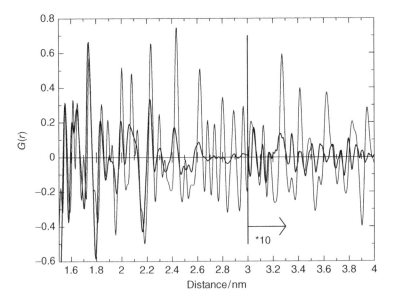

Figure 3.18 Intermediate distances of the ZnSe nanoparticle and bulk $G(r)$. The bulk $G(r)$ is scaled to match the nanoparticle $G(r)$. Above 3 nm, the nanoparticle $G(r)$ is multiplied by 10. Note that the very weak peaks in the nanoparticle $G(r)$ are synchronised with the bulk $G(r)$ up to distances of about 3.5 nm, the average nanoparticle diameter.

at distances of 2.1 and 2.3 nm are much wider for the nanoparticle than for the bulk structure. As the PDF is a measure of distances between atom pairs, we have the surprising result that in the nanoparticle structure the distance between atom pairs is well defined for pairs separated by 2.0 and 2.2 nm, while the intermediate distance at 2.1 nm is not well defined at all. This apparent contradiction allows information to be obtained regarding the defect structure of the nanoparticles. First, remember that the PDF is a projection of all interatomic distances in one dimension. Thus similar distances may be the projection of very different vectors in the crystal structure. A plausible explanation for the different widths therefore requires that the nanoparticle structure has different degrees of order in different directions. The simplest model one can test in this particular case contains stacking faults between the wurtzite and zincblende structures. This refinement improves the fit a little.

Slight differences between the experimental and the calculated PDF still exist with this model, and the next step would be to try out different shapes. Figure 3.19 shows calculated PDFs of extreme cases for particles with spherical, plate-like, and cylindrical shapes and for a particle terminated by the polyhedron shaped by the six symmetrically equivalent (110) faces and (100) and (00$\bar{1}$) planes. In order to focus on the effect that the shape has on the PDF, all particles here consist of the perfect wurtzite structure and have approximately the same volume. In comparison to the spherical particle, the peaks for the cylindrical particle drop in

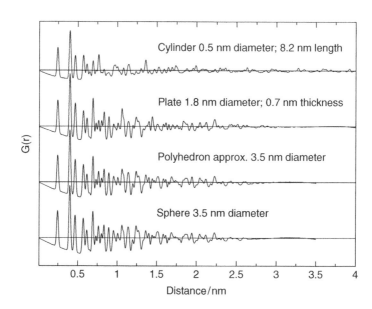

Figure 3.19   Influence of particle shape of ZnSe on calculated $G(r)$.

height more rapidly with increasing distance but extend with significant height to much longer distances. These longer distances correspond to interatomic distances along the long cylinder axis, which is much longer than the diameter of the sphere. The differences between the spherical and the plate-like particle are not as pronounced. For distances that correspond to approximately half the diameter, some individual differences exist between the two shapes. Towards the diameter, the peak heights drop off faster at greater distances for the plate-like particle. The differences between the 3 nm sphere and the particle terminated by (*hkl*) plane are insignificantly small. These shapes are too similar to be distinguished.

Let's go back to the comparison in Figure 3.16 between the PDFs of nanocrystalline and bulk ZnSe, looking at the first nanometre. For the PDF of nanocrystalline ZnSe, the peak height changes as a function of distance due to the finite size. The width of the first peak is identical in the nanocrystalline *versus* the bulk sample. Longer distances have an increasingly larger width for the nanocrystalline sample, in contrast to the bulk sample, for which the distances remain smaller and do not increase over the whole data range. In part, this increase is due to the high stacking fault density of the nanocrystalline sample. Both samples do show that the width of the first peak is significantly smaller than that of the second and further peaks, however. This change in the width of the interatomic distance distribution can in part be attributed to correlated displacements of the atoms away from the average position. The effect can be accounted for by a distance-dependent width, as given by Equation 3.27.

For our current example, the ZnSe nanoparticle, we can now test a final model. This model is an ellipsoidally shaped particle with rotational symmetry around the *c*-axis. This symmetry fits well to the hexagonal symmetry of an individual layer. Stacking faults are allowed to alternate between locally hexagonal and cubic sequences. The refinement improves again, and the final fit is shown in Figure 3.20, with refined parameters listed in Table 3.1.

3.5.3   Decorated ZnO Nanoparticle

The next example is illustrated to underline the ability to determine local structure of X-rays *versus* neutrons. The sample used for this example is nanocrystalline ZnO. ZnO crystallises in the wurtzite structure, and so too does nanocrystalline ZnO. This particular sample was synthesised using a sol–gel procedure by coprecipitation with citric acid. The citric acid serves to stabilise the small finite size. Figure 3.21

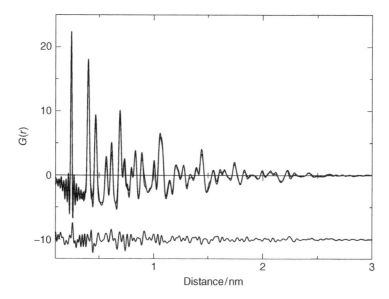

Figure 3.20 Final refinement of the structural model for the ZnSe nanoparticle. The model is an ellipsoidally shaped nanoparticle with growth faults. The peak width is corrected with r-dependence according to Equation 3.27.

Table 3.1 Refined parameters for the ZnSe nanoparticle. An ellipsoidal particle had been modelled with stacking faults that alternate between the wurtzite and zincblende structure types.

Parameter	Value
a-lattice parameter	0.3987 nm
c-lattice parameter	0.6492 nm
$z(Zn)$	0.3688
ADP	0.0083
Stacking fault probability	0.71
Diameter in (ab) plane	2.9 nm
Diameter along [001]	3.9 nm

shows the X-ray PDF based on a measurement with 100 keV X-rays at DESY, Germany. The local structure can be described by an ellipsoidal nanoparticle with diameters of 3.4 nm and a high stacking fault probability. No fundamental difference exists between this and the analysis of the ZnSe particles in the previous section.

Figure 3.22 shows the PDF obtained from neutron diffraction data. These data were collected at beam line NPDF, Lujan Center, Los Alamos National Laboratory, NM, USA, at a temperature of 30 K. Both sets of

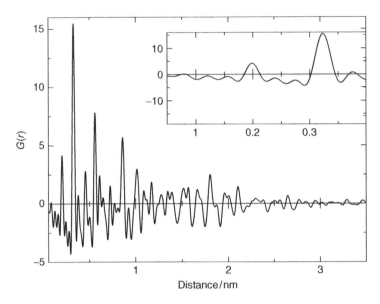

Figure 3.21 Experimental X-ray $G(r)$ of citric acid-stabilised ZnO. The particle diameter is about 3.5 nm and the $G(r)$ can be modelled very well with an equivalent of the ZnSe model. Data collected at BW5, DESY, at room temperature and with a wavelength of 0.1 nm.

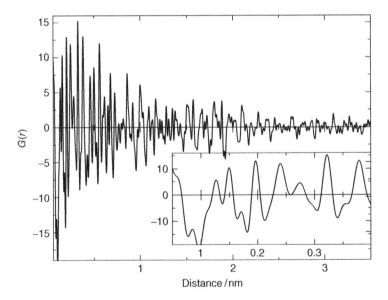

Figure 3.22 Experimental neutron $G(r)$ of citric acid-stabilised ZnO. Data collected at NPDF, Los Alamos National Laboratory at $T = 30\,K$. In addition to the ZnO distances, peaks at 0.12, 0.15, 0.24 and 0.274 nm are observed.

data exhibit peaks up to interatomic distances around 3.5 nm. The striking difference is observed at short interatomic distances. In this range, additional distances are observed in the neutron diffraction-based PDF at 0.13, 0.15, 0.24 and 0.27 nm. The first three distances correspond to the C–O, C–C and carbon next-neighbour carbon distances typically observed in organic molecules. The 0.27 nm distance, however, is not a prominent interatomic distance within the citric acid molecule. Within the molecule, several distances in the range 0.26–0.3.0 nm exist, though these are broadly distributed. A 0.27 nm distance is observed in several crystalline Zn metal–organic compounds that include carboxyl groups. In these crystal structures, one of the two oxygen atoms builds a 0.198 nm distance to a Zn atom, while the other is found at a distance of approximately 0.274 nm from adjacent Zn atoms. Thus, the observation of this additional distance that is not part of the ZnO core structure or of the stabilising organic shell shows that the organic molecule does actually form a layer on the surface of the ZnO core structure. For two more examples of the determination of organic molecules on the surface of nanoparticles, see references [41,42].

3.6 COMPLEMENTARY TECHNIQUES

Several complementary techniques exist that are suitable for investigating the local structures of materials. A very brief overview is presented here. For further details, the reader is referred to the cited literature.

The most obvious and immediately available complementary technique is the interpretation of the powder diffraction pattern itself.[43] As stressed above, the powder pattern and the PDF contain the same information, even though it is encoded very differently. The diffuse intensity and the Bragg reflection profile can also be interpreted with respect to the local structure. A powder pattern is usually interpreted without further manipulation, while the PDF is obtained from the normalised structure factor multiplied at each data point by Q. For this reason, the intensity of the initial powder pattern is relatively higher in the low-Q range and the interpretation tends to emphasise the Bragg intensity much more than the diffuse scattering or the high-Q part, which is often neglected as it seems not to carry any information. This leads to a stronger emphasis on the long-range order effects compared to the PDF, which stresses the local-order part of the structure.

Few PDF data have been obtained to date by measurements at an absorption edge. The vast majority of data are collected with high-energy X-ray diffraction techniques, and no immediate chemical information can be obtained from a PDF. This chemical information is the main strength of X-ray absorption techniques such as X-ray absorption near-edge spectroscopy (XANES) and especially extended X-ray absorption fine structure (EXAFS).[44] EXAFS data are obtained by measuring the absorption of a sample as a function of primary beam energy. At energies above the absorption edge, the absorption coefficient of the element will not decrease as a smooth function but shows oscillations. These oscillations around the smooth function are subject to a Fourier transform and yield a pattern in direct space. In analogy to the PDF, the peak positions and the peak width can be interpreted with respect to the immediate environment around the element at whose absorption edge the date were measured. The oscillations of the absorption curve are due to the interference between the photo electron emitted by the absorbing atom and the electron wave back-scattered by the surrounding atoms. As the range of the photo electron wave and of the back-scattered wave is limited, EXAFS data usually yield data on the first or second coordination shell only, in contrast to PDF data. The main advantage of EXAFS data is their chemical insight, as they specifically probe the environment of just one atom type.

High-resolution transmission electron microscopy (HRTEM), especially if taken at an aberration-corrected microscope, is another powerful complementary technique.[45,46] As the HRTEM image is obtained by recombination of the diffracted beams, the image does not show the crystal structure but instead the inverse Fourier transform of the scattered waves. As the electron scattering process is very intense, dynamical scattering theory applies even for thin samples. While this means that the HRTEM picture has to be carefully interpreted, as the white spots may actually be located at positions between the atoms, it is sufficient for this brief comparison to state that the HRTEM image is closely related to the projection of the crystal structure along the incoming electron beam. Thus one can see the crystal structure at least in two dimensions. Defects such as stacking faults, dislocations and the size and shape of nanosized crystals are immediately visible. With aberration-corrected microscopes, even the precise atom coordinates (including site substitution) can be detected. This extra dimension is a big advantage over the PDF data obtained from powder patterns, which produce a projection of the interatomic distances into one dimension. Compared to PDF experiments, sample preparation for an HRTEM experiment is much more involved, and one must keep in mind that many crystal structures are subject to serious beam damage by the high-energy electron beam.

In a nuclear magnetic resonance (NMR) experiment,[47,48] the nuclei of specific isotopes absorb and reemit energy while subjected to a high magnetic field. The associated resonant frequency depends on the isotope and the external magnetic field. For structural analysis, it is most relevant that the exact frequency depends on the local environment of the resonant isotope, which enables determination of the very local environment around specific isotopes. The isotope must have a non-zero nuclear spin; prominent examples are ^{1}H, ^{13}C and ^{15}N. The NMR technique is a very local probe and is sensitive to the immediate surroundings of the isotope.

REFERENCES

[1] F. Frey, *Acta Crystallogr. Sect. B*, **51**, 592 (1995).
[2] V. Nield and D. A. Keen, *Diffuse Neutron Scattering from Crystalline Materials*, Clarendon Press, Oxford, 2001.
[3] R. Welberry, *Diffuse Scattering and Models of Disorder*, Oxford University Press, Oxford, 2004.
[4] J. M. Cowley, *Diffraction Physics*, North Holland, Amsterdam, 1995.
[5] T. Egami and S. J. L Billinge, *Underneath the Bragg Peaks*, Pergamon, Amsterdam, 2003.
[6] S. J. Billinge and M. G. Kantzidis, *Chem. Comm.*, 749 (2004).
[7] S. J. Billinge and I. Levin, *Science*, **316**, 561 (2007).
[8] T. Proffen and H. Kim, *J. Mater. Chem.*, **19**, 5078 (2009).
[9] T. Weber and A. Simonov, *Z. Krist.*, **227**, 238 (2012).
[10] P. Debye, *Ann. Phys.*, **46**, 809 (1915).
[11] B. D. Hall, *J. Appl. Phys.*, **87**, 1666 (2000).
[12] A. Cervellino, C. Giannini and A.Guagliardi, *J. Appl. Cryst.*, **36**, 1148 (2003).
[13] C. Giacovazzo, H. L. Monaco, G. Artioli, D. Viterbo, M. Milanesio, G. Gilli, P. Gilli, G. Zanotti and G. Ferraris, *Fundamentals of Crystallography*, Oxford University Press, Oxford, 2011.
[14] R.B. Neder and T. Proffen, *Diffuse Scattering and Structure Simulation*, Oxford University Press, Oxford, 2008.
[15] V. Petkov, S. Shastri, B. Wanjala, R. Loukrakpam, J. Luo and C.-Z. Zhong, *Z. Krist.*, **227**, 212 (2012).
[16] P. J. Chupas, X. Qiu, J. C. Hanson, P. L. Lee, C. P. Grey and S. J. L. Billinge, *J. Appl. Crystallogr.*, **36**, 1342 (2003).
[17] P. J. Chupas, K. W. Chapman, G. Jennings, P. I. Lee and C. P. Grey, *J. Am. Ceram. Soc.*, **129**, 13822 (2007).
[18] D. A. Keen, *J. Appl. Crystallogr.*, **34**, 172 (2001).
[19] P. Juhás, T. Davies, C. L. Farrow and S. J. Billinge, *J. Appl. Crystallogr.*, **46**, 560 (2013).
[20] P. F. Peterson, M. Gutmann, T. Proffen and S. J. Billinge, *J. Appl. Crystallogr.*, **33**, 1192 (2000).
[21] K. Page, C. E. White, E. G. Estell, R. B. Neder, A. Llobet and T. Proffen, *J. Appl. Crystallogr.*, **44**, 532 (2011).
[22] R. B. Neder and V. I. Korsunskiy, *J. Phys.: Condens. Matter*, **17**, S125 (2005).

[23] R. B. Neder and V. I. Korsunski, *J. Appl. Crystallogr.*, **38**, 1020 (2005).

[24] R. C. Howell, T. Proffen and S. D. Conradson, *Phys. Rev. B*, **73**, 094107 (2006).

[25] K. Kodama, S. Iikubo, T. Taguchi and S. Shamnoto, *Acta Crystallogr. Sect. A*, **64**, 44 (2006).

[26] V. I. Korsunskiy, R. B. Neder, A. Hofmann, S. Dembski, C. Graf and E. Rühl, *J. Appl. Cryst.*, **40**, 975 (2007).

[27] C. Kumpf, R. B. Neder, F. Niederdraenk, P. Luczak, A. Stahl, M. Scheuermann, S. Joshi, S. K. Kulkarni, C. Barglik-Chory, C. Heske and E. Umbach, *J. Chem. Phys.*, **123**, 224707 (2005).

[28] R. B. Neder, V. I. Korsunskiy, C. Chory, G. Müller, A. Hofmann, S. Dembski, C. Graf and E. Rühl, *Phys. Stat. Sol. c*, **4**, 3233 (2007).

[29] F. Niederdraenk, K. Seufert, P. Luczak, S. K. Kulkarni, C. Chory, R. B. Neder and C. Kumpf, *Phys. Stat. Sol. c*, **4**, 3234 (2007).

[30] C. Chory, R. B. Neder, V. I. Korsunskiy, F. Niederdraenk, Ch. Kumpf, E. Umbach, M. Schumm, M. Lentze, J. Geurts, G. Astakhov, W. Ossau and G. Müller, *Phys. Stat. Sol. c*, **4**, 3260 (2007).

[31] F. Niederdraenk, K. Seufert, A. Stahl, R. S. Bhalerao-Panajkar, S. Marathe, S. K. Kulkarni, R. B. Neder and C. Kumpf, *Phys. Chem. Chem. Phys.*, **13**, 498 (2011).

[32] C. L. Farrow and S. J. Billinge, *Acta Crystallogr. Sect. A*, **63**, 232 (2009).

[33] O. Gereben, P. Jovari, L. Temleitner and L. Pusztai, *J. Optoel. Advanced Mater.*, **9**, 3021 (2007).

[34] M. G. Tucker, D. A. Keen, M. T. Dove, A. L. Goodwin and Q. Hui, *J. Phys. Cond. Matter*, **19**, 335218 (2007).

[35] R. L. McGreevy, *J. Phys. Cond. Matter*, **13**, R877 (2001).

[36] M. G. Tucker, M. T. Dove and D. A. Keen, *J. Appl. Crystallogr.*, **34**, 638 (2001).

[37] P. Scardi and M. Leoni, *Acta Crystallogr. Sect. A*, **58**, 190 (2002).

[38] M. Leoni, R. Di Maggio, S. Polizzi and P. Scardi, *J. Am. Ceram. Soc.*, **87**, 1133 (2004).

[39] P. Juhás, D. M. Cherba, P. M. Duxbury, W. F. Punch and S. J. L. Billinge, *Nature*, **440**, 655 (2006).

[40] P. Juhás, L. Granlund, P. M. Duxbury, W. F. Punch and S. J. L. Billinge, *Acta Crystallogr. Sect. A*, **64**, 631 (2008).

[41] K. Page, T. C. Hood, T. Proffen and R. B. Neder, *J. Appl. Crystallogr.*, **44**, 327 (2011).

[42] K. Page, T. Proffen, M. Niederberger and S. Seshadri, *Chem. Mater.*, **22**, 4386 (2010).

[43] R. Dinnebier and S. J. L. Billinge, *Powder Diffraction: Theory and Practice*, Royal Society of Chemistry, Cambridge, 2008.

[44] G. Brunker, *Introduction to XAFS: A Practical Guide to X-Ray Absorption Fine Structure Spectroscopy*, Cambridge University Press, Cambridge, 2010.

[45] B. Fultz and J. Howe, *Transmission Electron Microscopy and Diffractometry of Materials*, Springer, Berlin, 2012.

[46] R. Erni, *Aberration-Corrected Imaging in Transmission Electron Microscopy: An Introduction*, Imperial College Press, London, 2010.

[47] J. Keeler, *Understanding NMR Spectroscopy*, John Wiley, New York, 2010.

[48] H. Günther, *NMR Spectroscopy: Basic Principles, Concepts and Applications in Chemistry*, Wiley-VCH, 2013.

4

Electron Crystallography

Lu Han[a], Keiichi Miyasaka[b] and Osamu Terasaki[b,c]

[a] *School of Chemistry and Chemical Technology, State Key Laboratory of Composite Materials, Shanghai Jiao Tong University, Shanghai, China*
[b] *Graduate School of EEWS, WCU Energy Science & Engineering, KAIST, Daejeon, Republic of Korea*
[c] *Department of Materials & Environmental Chemistry, EXSELENT, Stockholm University, Stockholm, Sweden*
Email: terasaki@kaist.ac.kr & terasaki@mmk.su.se

4.1 INTRODUCTION

An optical microscope is an instrument that produces enlarged images of objects too small to be seen by the naked eye. Figure 4.1 shows the ray diagram of an optical microscope. Object distance, u, image distance, v, and focal length, f, are related as follows:

$$1/f = 1/u + 1/v \qquad (4.1)$$

Magnification M equals v/u and depends on f and either u or v, allowing us to magnify ($M \geq 1$) or demagnify ($M \leq 1$) an object. The resolution, δ, of an optical microscope is limited by the wavelength of visible light ($\lambda \sim 550$ nm for green light) and the numerical aperture, NA, through Rayleigh's criterion ($\delta = 0.61\lambda/NA$). This can achieve resolution of ~ 100 nm with the insertion of an index-matching material with a high refractive index between the atmosphere and the lens glass.[1] However, with this resolution it is not possible to solve a crystal structure, as the distance between neighbouring atoms in a crystal is normally 1–2 Å.

Structure from Diffraction Methods, First Edition. Edited by Duncan W. Bruce, Dermot O'Hare and Richard I. Walton.
© 2014 John Wiley & Sons, Ltd. Published 2014 by John Wiley & Sons, Ltd.

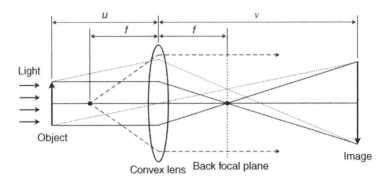

Figure 4.1   Ray diagram of an optical microscope.

In the early 1900s, a significant innovation was made when electrons were used instead of light to generate images. Since electrons have remarkably shorter wavelengths than photons, electron microscopes can achieve much higher magnification and resolution. Ernst Ruska *et al.* developed the first electron microscope, a transmission electron microscope (TEM), in 1931 and went on to win the Nobel Prize in Physics in 1986. The principle of a TEM is similar to that of an optical microscope, with electromagnetic lens being made convex through Lorentz force and used to focus the negatively charged electrons. When a point electron source is placed at the fore focal plane ($u = f$), we get a parallel electron beam ($v = \infty$), while when the source is placed at infinity ($u = \infty$) we get a convergent electron beam at the back focal plane ($v = f$).

TEMs can provide four-dimensional (4D) information, *i.e.* real-space (microscopy), reciprocal-space (diffractometry), energy-space (spectroscopy) and time-space (from video rate to femtoseconds). We define **electron crystallography** (EC) as crystallography that uses electrons as a probe through microscopy and/or diffractometry, whereas X-ray crystallography is based on diffraction. TEM can provide an electron diffraction (ED) pattern on the back focal plane of the objective lens and can supply real-space information through images on the image plane. The phase information of the **crystal structure factor** (CSF) is lost in the diffraction pattern (kinematical) but preserved in the images, so analysis of the images has a very important role in structural determination. The history of EC is well described in Chapter 1 of the recently published *Electron Crystallography: Electron Microscopy and Electron Diffraction*.[2] Further information on structural characterisation by EC can be found in references [3–9].

4.2   CRYSTAL DESCRIPTION

Let's define a crystal as an object with translation symmetry in a space of n-dimension (nD; $1 \leq n \leq 3$) and positional vector defined in the space.

4.2.1   Fourier Transformation and Related Functions

Fourier transforms (FTs) $F(q)$, $G(q)$ and $H(q)$ are defined for gentle functions of $f(r)$, $g(r)$ and $h(r)$ as follows:

$$\mathcal{F}[f(r)] = F(q) \equiv \frac{1}{\sqrt{(2\pi)^n}} \int f(r) e^{-iq \cdot r} d^n r \qquad (4.2)$$

$$\mathcal{F}[g(r)] = G(q)$$

$$\mathcal{F}[h(r)] = H(q)$$

The inverse Fourier transforms of $F(q)$, $G(q)$ and $H(q)$, $\mathcal{F}^{-1}[F(q)]$, $\mathcal{F}^{-1}[G(q)]$ and $\mathcal{F}^{-1}[H(q)]$, are as follows:

$$\mathcal{F}^{-1}[F(q)] = \frac{1}{\sqrt{(2\pi)^n}} \int F(q) e^{iq \cdot r} d^n q = f(r) \qquad (4.3)$$

$$\mathcal{F}^{-1}[G(q)] = g(r)$$

$$\mathcal{F}^{-1}[H(q)] = h(r)$$

The **convolution**, $h(r)$, of functions $f(r)$ and $g(r)$ is defined using the symbol $*$ as:

$$h(r) = f(r) * g(r) \equiv \int f(r')g(r - r') d^n r' \qquad (4.4)$$

Fourier transformation of $h(r)$ is then given by:

$$\mathcal{F}[h(r)] = \mathcal{F}[f(r) * g(r)] = \mathcal{F}[f(r)] \cdot \mathcal{F}[g(r)] = F(q) \cdot G(q) = H(q) \qquad (4.5)$$

Another important operation, called the **correlation operation** of $f(r)$ and $g(r)$, is defined using the symbol $\otimes$ as:

$$f(r) \otimes g(r) \equiv \int f(r')g(r + r') d^n r' \qquad (4.6)$$

If $g(r) = f(-r)$ then the correlation is an **autocorrelation**:

$$\mathcal{F}\,[f(r) \otimes f(-r)] = \mathcal{F}\,[f(r)] \cdot \mathcal{F}\,[f(-r)] = F(q) \cdot F(q)^* = |F(q)|^2 \quad (4.7)$$

4.2.2   Lattices

An nD periodic lattice $L(r)$ is given by an infinite array of points in space using Dirac delta function, $L(r) = \sum_{r_n} \delta(r - r_n)$, where any lattice point r_n is a centre of symmetry and is given by a linear combination of primitive translational vectors, $\{a_i\}$, $1 \leq i \leq n$, as:

$$r_n = \sum_{i=1}^{n} n_i a_i \tag{4.8}$$

where n_i $(1 \leq i \leq n)$ are arbitrary integers.

Each lattice point has identical surroundings to all others. The primitive vectors of reciprocal lattice $\{b_j\}$, $1 \leq j \leq n$, are defined corresponding to $\{a_i\}$ as:

$$a_i \cdot b_j = 2\pi\delta_{ij} \quad (\text{mod } n) \tag{4.9}$$

where $i, j = 1, 2, 3$ and δ_{ij} represent the Kronecker delta. Reciprocal lattice points are given in b_j space by reciprocal vectors q_h with integers of h_k, $1 \leq k \leq n$.

$$q_h = \sum_{k=1}^{n} h_k b_k \tag{4.10}$$

For a 1D periodic lattice, $r_1 = n_1 a_1$ and Equation 4.9 becomes:

$$a_1 \cdot b = 2\pi \cdot \text{integer} \tag{4.11}$$

This gives a set of planes of b with equal spacing $2\pi/|a_1|$ perpendicular to a_1. A 2D lattice (planar lattice) is described by a convolution of two independent 1D lattices, a_1 and a_2, and hence the Fourier transform of a 2D lattice is given by the product of the Fourier transforms of the original 1D lattices, as shown in Figure 4.2 (following Figure A8-1 in *Space Groups for Solid State Scientists*, second edition[10]).

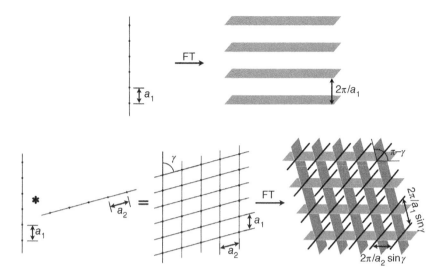

Figure 4.2 Schematic drawing of the 1D and 2D lattices and the corresponding reciprocal lattices.

4.2.3   Crystals and Crystal Structure Factors

For simplicity, let's examine the 3D ($n = 3$) case.

Any position r in the crystal is given by the origin of the mth unit cell, r_m, and coordinates (x, y, z) within the cell as:

$$r = r_m + xa_1 + ya_2 + za_3 \qquad (4.12)$$

where $0 \leq x, y, z \leq 1$.

Each reciprocal lattice point hkl represents a set of lattice planes in real space, perpendicular to the reciprocal lattice vector $q_{hkl} = hb_1 + kb_2 + lb_3$. The interplanar spacing $d_{hkl} = 2\pi/|q_{hkl}|$.

A crystal structure $C(r)$ is given by atomic distribution within a basis $B(r)$ and underlying the periodic lattice $L(r)$. A crystal can take any size and shape, given by $Z(r) = 1$ when r is inside the crystal and $Z(r) = 0$ otherwise. The crystal $C(r)$ and CSF $F(q_{hkl})$ are then defined by the following two equations:

$$C(r) = \{B(r) * L(r)\} \cdot Z(r) = B(r) * \{L(r) \cdot Z(r)\} \qquad (4.13)$$

and:

$$F(q_{hkl}) = \mathscr{F}\,[C(r)] = \mathscr{F}\,[B(r)] \cdot [\mathscr{F}\,[L(r)] * \mathscr{F}\,[Z(r)]] \qquad (4.14)$$

In most textbooks, for simplicity the crystal $Z(r)$ is given by the number of the unit cell, N_1, N_2, N_3, which is a maximum number of n_1, n_s, n_3, along $a_i, i = 1, 2, 3$ as $Z(r) = 1$ for $n_1 \leq N_1, n_2 \leq N_2, n_3 \leq N_3; Z(r) = 0$ otherwise. $\mathscr{F}\,[Z(r)]$ is then given by the so-called Laue function or shape (geometrical) function as:

$$\mathscr{F}\,[Z(r)] = \left\{ \frac{\sin \pi N_1 h}{\sin \pi h} \right\} \cdot \left\{ \frac{\sin \pi N_2 k}{\sin \pi k} \right\} \cdot \left\{ \frac{\sin \pi N_3 l}{\sin \pi l} \right\} \qquad (4.15)$$

The advantage of using the general equation over the Laue function will be shown later for a thin single crystal with vicinal surface.

Crystallographic **point-group** (**PG**) symmetries require one, four (oblique, rectangular, square and hexagonal) and seven (triclinic, monoclinic, orthorhombic, tetragonal, trigonal, hexagonal and cubic) crystal systems for 1D, 2D and 3D systems, respectively. A Bravais lattice is a distinct lattice type; there are one, five and fourteen Bravais lattices for 1D, 2D and 3D systems, respectively. The combination of 14 Bravais lattices, 32 crystallography PGs and the screw axis and glide plane symmetry operations results in a total of 230 different **space groups** (**SGs**), describing all possible crystal symmetries.

4.2.4 Simple Description of Babinet's Principle

A special case is the porous solid. Let's take a simple 1D model system in order to explain Babinet's principle, which is useful to understanding 'contrast scattering' in porous nanocrystals. Babinet's principle states that the diffraction pattern from an aperture, $a(x)$, is the same as the pattern from an opaque object of the same shape, $b(x) \equiv 1 - a(x)$, illuminated in the same manner.

Let us take a rectangular function, $f(x) = 1$, for, $-L/2 \leq x \leq L/2$ and $f(x) = 0$ otherwise (Figure 4.3a). Using $\mathrm{sinc}(x) \equiv \sin(x)/x$ (shown in the inset of Figure 4.3):

$$\mathscr{F}\,[f(r)] = F(q) = (1/\sqrt{2\pi}) \int_{-L/2}^{L/2} e^{-iqx}\,dx = (L/\sqrt{2\pi}) \cdot \mathrm{sinc}(qL/2)$$
$$(4.16)$$

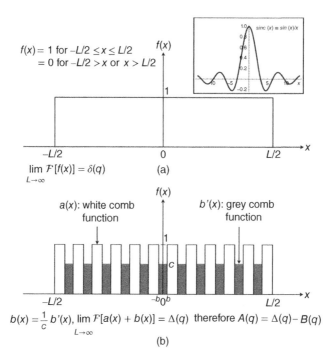

Figure 4.3 Schematic drawing of Babinet's principle for the diffraction pattern from (a) an aperture and (b) a periodic comb function.

It is noteworthy that integral representation for a δ-function is expressed as:

$$\delta(x) = 1/\sqrt{2\pi} \int_{-\infty}^{\infty} 1/\sqrt{2\pi}e^{-iqx}dq$$

$$\text{therefore } \mathscr{F}[\delta(x)] \equiv \Delta(q) = 1/\sqrt{2\pi}e^{-iqx} \tag{4.17}$$

Therefore:

$$\mathscr{F}[a(x) + b(x)] = \Delta(q) \tag{4.18}$$

and:

$$\mathscr{F}[a(x)] \equiv A(q) = \Delta(q) - \mathscr{F}[b(x)] \equiv \Delta(q) - B(q) \tag{4.19}$$

Except for the origin of reciprocal space, the Fourier transforms of arbitrary shapes of an aperture $a(x)$ and of the conjugate of the same aperture $b(x)$ are the same in magnitude but opposite in sign. Therefore, both will produce the same diffraction pattern (a real 3D example will be shown later using MCM-48 and its carbon replica (Figure 4.29)).

Taking periodic comb functions $a(x)$ and $b'(x)$ with $a(x) + (1/c)$ $b'(x) = 1$, the same argument is valid (Figure 4.3b).

4.3   ELECTRON MICROSCOPY

4.3.1   Interaction between Electrons and Matter

Two different types of electron illumination – **parallel** and **focused** (convergent) – are used to obtain real and wave-number space information. These produce, respectively, selected area electron diffraction (SAED) and convergent beam electron diffraction (CBED) in diffraction mode, and conventional transmission electron microscope (CTEM) images and scanning transmission electron microscope (STEM) images in image mode.

Electrons are negatively charged particles that interact strongly with the atoms in a crystal *via* their electrostatic potentials, $V(r)$, of the Coulomb forces. Electrons are scattered elastically or inelastically, and various signals can be produced through the interaction that occurs when a high-energy electron is incident on a specimen (Figure 4.4). Elastically scattered electrons and backscattered electrons (BSEs) are formed by elastic scattering, while characteristic X-rays, secondary electrons (SEs) and Auger electrons (AEs) are induced by inelastic scattering processes. A STEM high-angle annular dark field (HAADF) image is formed mainly by inelastically scattered electrons with lattice

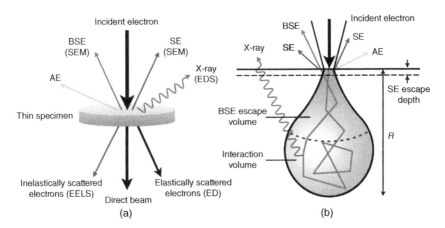

Figure 4.4   Signals generated using (a) a high-energy electron beam with a thin specimen and (b) a focused beam with a bulk specimen.

vibration, called thermal diffuse scattering (TDS), but energy loss is too small to be detected ($\sim$0.03 eV at room temperature). It is worth noting that a thin specimen allows incident electrons to be scattered in both forward and backward directions, while with a bulk specimen only backscattering is available. When a focused beam is incident on a bulk specimen, the penetration depth, R(nm), can be estimated as follows:[11]

$$R = 27.6 \, \frac{AE^{5/3}}{\rho Z^{8/9}} \tag{4.20}$$

where A = mass number (g mol^{-1}), E = energy of impact electron (keV), ρ = density (g cm^{-1}), Z = atomic number.

Each of these signals can be detected by specialised detectors and used for different purposes. The direct beam and the elastically scattered electrons are of special interest in TEM, the coherence of which forms ED patterns. In scanning electron microscopy (SEM), SEs and BSEs are conventionally defined as electrons with kinetic energy smaller and larger than $\sim$50 eV, respectively. The number of SEs emitted from the specimen surface depends greatly on the incident angle, which is employed in surface topology observation in SEM. BSEs are high-energy electrons that originate from interaction with the nucleus and are backscattered out of the specimen by elastic scattering (Rutherford or Mott scattering). They may undergo inelastic collisions, but all have an emitted energy of over 50 eV, by definition. They provide compositional information on the specimen. The chemical and elemental characterisations of a sample can be analysed by energy-dispersive X-ray spectroscopy (EDS) using characteristic X-rays. The inelastic scattered electrons can also provide information on chemical composition, bonding, surface properties *etc.* *via* electron energy loss spectroscopy (EELS). EELS tends to show advantages at relatively low atomic numbers with thin specimens, while EDS is suitable for the heavier elements.

4.3.2   Scanning Electron Microscopy

SEM is a powerful technique by which to observe both the topological features of and elemental information on a specimen by scanning it with a focused beam of electrons. The ray diagram of a typical scanning electron microscope is shown in Figure 4.5. An electron beam is emitted from an electron gun, which typically has an energy ranging from 100 eV to 30 keV. There are two types of electron gun: (i) the thermionic, tungsten hairpin filament and LaB$_6$ single-crystal and (ii) field emission gun (FEG).

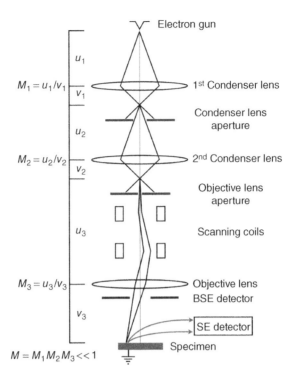

Figure 4.5 Ray diagram of a scanning electron microscope. Each lens demagnifies by M_i, and as a whole the scanning electron microscope demagnifies by M.

The Schottky emission cathode is categorised between the two because it also includes thermionic emission. The electron beam is demagnified by one or two condenser lenses and an objective lens into a spot with subnanometer diameter. The beam is deflected in the x-and y-axes by pairs of scanning coils or deflectors so that it scans over a certain area of the sample surface. The signals are then emitted from each point of the scanning area, collected by corresponding detectors, converted into electric signals and amplified.

The types of signal produced by a scanning electron microscope include SE, BSE, characteristic X-rays, light (cathodoluminescence, CL) *etc*. Most SEs are produced inside interaction volumes, and SEs created only near the surface (escape depth < several nanometres) can escape from the surface to vacuum. The dependence of SE yield on the angle of the direction of incident electrons to the surface normal, φ, is given by $1/\cos(\varphi)$ and thus SE signals are considered to be most suitable for observing surface topography. Figure 4.6 shows the number of electrons

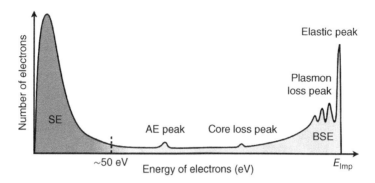

Figure 4.6 Energy distribution of SEs and BSEs for electrons with impact energy, E_{Imp}. Electrons with energy below and above 50 eV are classified as SEs and BSEs, respectively.

emitted from a sample as a function of their kinetic energy. Incident electrons (or BSEs) are scattered inelastically. Atomic electrons then gain energy; electrostatically weakly bound electrons such as valence and conduction electrons are released from atomic confinement on gaining even a small amount of energy.

SEM has some advantages over TEM, as it has a larger depth of field for the determination of crystal morphology and fine surface structures, such that surface topology can be observed as different contrasts in the image through a relatively simple experiment.

First let's discuss crystal morphology.

There are two typical crystal morphologies, known as equilibrium and growth forms. In the case of equilibrium form, the crystal and its surroundings are thermodynamically in equilibrium, and morphology is governed by minimum surface energy under constant volume and is well described by Wulff construction. The growth form is governed by anisotropy of growth rates in different crystallographic directions and is strongly dependent on growth conditions. In both cases, however, crystal morphologies should be commensurate with the PG symmetry of the crystal.

Let us take silica mesoporous crystals as an example. Figure 4.7a–d shows the typical morphologies of 2D-hexagonal $p6mm$[12], 3D-hexagonal $P6_3/mmc$[13] and 3D-cubic MCM-48 ($Ia\bar{3}d$)[14] and SBA-1 ($Pm\bar{3}n$)[15] mesoporous crystals, respectively. Coresponding PG symmetries, such as $6mm, 6/mmm$ and $m\bar{3}m$, and schematic drawings of morphologies with surface indices are also shown. In the case of $p6mm$, the indices of the external surface are not determined uniquely from

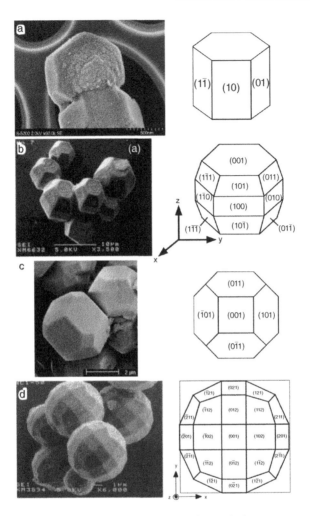

Figure 4.7 Typical SEM images showing crystal morphologies. (a) *p6mm*[*6mm*]. Reprinted with permission from [12] Copyright (2003) Wiley-VCH. (b) *P6₃/ mmc*[*6/mmm*]. Reprinted with permission from [13] Copyright (2002) American Chemical Society. (c) *Ia3̄d*[*m3̄m*]. Reprinted with permission from [14] Copyright (1998) Royal Society of Chemistry. (d) *Pm3̄n*[*m3̄m*]. Reprinted with permission from [15] Copyright (2001) American Chemical Society.

the SEM image but from a combination of a TEM image and the ED pattern.

Using low-impact energy electrons, we are able to observe SEM images from electrically insulating materials without any coatings, because the total electron yield (ratio of number of emitted electrons (sum of

BSEs and SEs) to number of incident electrons) becomes close to 1. On overcoming the resolution problem at low energy, surface fine structures include not only growth steps but also channel and cage openings, as well as their arrangement at the external surface have been observed.[16] For example, the channel openings, growth steps and heights are observed in high-resolution scanning electron microscopy (HRSEM) images of SBA-16 taken almost along the threefold [111] axis.[17] Figure 4.8b is an enlarged image of Figure 4.8a, allowing us to clearly see layer-by-layer growth steps in various zeolites and mesoporous silicas (*e.g.* SBA-16), micropores in the silica wall of SBA-15 and so on.

Now let's discuss atomic number-dependent contrast.

As the energy of incident electrons is quite high compared to that of electrons inside a sample, the two can be distinguished. Therefore, BSEs are incident electrons that are ejected from a sample without too much loss of energy, as a result of several scatterings or phonon (extremely small energy loss)/plasmon excitations. As the cross-section for high-angle elastic scattering (Rutherford scattering) is proportional to Z^2, we can obtain strong atomic number-dependent contrast in BSE images. BSEs come from approximately the top half of the interaction volume (Figure 4.4b and Equation 4.20), some tens or hundreds of nanometres depending on the incident electron energy and atomic number of specimen Z, while SEs come from the surface. This can be seen in Figure 4.9, where $Au@TiO_2$ rattle-spheres[18] are shown by SEs and BSEs. Both SE and BSE images allow the observation of gold particles on the surface of the spheres, but SE images provide only the surface morphology while BSE signals are able to distinguish the gold particles from TiO_2, as they are heavy and thus very bright in contrast. SE gives

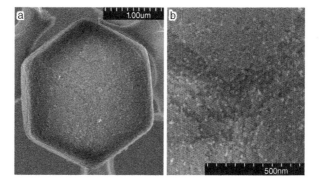

Figure 4.8 (a) HRSEM image of SBA-16. (b) Enlargement of (a). Reprinted with permission from [17] Copyright (2004) Elsevier Ltd.

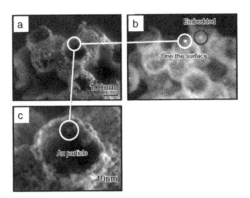

Figure 4.9 (a, c) SE image and (b) BSE image of an Au@TiO$_2$ sample: titanium nanoparticles creating interconnected macrospherical cavities, each encapsulating a gold nanoparticle. Reprinted with permission from [16] Copyright (2009) JEOL Ltd.

mainly topographical information, so both Au nanoparticles and TiO$_2$ particles of the sphere shell give a similar brightness. On the other hand, BSE's brightness is strongly dependent on atomic number, Z, so we can distinguish Au from TiO$_2$ particles; an Au particle can even be observed when encapsulated within the TiO$_2$ shell if it can be incorporated in the interaction volume of the electron probe. Furthermore, by comparing SE and BSE images, we can judge whether an Au nanoparticle is on the surface of the shell or behind it.

4.3.3 Transmission Electron Microscopy

TEM is a powerful tool for obtaining internal structural information at length scales from several hundreds of nanometres to sub-ångström, allowing crystal structures to be solved.

In CTEM, we use parallel electron beam illumination, plane wave, to obtain ED patterns or TEM images at accelerating voltages of 30–1250 kV. The ray diagrams of the formation of SAED and TEM images are schematically illustrated in Figure 4.10.

When an electron beam passes through a specimen, it interacts with the specimen to form an exit electron wave at the exit surface. The exit wave is transformed by an objective lens into a diffraction pattern at its back focal plane and to an image at the image plane. Following Equation 4.1, by tuning the focal length, f, through the current of the intermediate lens we can obtain information at different positions (back focal plane

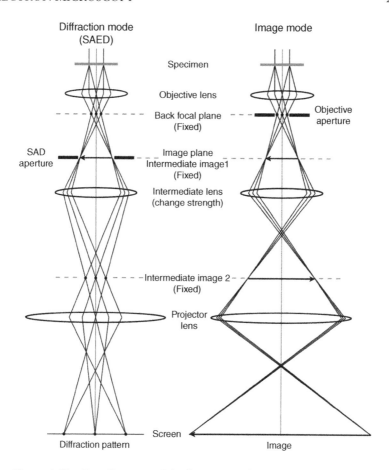

Figure 4.10   Ray diagrams of the formation of SAED and TEM images.

(SAED) or image plane (image)) with different values of u and bring it to a single position for use by the projector lens with fixed value of v in Equation 4.1. The corresponding information can then be projected onto a fluorescent screen or charge-coupled device (CCD) by the projector lens. An objective aperture is placed at the back focal plane of the objective lens and used in imaging mode, which selects some of the diffracted beams in order to form an image in the image plane, while a selected area diffraction (SAD) aperture is inserted at the image plane of the objective lens, which selects an area of interest of the specimen in order to form SAED. By using convergent beam illumination, STEM and CBED experiments can be carried out.

4.4   ELECTRON DIFFRACTION

4.4.1   X-Rays (Photons) *versus* Electrons

X-ray crystallography, using X-ray diffraction (XRD) with either a powder or a single-crystal diffractometer, is the most precise and most popular method of structural analysis. However, many crystals are synthesised only in powder form and quite often in small amounts, which is of no use for single-crystal XRD analysis. Structure determination by powder XRD can be used as an alternative, but it too has its limitations: reflections with similar scattering angles can overlap, causing difficulty in finding unit cell and ambiguity in their relative intensities, and a sample can contain defects, giving rise to a poorly resolved powder XRD profile with insufficient structural information. In that case, EC is the only possible technique by which to solve/characterise the crystal structure.

The interaction of X-ray photons (electromagnetic waves) and electrons (matter waves) with atoms and material will be discussed briefly, before we give equations for diffraction intensity and make comparisons between them.

X-rays are electromagnetic waves whose motions are described by Maxwell's equation. At the same time, they have particle aspects as well, so that *e.g.* Compton scattering can be described as photon–electron collision. For X-rays, the relationships among energy, E_x, wavelength, λ_x, and de Broglie momentum, p_x, are governed by the following equations:

$$E_x = h\nu_x = hc/\lambda_x \text{ or } \lambda_x(\text{Å} = 10^{-10}\text{m}) = 12.4/E_x(\text{keV}) \qquad (4.21)$$

and:

$$p_x = h/\lambda_x \qquad (4.22)$$

where h is Planck's constant, $h = 6.626 \times 10^{-34}\,\text{J s}^{-1}$.

The energy of X-rays with a wavelength of 1.54 Å (~Cu Kα) is given by:

$$E_x = 8.27(\text{keV}) = 1.325 \times 10^{-15}(\text{J}) \qquad (4.23)$$

Electrons too are particles that can be described as a matter wave. Their motions are described by the Shrödinger equation. According to Fujiwara, we can use the Shrödinger equation for high-energy electrons instead of the Dirac equation following relativistic corrections for electron mass and energy.[19]

Electrons accelerated by $E(\text{keV})$ have momentum p_e, kinetic energy E_e and de Broige wavelength λ_e, with the following relationships:

$$p_e = mv, \; E_e = p_e^2/(2m) \tag{4.24}$$

$$\lambda_e = h \, / \, p_e = h \, / \, \sqrt{2mE_e} = 0.3873/\sqrt{E_e(\text{keV})} \tag{4.25}$$

The dependence of wavelength on energy is different for X-rays *versus* electrons. The electron energy corresponding to a wavelength of $\lambda_e = 1.5 \, \text{Å}$ (X-ray, $\sim$ Cu $K\alpha$) is $E_e = 66.7 \, \text{eV}$, and the wavelength of 200 keV electrons is $\lambda_e = 0.0251 \, \text{Å}$ (after relativistic corrections).

The conditions for constructive interference by a crystal are given using lattice spacing d_{hkl} and scattering angle $2\theta_{hkl}$ by the Bragg equation, $2d_{hkl} \sin \theta_{hkl} = \lambda$. These conditions can be illustrated by another representative Laue condition and Ewald construction as shown in Figure 4.11 for Au thin single-crystal film. The d_{200} of Au is 2 Å, so the reciprocal lattice point corresponding to 200 reflection of Au is $(2\,\text{Å})^{-1} = 0.5 \, \text{Å}^{-1}$. The radius of the Ewald sphere, $1/\lambda_e$, for electrons with energy 200 keV is *ca* $40 \, \text{Å}^{-1}$, and that for X-rays, Cu $K\alpha$, is *ca* $0.66 \, \text{Å}^{-1}$.

The Ewald sphere for the electrons is so close to a plane that many reflections will be excited simultaneously. This difference – in how many reflections are simultaneously excited when electrons are incident along a zone axis through a different radius of Ewald sphere for electrons *versus* X-rays – makes electron scattering more dynamic than X-ray scattering. Furthermore there is a large difference in scattering power. Reciprocal lattice points that are elongated by $2/t$ **perpendicular to the crystal surface** become the reciprocal-lattice rods (rel-rods for short) through the effect of the shape function previously described; intensity distributions are given by $\text{sinc}^2(x)$, as shown in Figure 4.11 (top right).

4.4.2   Scattering Power of an Atom

Electrons are scattered through interaction with the electrostatic potential of the electrons and nuclei that make up the constituent atoms of a solid. Let us start with the interaction of atoms with X-rays and electrons, and compare the two. The strength of the interaction – the scattering power of an atom – is described by the **atomic scattering factor**. An X-ray is an electromagnetic wave (transverse), so we require a polarisation vector. An incident X-ray can be described by:

$$E_i = \varepsilon_0 \exp i[(k_0 \cdot r) - \omega t] \text{ and } \varepsilon_0 \cdot k_0 = 0 \tag{4.26}$$

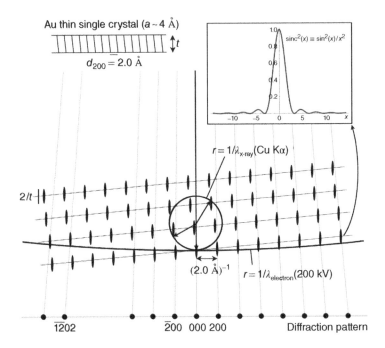

Figure 4.11 Schematic drawing of the Laue condition and Ewald construction for both X-rays (Cu Kα) and electrons (200 kV).

Using Thomson scattering for X-rays in a stationary state:

$$E_f(r) = [\varepsilon_0 \exp i[k_0 \cdot r] + \{-r_e(\varepsilon_0 \cdot \varepsilon_f)f^X(K)\}\varepsilon_f \exp i[|k_f| \cdot r]]/r \quad (4.27)$$

where scattering vector $K \equiv k_f - k_0$, $|K| = 4\pi \sin\theta/\lambda[\text{Å}^{-1}]$, ε_0 and ε_f are unit polarisation vectors of the electric field for the incident and scattered X-rays and the Thomson scattering length or classical electron radius $r_e = e^2/4\pi\varepsilon_0 mc^2 = 2.8179 \times 10^{-5}[\text{Å}]$. The first term represents the incident beam and the second term the scattered beam. Therefore, scattered intensity is given by:

$$I_{scat}^{x-ray}(k_f) = |\{-r_e(\varepsilon_0 \cdot \varepsilon_f)f^X(K)\}\varepsilon_f \exp i[|k_f| \cdot r]/r|^2 = Pr_e^2 f^X(K)^2/r^2 \quad (4.28)$$

where polarisation factor $P = (\varepsilon_0 \cdot \varepsilon_f)^2$.

For X-rays, the scattering factor of an atom, $f^X(K)$, is given in units of Thomson scattering length as:

$$f^X(K) = \int \rho_{\text{atom}}(r)e^{-iKr}d^3r \quad (4.29)$$

where $\rho_{atom}(r)$ is an electron charge distribution of the atom.

For electrons, based on the first Born approximation:

$$\Psi(r) = \exp i[k_0 \cdot r] - f^{el}(K) \exp i[|k_f| \cdot r]/r \tag{4.30}$$

The scattering factor for electrons is given in units of [Å] as:

$$f^{el}(K) = \int V_{atom}(r)e^{-iKr}d^3r = \frac{2\pi me}{h^2}\frac{[Z - f^X(K)]}{K^2} \tag{4.31}$$

where $V_{atom}(r)$ is an electrostatic potential distribution of the atom.

The scattering power of an electron is approximately 10^4 times greater than that of an X-ray (e.g. for Fe, $f^{el}(K = 0) = 7.4 \times 10^{-8}$ cm and $f^X(K = 0) = 7.3 \times 10^{-12}$ cm). This indicates that two aspects of electron scattering will be distinct from those of X-ray scattering: (i) the dynamic scattering effect will be serious and (ii) much smaller scattering objects (ca 10^{-8} times smaller) will have enough scattering intensity to be studied using electrons. It will be shown later that we can obtain single-crystal ED patterns or Fourier diffractograms (FDs) from Fourier transformation of high-resolution transmission electron microscopy (HRTEM) images taken from small and thin crystalline areas. TEM is therefore very powerful for the structural study of crystals smaller than 1 μm in size.

4.4.3   Crystal Structure and Electron Diffraction

A crystal is a 3D periodic array of unit cells, each of which contains a group of atoms or cavities. Therefore, in order to solve a crystal structure, its atomic positions must be determined, i.e. the distribution of scattering density – electron density, $\rho_{cryst}(r)$, for X-rays or electrostatic potential distribution, $V_{cryst}(r)$, for electrons – in a unit cell. $V_{cryst}(r)$ or $\rho_{cryst}(r)$ is obtained through a scattering experiment via the CSF, $F(h)$, for h reflection ($h = hb_1 + kb_2 + lb_3$, where h, k and l are a set of integers called reflection indices). $F(h)$ is the scattering amplitude and is given by the Fourier coefficient of $V_{cryst}(r)$ or $\rho_{cryst}(r)$. Hereafter we focus on electron microscopy (EM), so only $V_{cryst}(r)$ should be used:

$$F(h) = \mathscr{F}\{V_{cryst}(r)\} = 1/\sqrt{(2\pi)^3}\int V_{cryst}(r)e^{-ih\cdot r}d^3r = |F(h)|e^{-i\theta(h)}$$
$$\tag{4.32}$$

where $\theta(h)$ is the phase of CSF for h reflection, $\theta(h)$ is a function of the coordinates of the origin and $F(h)$ is complex in general.

Only absolute values, *i.e.* moduli $|F(h)|$, can be obtained from diffraction intensity, $I(h)$, for reflection h as given:

$$I(h) = \mathscr{F}\{V_{cryst}(r) \otimes V_{cryst}(-r)\} = F(h) \cdot F(h)^* = |F(h)|^2 \qquad (4.33)$$

It is clear from Equation 4.33 that phase information, $\theta(h)$ in Equation 4.32, which is very important for the structure determination, disappears in intensity, $I(h)$. The positions of h at which $I(h)$ is not equal to 0 give information on a crystal system, unit cell parameters and possible SGs. It is to be noted that the SG itself is not a structure solution but gives symmetry relations for an arrangement of the groups of atoms/cavities in the unit cell. Once we have obtained the phase relationships of CSF, $\theta(h)$, the structure $V_{cryst}(r)$ can be determined straightforwardly by inverse Fourier transform as:

$$V_{cryst}(r) = \mathscr{F}^{-1}[F(h)] = 1/\sqrt{(2\pi)^3}\int F(h)e^{ih\cdot r}d^3h \qquad (4.34)$$

or by Fourier sum as:

$$V_{cryst}(r) = V_{cryst}(x,y,z) = 1/\sqrt{(2\pi)^3}\sum_{h,k,l} F(h,k,l)e^{2\pi i(hx+ky+lz)} \qquad (4.35)$$

The simplest assumption by which to describe the exit electron wave field at the exit surface is that a crystal will change only in phase during the electron scattering process; this is called the **phase object approximation (POA)**. The electron wave field at the exit surface of the specimen with thickness t, *i.e.* the exit wave function, is then given by:

$$\phi_t(x,y) = \exp[i\sigma t \cdot V_p(x,y)] \qquad (4.36)$$

If we can take **weak phase object approximation (WPOA)** $(i\sigma t \cdot V_p(x,y) << 1)$ then:

$$\phi_t(x,y) = 1 + i\sigma t \cdot V_p(x,y) \qquad (4.37)$$

where the projected potential for a crystal thickness t is given by:

$$t \cdot V_p(x,y) = \int_0^t V(x,y,z)dz \qquad (4.38)$$

and the interaction parameter $\sigma = \pi/(\lambda E)$ is equal to $0.00729\,[V^{-1}\,nm^{-1}]$ at accelerating voltage $E = 200$ kV.

At the back focal plane of the object lens, we encounter Fourier transform of the exit wave:

$$F_h(u,v) = \mathscr{F}\{\phi_t(x,y)\} = \Delta(u,v) + i\sigma t \cdot \mathscr{F}\{V_p(x,y)\}$$

$$= \Delta(u,v) + i\sigma t \cdot V_h(u,v) \tag{4.39}$$

where u and v are coordinates in reciprocal space; the first term corresponds to the transmitted beam and the second to the scattered beams:

$$F_h(u,v) \equiv F(h) = (2\pi me/h^2)V_h(u,v) \tag{4.40}$$

where h, m and e are Planck's constant, electron mass and electron charge. The ED pattern:

$$I_h(u,v) = F_h(u,v) \cdot F_h(u,v)^* = |F_h(u,v)|^2 \tag{4.41}$$

is observed at the back focal plane.

4.4.4 Relationship between Real and Reciprocal Space

There are different types of ED, as shown in Figure 4.12. A polycrystalline specimen with randomly orientated crystals normally gives rise to ring patterns, called Debye–Scherrer rings, while single crystals show patterns with sharp spots. If a large convergent electron beam is used instead of a parallel incident beam, a CBED pattern can be observed, showing discs containing detailed information about the symmetry relationships, which will be shown later.

As we have discussed, reciprocal space is the Fourier transform of real space, each reflection hkl in reciprocal space representing a set of parallel lattice planes in real space, which are perpendicular to the reciprocal lattice vector, q_{hkl}; the interplanar spacing, d_{hkl}, is inversely proportional to the length of q_{hkl}. Thus, the longer the unit cell parameters a_1, a_2, a_3, the shorter the reciprocal lattice vectors b_1, b_2, b_3. b_1 is perpendicular to the $(a_2 a_3)$ plane, b_2 to the $(a_3 a_1)$ plane and b_3 to the $(a_1 a_2)$ plane. This relationship means that the symmetry in real and reciprocal space is coherent. Figure 4.13 shows the relationship between the crystal structure in real space and the ED in reciprocal space. The basis of the crystal structure, a collection of atoms ($B(r)$ in Equation 4.13), is shown. It is worth noting that a diffraction pattern (kinematical) always shows inversion symmetry with 1 of the 11 possible centrosymmetric

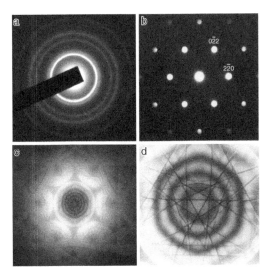

Figure 4.12 Different types of ED pattern. (a) Polycrystalline Au nanoparticles. (b–d) CBED patterns taken along the [111] axis of an Si crystal with different 2α angles: (b) $2\alpha \approx 0$, corresponding to the SAED pattern; (c) the whole CBED pattern, with $2\alpha = 12$ mrad; (d) enlarged central disc of (c), showing contours and dark lines – only threefold symmetry can be observed, rather than the sixfold symmetry in the SAED pattern of (b).

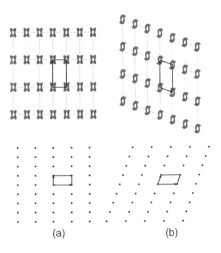

Figure 4.13 Relationship between the crystal structure in real space and the ED in reciprocal space of (a) an orthorhombic unit cell and (b) a monoclinic unit cell.

crystallographic PGs (called Laue classes), which will be discussed in the next section as Friedel's law.

It is worth noting that the position of each diffraction spot is determined by the size and type of the unit cell in real space; in other words, the position of the diffraction spot is exactly the same for different structures with identical unit cell dimensions. However, the atomic position inside the unit cell generates different amplitudes and phases from the diffraction spots in reciprocal space. The symmetry of the crystal makes all symmetry-related reflections have the same amplitude, but the phases are relatively complicated. It is also worth noting that systematic absences (called **systematic extinctions**) in *hkl* reflections arise when translational symmetry elements such as lattice centring or non-translational symmetry elements such as screw axes and glide planes are present. One example is shown in Figure 4.14; systematic extinction can be clearly observed in Figure 4.14b, caused by the existence of the glide planes.

4.4.5 Friedel's Law and Phase Restriction

Friedel's law comes from Equation 4.7 and is very important in kinematical diffraction.

$$F(hkl) = F^*(\overline{hkl}) \tag{4.42}$$

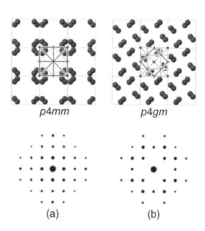

p4mm   p4gm

(a)   (b)

Figure 4.14 Two different crystals with the same unit cell dimensions but different atomic positions. The plane groups are (a) *p4mm* and (b) *p4gm*, where systematic extinctions *h*0: *h* = 2*n* can be seen.

where F^* is the complex conjugate of F. From the definition of CSF $F(hkl)$ for hkl reflection and Equation 4.42, the amplitudes of the two reflections are identical, as:

$$|F(hkl)| = |F(\overline{hkl})| \qquad (4.43)$$

The pair of reflections (hkl) and $(\overline{hkl})$ is called Friedel's pair. Its phases are related, as:

$$\theta(hkl) = -\theta(\overline{hkl}) \qquad (4.44)$$

The diffraction patterns are always centrosymmetric according to Friedel's law. For centrosymmetric structures, the phase shift between (hkl) and $(\overline{hkl})$ is zero:

$$\theta(hkl) = \theta(\overline{hkl}) \qquad (4.45)$$

Therefore, the phase must be both θ *and* $-\theta$, suggesting that the phase of all reflections in centrosymmetric SGs must be only 0 or π, taking the inversion centre as the origin. This phase restriction can greatly simplify the structural solution, as only two possible phase values need to be considered; however, it is still a tremendous amount of work to determine the phase values, as there are 2^n combinations for n reflections. Therefore, various methods, including direct methods, charge flipping, Patterson map *etc.*, must be applied in order to identify the phase relationships. Later we will show the advantages of using HRTEM images, which retain the phase information, allowing it to be extracted directly.

4.4.6 Information on the 0th, 1st and Higher-Order Laue Zone

A zone axis $[uvw]$ is a lattice row parallel to the intersection of two (or more) families of lattices planes, it is the direction in real space from which an ED pattern is taken. This can be written as:

$$[uvw] = ua_1 + va_2 + wa_3 \qquad (4.46)$$

where u, v and w are integers. The zone axis is parallel (or almost parallel) to the incident electron beam and the reciprocal lattice planes observed in the ED patterns can be called zones; these are perpendicular to the zone axis. All reflections on the zones obey the Laue equation:

$$uh + vk + wl = n \qquad (4.47)$$

where n is an integer.

- $n = 0$, **0th-order Laue zone (ZOLZ):** The ZOLZ contains the reciprocal lattice plane, which passes through the reciprocal lattice origin – the 000 reflection. The number of reflections that can be observed in the ZOLZ is dependent upon the radius of the Ewald sphere. The ZOLZ gives projected structural information, which is very important, and normally only the ZOLZ can be observed from an SAED pattern.
- $n \neq 0$, **higher-order Laue zone (HOLZ):** The Ewald sphere may also cut the reciprocal lattice next to the ZOLZ with $n = 1$ or $n = 2$, allowing the information from the 1st- and 2nd-order Laue zone (FOLZ and SOLZ) to be revealed. Reflections of the HOLZ appear outside the ZOLZ, separated by concentric circles. The distance between the HOLZ and the reciprocal lattice origin, 000, can provide information on the lattice planes perpendicular to the zone axis.

A schematic drawing of a crystal taken from zone axis [001] is shown in Figure 4.15. From the geometrical relationship, obviously:

$$d_{hk0} = ML\lambda/D_{hk0} \qquad (4.48)$$

where L is the camera length of the TEM, M is the magnification of the film and D_{hk0} is the distance between reflection 000 and $hk0$ on the film.

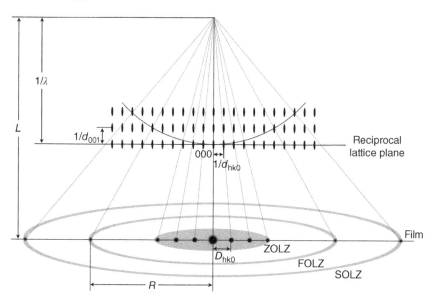

Figure 4.15 Schematic drawing of the geometry of the ED, ZOLZ and HOLZ of a crystal taken from the [001] zone axis.

The reflections in the ZOLZ have the relation:

$$0 \cdot h + 0 \cdot k + 1 \cdot l = 0 \qquad (4.49)$$

which means that all the reflections of the ZOLZ are $hk0$. Obviously, all the reflections in the FOLZ and SOLZ are $hk1$ and $hk2$, respectively.

Note that the radius of the FOLZ, R, represents the information from the lattice spacing perpendicular to the zone axis, from which d_{001} can be calculated as:

$$d_{001} = 2M^2 L^2 \lambda / R^2 \qquad (4.50)$$

4.4.7 Determining Unit Cell Dimensions and Crystal Symmetry

From a set of ED patterns, it is possible to obtain three primitive translation vectors in reciprocal space.

However, it is hard to judge the symmetry of the crystal from an independent ED pattern and it is not so easy to obtain the two main zone axes from one crystal due to the hardware limitation of TEM, in which the tilting angle is relatively small, especially for TEM designed for high resolution. To solve this problem, a series of ED patterns with one common axis and different tilting angles can be collected. As each ED pattern is the cross-section of the corresponding reciprocal lattice planes, the 3D reciprocal lattice can be reconstructed by combining these tilting series and thus the 3D information from the unit cell dimensions – the extinction conditions – can be revealed. Figure 4.16 shows one tilting series from a crystal with orthorhombic structure, from which the unit cell parameter can be calculated. Recently, Akaogi *et al.* discussed a method for determining cell parameters (a, b, c, α, β and γ) from a CBED pattern.[20]

Following determination of the unit cell parameters and indexing of the ED patterns, the systematic extinctions of each ED pattern can be obtained. Then the reflection conditions can be summarised and possible SGs of the crystal can be found by checking International Tables for Crystallography, Volume A: Space-Group Symmetry. Laue class and PG information can be obtained from the intensity distribution of ED patterns and crystal morphology, respectively, which will help in the unique determination of an SG.

However, in order to solve the structure, it is necessary to have information on both amplitudes and phases for each reflection

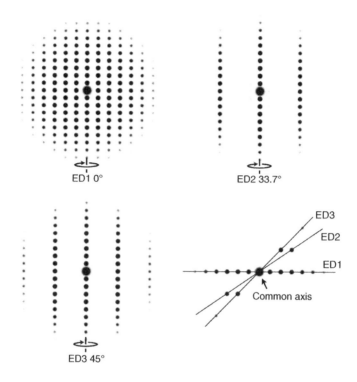

Figure 4.16 Determination of the unit cell from a tilting series of ED patterns.

in order to reconstruct the 3D crystal structure and solve the atomic positions.

4.4.8 Convergent Beam Electron Diffraction

CBED is the oldest ED technique, developed by Kossel and Möllenstedt in 1939, prior to the development of SAED by LePoole in 1947. In CBED, the Ewald sphere shown in Figure 4.17 covers a certain volume with cylindrical symmetry, depending on the convergent angle, α. The diffracted beam forms a disc of CBED pattern and the intensity at each point in the disc is produced from an independent beam of blue or red incidence. In other words, we see diffraction intensity distribution as a function of different conditions of incidence. CBED can determine PG and SG from nanocrystals. Because of its dynamical diffraction features,

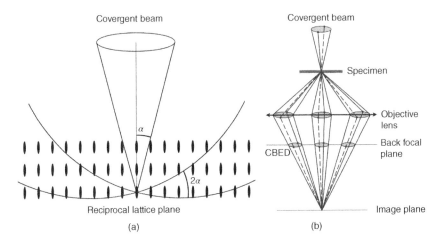

Figure 4.17 Schematic drawings of CBED. (a) Ewald spheres can scan reciprocal space with an effective thickness of twice the convergent angle, α. (b) Ray diagram specifying incidences marked by solid and dashed lines.

it can distinguish all 32 PGs, while kinematical diffraction distinguishes only 11 centrosymmetric PGs (called Laue classes, following Friedel's rule). A recent paper by Tsuda *et al.* on $FeCr_2O_4$ provides a very good example, showing both SG and atomic parameter refinements.[21] Although dynamical diffraction effects are important to CBED, however, we want to stress that a kinematical CBED approach to nanostructure determination is also very useful and powerful, as shown by McKeown and Spence.[22]

Figure 4.18 shows the CBED pattern of GaAs crystal (SG $F\bar{4}3m$), taken from [111]. It apparently shows *6mm* symmetry and sixfold

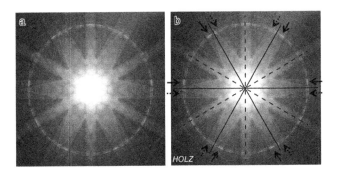

Figure 4.18 CBED pattern of GaAs crystal (SG: $F\bar{4}3m$), taken with [111] incidence. Kindly provided by Dr Tadahiro Yokosawa.

rotation symmetry with two independent mirrors drawn by solid and dashed lines, but if we look carefully at the HOLZ pattern, fine details (marked by dotted and solid arrows) are different. Therefore, the solid lines are not mirror planes but the dashed lines are, and symmetry becomes $3m$.

4.5   IMAGING

4.5.1   Crystal Structure and TEM Images

At the image plane, the wave function $\Phi(x, y)$ is modified by the objective aperture function, $A(u, v)$, the objective lens characteristics envelope function, $Env(u, v)$ and the contrast transfer functions, $CTF(u, v)$, which show the transfer ability of the objective lens in structural details, as:

$$\Phi(x, y) = \mathcal{F}\{F_h(u, v) \cdot A(u, v) \cdot Env(u, v) \cdot CTF(u, v)\} \qquad (4.51)$$

$$CTF(u, v) = \sin \chi(u, v) \qquad (4.52)$$

$$\chi(u, v) = \pi \left\{ C_s \lambda^3 \left(u^2 + v^2\right)^2 / 4 - \Delta f \lambda \left(u^2 + v^2\right) / 2 \right\} \qquad (4.53)$$

where C_s and Δf are a spherical aberration coefficient of an objective lens and the defocus value, respectively. $CTF = -1$ for a wide range of (u, v) would be ideal, but it is a complex function of C_s, Δf, u and v.

The observed image, $I(x, y)$, is given as:

$$I(x, y) = \Phi(x, y) \cdot \Phi(x, y)^* = 1 - 2\sigma t \cdot V_p(-x, -y) * \mathcal{F}\{\sin \chi(u, v)\} \qquad (4.54)$$

where $*$ is the convolution operation.

The FD obtained from the HRTEM image, $I_{image}(h)$ (by choosing a thin region), is:

$$I_{image}(h) = \mathcal{F}^{-1}\{I(x, y)\} = 2\sigma t \cdot \{F(h) / (2\pi me / h^2)\} \sin \chi(u, v) \qquad (4.55)$$

Therefore, $I_{image}(h)$ is proportional to the CSF $F(h)$ and thickness t multiplied by the CTF. So the CSFs can be obtained through Fourier transformation of the HRTEM image after recovery from the CTF effect, which is a function of C_s and Δf (Equations 4.52 and 4.53).

HRTEM images should be taken from thin areas at good resolution and with good signal-to-noise ratio (SNR) in order to (i) fulfil the condition of WPOA and (ii) obtain the genuine extinction rule for the space group determination through the FD of the HRTEM image.

4.5.2   Image Resolution

The theoretical resolution of an optical microscope with perfect lens R_{th} is given as:

$$R_{th} = 1.22\lambda/\alpha \qquad (4.56)$$

where α and λ are the angular aperture size and the wavelength of light, respectively. Of course, the resolution required strongly depends on what kind of structural information we are interested in. The resolution of an optical microscope is far below that required to resolve a structure at the atomic scale, as (i) for the visible light range the wavelength is 3000–$8000\,\text{Å}$ and of the order of 1 (most textbooks use semi-angle $\alpha/2$) and (ii) for the X-ray range we cannot create a reasonable lens, although it is in the range of atomic resolution.

On the other hand, since an electron has charge we can create an electromagnetic lens through Lorentz force, and hence an electron microscope. The wavelength of electrons at $200\,\text{kV}$ is $0.0251\,\text{Å}$, which is small enough to resolve structure at the atomic scale if we can keep α to the order of 1. In the case of an electromagnetic lens, the lens is not perfect and has various kinds of aberration. A point source will be imaged not as a point but as a disc by a lens with spherical aberration. The disc diameter of confusion, R_{sph}, is given by $0.5C_s(\alpha/2)^3$, where C_s is a spherical aberration constant of the lens. Taking the effect of the aberration into account, the resolution can be expressed as:

$$R = \sqrt{R_{th}^2 + R_{sph}^2} \qquad (4.57)$$

since R_{th} and R_{sph} are independent of each other. R takes a minimum value of:

$$R_{min} = 0.9\lambda^{1/4}C_s^{1/4} \qquad (4.58)$$

at optimum angular aperture size:

$$\alpha_{optimum} = 1.54(\lambda/C_s)^{1/4} \qquad (4.59)$$

In other words, an effective value of $\alpha_{optimum}$ in Equation 4.59 for the electromagnetic lens is mainly given by spherical aberration as $1.54(\lambda/C_s)^{1/4}$, and is of the order of 10^{-3}. The attainable resolution, R_{min}, is in the range of 1.8–$2.5\,\text{Å}$ for most EMs with accelerating voltage of 200 and $300\,\text{kV}$. The recent development of EM with C_s collector(s) is of enormous value in enhancing resolution, allowing the electron beam diameter on a specimen to be smaller than $1\,\text{Å}$ with a C_s

(beam) corrector STEM. However, some crystals, such as zeolites and metal–organic frameworks (MOFs), are so electron beam-sensitive that electron beam damage is a limiting factor for better resolution.

4.5.3 Limitation of Structural Resolution

The structural resolution obtained through HRTEM is given by $[|b|_{max}]^{-1}$, where $|b|_{max}$ is the magnitude of maximum scattering vectors that can effectively contribute to image formation (the Fourier transform). For a centrosymmetric crystal, we can make $F(b)$ real by taking an origin at an inversion centre; $\theta(b)$ will be either $0(+)$ or $\pi(-)$.

Figure 4.19 explains the above situation. A plane wave incident on a thin single crystal at the object plane with crystal thickness t and (hkl) lattice plane is not parallel to the incident electron beam but inclined. All points on the exit surface become sources of spherical waves, and interference of the waves forms b diffraction spots with intensities, $I(b)$,

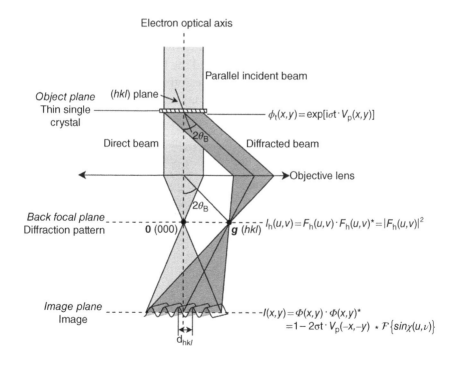

Figure 4.19   Ray diagram of TEM.

at infinite distance from the crystal (Fraunhofer diffraction). An objective lens bring h diffraction spots to the back focal plane. We can collect $I(h)$ *via* a diffraction experiment and thereby a data set of $|F(h)|^2$. By assuming $\theta(h)$, we obtain a structural solution through an inverse Fourier transform. Using TEM, we can observe both an ED pattern and an image from the same area merely by changing the currents of an intermediate lens in order to transfer information from the back focal plane or image plane to a detector, as shown in Figure 4.10. The advantages and drawbacks of observing HRTEM images are discussed in the next section.

4.5.4 Electrostatic Potential and Structure Factors

As we have discussed, FD obtained from the HRTEM image, $I_{\text{image}}(h)$, is proportional to the CSF $F(h)$ and thickness t multiplied by the CTF, and the CSFs can be obtained through Fourier transformation of the HRTEM image following CTF correction. The projected potential corresponds to the coherence of the diffracted beams, each of which is a set of Bragg planes with different amplitudes and phases. The Fourier synthesis of electrostatic potential, $V_{\text{cryst}}(r)$, can be understood through the summation of these Bragg planes. The HRTEM image of the cubic mesoporous crystal SBA-6 (SG: $Pm\overline{3}n$) and the FDs taken from the [001] axis are shown in Figure 4.20,[23] while the amplitudes and phases of $hk0$ reflections extracted from the HRTEM image are listed in Table 4.1.

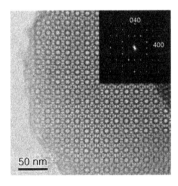

Figure 4.20   HRTEM image of SBA-6 (SG $Pm\overline{3}n$) and the FD taken along the [001] axis. Data taken from [23].

Table 4.1 Amplitudes and phases of $hk0$
reflections extracted from the HRTEM
image.

h	k	l	Amplitude	Phase
2	1	0	100.0	0
2	0	0	81.4	0
4	0	0	68.4	π
3	2	0	35.2	π
4	2	0	25.3	π
4	1	0	21.7	0
3	3	0	19.2	0
4	4	0	15.1	0
2	2	0	11.4	π
1	1	0	11.4	0
4	3	0	5.6	π
5	2	0	5.2	0
3	1	0	5.0	π
5	3	0	3.8	π
5	4	0	2.6	π
6	1	0	2.2	π
6	2	0	1.7	π
6	3	0	1.6	0
6	0	0	1.6	π
0	6	0	1.6	π
5	5	0	1.2	0
4	6	0	1.1	0
5	1	0	0.8	π
7	1	0	0.8	0

The Fourier synthesis of the projected potential map is shown in Figure 4.21. The d -spacing of each Bragg plane is equal to the periodicity of the corresponding diffracted beam. The reflection (200) is parallel to the [010] axis and repeats twice along the [100] axis, while (020) is parallel to the [100] axis. The coherence of (200) and (020) shows the 2D-square contrast with maximum value at their intersection points. (400) is half the wavelength of (200) and repeats four times along the [100] axis. However, the phase of (400) is π and we can see clearly from Figure 4.21 that the origin is shifted from white to black (anti-phase). The reflections (210) and (2$\bar{1}$0) show evidence of a rhombic lattice. When we take the symmetry into account, the sum of all {210} reflections shows a square lattice with $p4mm$ symmetry. The sum of {210} and {200} already shows similar contrast to the HRTEM image, but without the details. By totalling the five strongest reflections, the structure can be made easily

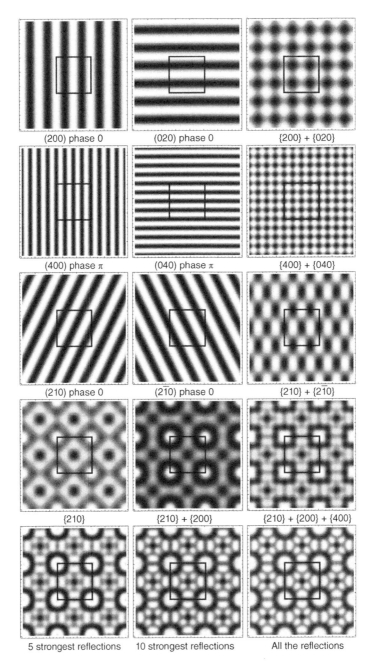

Figure 4.21 Fourier synthesis of the projected potential map of the mesoporous crystal SBA-6 (SG: $Pm\bar{3}n$) along the [001] axis.

recognisable, looking almost the same as the original HRTEM image. It is worth noting that strong reflections with large amplitudes contribute more to the potential, while the high-order reflections contribute more to the resolution.

4.5.5   Image Simulation

The idea of HRTEM image simulation comes from the fact that sometimes it is not possible to go back from the experimental HRTEM image to the original structure. Thus one assumed structure can be made to simulate the image under various conditions and compared with the experimental one, then modified and compared again *etc*. This technique is especially useful for the interpretation of defect structures as the EC method is invalid for the reconstruction of local defects and intergrowth.

The HRTEM image is sensitive to many parameters, including the crystal structure, unit cell dimensions, SG, atomic positions, alignment of the optical axis and beam divergence, alignment of the crystal zone axis, defocus value of the objective lens, thickness of the crystal *etc*.

The multislice method for modelling the contrast of images under different conditions was developed by Cowley and Moodie and is widely used.[24] A crystal of thickness t is split into many thin slices with thickness Δt normal to the incident electron beam. Δt can be so thin that the POA is valid for each slice. The projected potential of the first slice is calculated and then transferred to the second slice and so on. Then the final exit wave function is obtained by calculating the electron propagation of all slices.

Today, HRTEM image simulation of crystals is widely used for structural analysis, with several commercial software packages available.[25]

4.6   THE EC METHOD OF SOLVING CRYSTAL STRUCTURES

We have shown that EC has the advantages of allowing direct visualisation of the real space of structures and of retaining the crystal structure phase information. Recently, more and more nanomaterials have been synthesised; these can be formed by the self-assembly of surfactant micelles or polymers as templates or other building blocks such as biomolecules. However, such materials are normally not highly crystalline and some are amorphous on the atomistic scale, though ordered on the nanometre scale. The scattering in these materials arises from regular arrays of pores or template molecules. Structure determination by the XRD powder profile alone is very hard as the mesoscale ordering is very sensitive to both synthesis conditions and synthesis time, and the crystals contain local structural variations that give a few broad peaks and large damping in scattering intensities.

In this section, various materials solved by EC methods will be described in detail: (i) 1D atomistic crystal – a layer of MFI (zeolite framework code given by the International Zeolite Association) zeolite nanosheets; (ii) 2D mesoscale crystal – an MCM-41 (structure code given by Mobil) (plane group $p6mm$) and DNA self-assembly structure (plane group $p4mm$); and (iii) 3D mesoscale cystal – face centred-cubic (SG $Fm\bar{3}m$) and minimal surfaces (SG $Ia\bar{3}d$ and $Pn\bar{3}m$).

4.6.1   1D Structures

A good example of a 1D structure is the MFI zeolite nanoshteets. Ryoo *et al.* succeeded in synthesising hydrothermally ultrathin MFI nanosheets, grown in the (ac) plane with just a single unit cell's thickness (2.0 nm) along the b-axis, by using a di-quaternary ammonium-type surfactant, as shown in Figure 4.22a.[26] The layered lamellar structure was very uniform and the interlayer distance could be precisely controlled according to the hydrophobic chain length of the surfactant. Pillaring was carried out with tetraethoxysilane (TEOS) prior to the removal of the template in order to make mesopores within the interlamellar spaces. An HRTEM image of the layered MFI sheets synthesised by use of $C_nH_{2n+1}-N^+(CH_3)_2-C_6H_{12}-N^+(CH_3)_2-C_6H_{13}(n = 22)$ with pillars is shown in Figure 4.22b; the thickness of both the MFI layer and the mesopore layer is ~3 nm.[27] Up to fourth-order reflections corresponding to the interlayer distance were observed in the powder XRD pattern at small scattering-angle range due to the interlamellar structural coherence; interestingly, the accidental extinction of the second-order reflections can be clearly observed in the powder XRD patterns from the sheets synthesised with $n = 22$ but not with $n = 18$ (Figure 4.22c).

As we have seen with Equations 4.13 and 4.14, the accidental extinction comes not from the symmetry of the lattice through $\mathscr{F}[L(r)]$ but from the zero-cross of $\mathscr{F}[B(r)]$ in the 1D crystal. $B(r)$ does not bring

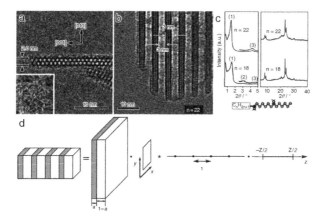

Figure 4.22   MFI zeolite nanosheets synthesised using $C_nH_{2n+1}-N^+(CH_3)_2-C_6H_{12}$ $-N^+(CH_3)_2-C_6H_{13}(n = 22)$. (a, b) HRTEM image. Reprinted with permission from [26] Copyright (2009) Nature Publishing. (c) Powder XRD patterns showing that when $n = 22$, the second-order reflection is missing due to accidental extinction. (d) Graphical representation of the crystal and symbolic expression of the CSF for the 1D system [26, 27]. Reprinted with permission from [27] Copyright (2010) Elsevier Ltd.

extra symmetry to the lattices. A graphical representation is given in Figure 4.22d.

For 1D crystals, $C(r)$ is reduced to:

$$C(r) = \{B(z) \cdot H(x,y)\} * \{L(z) \cdot S(z)\} \qquad (4.60)$$

where $B(z)$ is the basis (slab) with unit thickness corresponding to the lattice periodicity along z, $L(z)$ is the infinite 1D lattice function along z and $H(x,y)$ and $S(z)$ are the crystal size functions along the (xy) plane (2D) and z-axis (1D), respectively.

If the size of the crystal slab in the x- and y-directions and the number, n, of nanosheets along z are large enough then $F(k)$ will have a non-zero value only when the wave vector, k, is equal to $lc^*(l = \text{integer})$. Then:

$$\mathscr{F}[B(z)] = (A - B) \int_0^a e^{ikz} dz + B \int_0^1 e^{ikz} dz \qquad (4.61)$$

where A and B are scattering factors for slabs A and B, respectively.

Figure 4.23 shows the intensities of reflections for $k = lc^*$, obtained from the equations above, with $l = 2, 3, 4$ and 5 reflections normalised to the intensity of $l = 1$ for both the $B/A = 0$ and the $B/A = 0.5$ cases, since $F(k = c^*) \neq 0$. It is clear from the figure that the intensity of

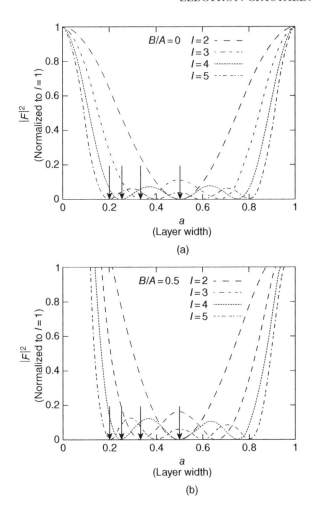

Figure 4.23 Relative XRD intensities $|F(k)|^2$ of lth-order reflections relative to the 1st-order reflection for (a) $B/A = 0.0$ and (b) $B/A = 0.5$. Reprinted with permission from [27] Copyright (2010) Elsevier Ltd.

nth-order reflection ($l = n$) extinguishes when the rational layer thickness reaches $1/n$. The intensity of the 2nd-order reflection of MFI nanosheets will disappear at $B/A = 0.5$.

Recently, Tsapatsis *et al.* published an article on the 'Direct Synthesis of Self-Pillared Zeolite Nanosheets' in *Science*.[28] We will leave this for a later article as we need to describe some structural details of intergrowth between two zeolite structures, MFI and MEL (zeolite framework code given by the International Zeolite Association) zeolite structures.

4.6.2   2D Structures

It is difficult to determine crystal structures solely from powder XRD data, as mentioned above, even if a crystal has a 2D structure such as *p6mm*.[29] A set of two TEM images with incidences parallel and perpendicular to the channel gives conclusive evidence of *p6mm* symmetry with a 1D-channel system. The situation is shown schematically in Figure 4.24. TEM images, in conjunction with simulation, enable the observation and discussion of (i) the 1D nature of the channels (Figure 4.24b,c) and (ii) their 2D hexagonal arrangement (*p6mm*), together with channel shape and wall thickness (Figure 4.24d). SBA-15 has *p6mm* symmetry, like MCM-41, although it also has channel connectivities through randomly arranged complementary pores inside the silica wall, unlike MCM-41. The existence of randomly arranged complementary pores can only be clarified at present by TEM, through observation of replicas such as a Pt nanonetwork structure;[30] otherwise, low-voltage HRSEM images must be used.

Kresge *et al.* carefully combined EM observations with powder XRD experiments in order to solve for the first time the structure

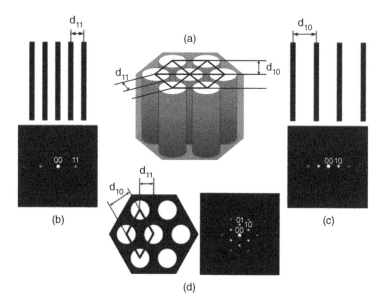

Figure 4.24   Schematic diagram of two HRTEM imaging modes from two principal directions for a 2D hexagonal *p6mm* structure and the corresponding HRTEM images and ED patterns.

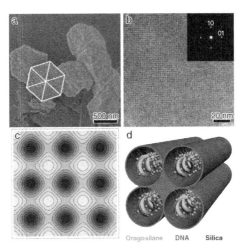

Figure 4.25 Morphology and structure of a DNA–silica complex with a 2D square *p4mm* symmetry. Reprinted with permission from [31] Copyright (2009) Wiley-VCH.

of MCM-41.[29] This clearly indicates the importance of TEM in the structural analysis of mesoporous crystals, although many papers since published comprise an unfortunate mix of possible (speculative) structure and structural solution.

Another example is shown in Figure 4.25. Crystals of a DNA–silica complex were synthesised by the self-assembly of DNA as a template. Interestingly, a rare 2D square structure possessed an inconsistent hexagonal morphology and appeared as the dominant structure. Using TEM, it was found that the hexagonal morphology of the crystal platelets with 2D square symmetry was composed of domain structures arranged at 60° to one another and that the 2D square structure was transformed from a 2D hexagonal structure. It is worth noting that the *p4mm* structure with small interaxial separation was formed at higher concentrations by the formation of an electrostatic 'zipper', with interactions between the negatively charged strands and positively charged grooves of opposing molecules.[31]

4.6.3 3D Structures

TEM observation is powerful, but a TEM image is essentially projected structural information about a specimen along the direction of incident electrons. Therefore, in order to build a 3D structure it is necessary to combine images observed from a number of different incidences.

A general approach for this is **tomography**, which will be discussed in Section 4.7.2. If the material is a crystalline, *i.e.* a periodic system, we can apply crystallography instead of tomography. Using crystallography, we can (i) dramatically reduce the number of images required for the reconstruction to just a few, depending on the crystal symmetry (the higher the crystal symmetry, the fewer images are required) and (ii) dramatically enhance the SNR, because structural information concentrates only on reciprocal lattice points.

There are two main approaches to obtaining a structural solution: one based on a 3D ED data set and is the other on HRTEM images (Figure 4.26). The biggest difference comes from the 'phase problem': the former approach assumes and finds the best possible solution, while the latter determines phases and obtains a unique solution (although resolution is limited).

Cowley[32] and Kuwabara[33] measured ED patterns in order to solve structures. Dorset and Hauptman[34] were the first to apply direct phase determination using 'triplet' and 'quartet' structure invariant for electron diffraction intensity data obtained from organic macromolecules. Dorset showed the power of EC for various crystals. Carlsson *et al.* [35] obtained a framework structure and localised the Fe atomic position of Fe-oxide incorporated in a cavity of Na−Y (FAU)

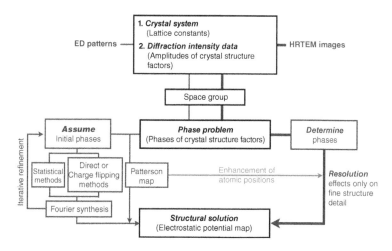

Figure 4.26 Illustration of the structural solution obtained through the use of ED patterns and HRTEM images.

zeolite. Dorset and Gilmore[36] further refined the crystal structure of Na–Y from 87 unique reflections observed by Carlsson *et al.*, based on a maximum-entropy and -likelihood approach. Wagner *et al.*[37] solved the unknown structure of large-pore, high-silica zeolite, SSZ-48, from ED data (326 unique reflections) *via* direct methods. Recent progress in collecting ED data sets has been made by precession and beam tilting, with many structural reports coming from ED pattern-based analysis, as will be discussed in Section 4.7.3. It is worth mentioning that Ohsuna *et al.*[38] cleverly introduced the Patterson map to enhance the positions of heavier atom (Si) pairs in HRTEM images without losing its 'unique solution' characteristic.

Now let us discuss 3D structure, taking MCM-48 with SG symmetry $Ia\bar{3}d$ (PG symmetry $m\bar{3}m$) as an example. HRTEM images of JEM-3010 at 300 kV along [100], [110] and [111] are shown in Figure 4.27, with corresponding FDs inserted at the top right of each. The following extinction conditions for $hkl : h + k + l = 2n$, $0kl : k, l = 2n$, $hhl : 2h + l = 4n$ and $h00 : h = 4n$ lead to $Ia\bar{3}d$ uniquely. This is a lucky case; quite often extinction conditions will lead to a few possible SGs. Additional information obtained from projected symmetry (*i.e.* plane groups from

Figure 4.27 HRTEM images and corresponding FDs of MCM-48, taken along the (a) [100], (b) [110] and (c) [111] axes.

HRTEM images) and/or PG symmetry from crystal morphology from SEM images will give a unique SG in most cases. For example, the plane groups for the images of MCM-48 taken along [100], [110] and [111] are *p4mm*, *c2mm* and *p6mm*, respectively, and point symmetry is $m\bar{3}m$.

Ohsuna has developed software, MesoPoreImage,[39] that simulates HRTEM images and discusses observed HRTEM images of MCM-48, AMS-10, SBA-16 and SBA-6.

Figure 4.28a shows simulated images of MCM-48 for the [111] incidence as a function of defocus value. The image changes from honeycomb-type to an equilateral triangular network at −3000 nm defocus, which can be explained by an important reflection changing its sign through *CTF* (Figure 4.28b). Generally speaking, A HRTEM image of mesoporous crystals is stable in contrast for a wide range of defocus value, but it should be borne in mind that the situation shown

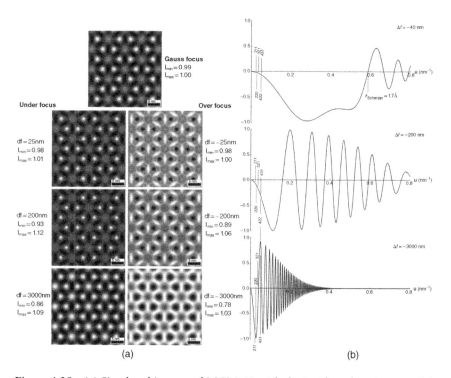

Figure 4.28 (a) Simulated images of MCM-48 with the incident direction parallel to the [111] direction. Defocus values and minimum and maximum intensities are indicated beside each image. Reprinted with permission from [39] Copyright (2011) Elsevier. (b) $\sin\chi(u)$ *versus u* curves modified by the damping envelope function for various defocus values with the positions of the reflections of MCM-48 for [111] incidence.

Figure 4.29 HRTEM images of mesoporous silica crystal with (a) $Ia\bar{3}d$ and (b) its carbon replica, taken along [111]. Corresponding FDs are inserted. Reprinted with permission from [17] Copyright (2004) Elsevier Ltd.

in the figure may happen. However, for atomistic crystals the *CTF* is so sensitive that it changes rapidly and interpretation of the image becomes very difficult.

In order to show the importance of phase information on CSFs in obtaining a structural solution, two HRTEM images are shown in Figure 4.29a,b: silica MCM-48 in (a) and a carbon network synthesised from the pores of MCM-48 (carbon replica) in (b). The two images give opposite contrasts but exactly the same diffraction patterns (see insets). This is because only the phase relationships among the *h* reflections are different between the two, as discussed above with reference to Babinet's theorem.

Recently, we have synthesised a crystal with face-centred cubic (FCC) $Fm\bar{3}m$ symmetry, showing both a well-resolved powder XRD profile and HRTEM images. Figure 4.30 shows SEM images of the morphologies of the FCC mesoporous crystals, which can be categorised into two types: plate and polyhedron. The plate type has a truncated triangular shape, while the polyhedron typically has an icosahedral shape, although deca-hedral, Wulff and D-Wulff polyhedrons can also be observed. However, its icosahedron or truncated triangle plate morphology is not consistent with the corresponding PG symmetry, $m\bar{3}m$. This phenomenon has been explained in terms of the peculiar occurrence of multiple twinning (as for Au nanoparticles) during the formation of the mesostructure.[40−42]

HRTEM images of these FCC mesoporous crystals show that a twin-ning boundary can occasionally be observed along the [110] axis, as shown with the corresponding ED pattern in Figure 4.31a. Figure 4.31b shows an HRTEM image and corresponding ED pattern taken from the vertex of a decahedron particle composed of five tetrahedrons, suggesting the formation of five twinning boundaries surrounding a fivefold centre. The electron crystallographic 3D reconstruction of the unit cell shows typical cage-type mesostructure, which highly spherical cavities arranged in an FCC manner. The FCC structure, which is also known as a cubic

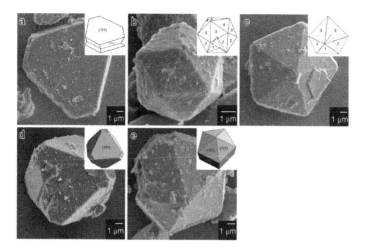

Figure 4.30 Typical SEM images and corresponding schematic drawings of calcined mesoporous crystals coated by gold. (a) Particle of plate type. Its vertices are often truncated. (b) Icosahedral shape. (c) Decahedral shape. (d) Particle showing the Wulff polyhedron. (e) Particle showing the D-Wulff polyhedron. Reprinted with permission from [40] Copyright (2006) Wiley-VCH.

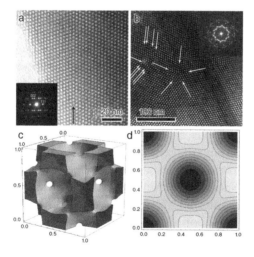

Figure 4.31 (a) HRTEM image taken along [110], with corresponding ED pattern (inset). A single twin boundary is occasionally observed, as indicated by the arrow. (b) HRTEM image taken from the vertex of a decahedron particle composed of five tetrahedrons. (c) Electron crystallographic 3D reconstruction of the calcined mesoporous crystal. (d) Contour plot sliced perpendicular to [001] at $z = 0$(a, c, d). Reprinted with permission from [40] Copyright (2006) Wiley-VCH; (b) Reprinted with permission from [41] Copyright (2007) Royal Society of Chemistry.

close-packed (CCP) structure, can be described as ABC stacking of hard spheres along the $<111>_{cubic}$ direction and has the highest packing density of perfect spheres, at ~0.74. The structural analogue, the hexagonal close-packed (HCP) structure with SG $P6_3/mmc$, can be described as AB stacking of layers along the $[001]_{hex}$ direction; it has the same packing density and can easily intergrow with the FCC structure. However, the sample shows overwhelming FCC over HCP stacking, and HCP stacking was observed only at the twin boundaries.

The exceptional morphologies can be explained by the formation of multiply-twinned particles (MTPs). The MTP model was first proposed by Ino in the 1960s for abnormal particles observed in thin gold films formed by evaporation on sodium chloride and potassium chloride in ultra-high vacuum (UHV).[43] For an FCC crystal, the surface energies of the low-index crystallographic facets follow an order of $\gamma\{111\} < \gamma\{100\} < \gamma\{110\}$, suggesting that an octahedral or tetrahedral shape should be formed by the FCC single crystal in order to maximise the exposure of {111} facets. A truncated octahedron (Wulff polyhedron) can be expected, with a nearly spherical shape with {111} and {100} facets, which make the smallest surface area, minimising the total interfacial free energy. On the other hand, MTPs with fivefold twinning feature can be also formed, including icosahedra and decahedra if the lowest total free energy is achieved by maximising the surface coverage with {111} facets. In these MTPs, tetrahedra surrounded by four {111} faces are accumulated one after another on a growth nucleus with {111}. It is worth noting that normally the MTPs are not capable of growth to a large size because of the internal elastic strain arising from angular misfits of tetrahedral single crystals ($70.5° × 5 = 352.5°$). However, the MTPs of the FCC mesoporous crystal show a remarkably large number of constituents (10^9) compared with those of FCC metals ($< 10^5$), suggesting that the surfactant micelles are more flexible before the completion of silica polymerisation than they would be in metal atoms, which can compensate for the internal strain energy during the formation of MTPs.

Very recently, mesoporous crystal hollow spheres have been synthesised with amino acid-derived anionic amphiphilic molecules in the presence of non-ionic surfactant.[44] The crystals show spherical morphologies; however, after cross-section polishing, which uses an accelerated argon beam to cut the material, unusual polyhedral hollows were revealed. The icosahedral hollow, in which each vertex is shared by five faces and has fivefold symmetry, was frequently observed. Other types of polyhedron can be also categorised, in which each vertex is shared by three or four faces (Figure 4.32).

Figure 4.32 Morphology of a hollow mesoporous crystal. (a) SEM image of the calcined sample, showing a spherical shape. (b–d) Cross-section of the hollow mesoporous crystal, showing various inner morphologies. Reprinted with permission from [44] Copyright (2011) American Chemical Society.

The HRTEM images show that the hollow mesoporous crystal is also formed by a similar multiply-twinned feature, and the twinning structures are observed at the boundary of the adjoined facets. Figure 4.33 show TEM images with different tilting angles of a mesoporous crystal with an icosahedral hollow. As shown in Figure 4.33b, the twinning boundary is observed along the [110] axis *via* the common (111) surface. By tilting the crystal, $\bar{3}$ symmetry is observed from the centre area of an inner facet, showing the typical <111> contrast. These results show that the inner facets of the hollow are {111}. A decahedral hollow with a fivefold centre, Wulff polyhedron single crystal and irregular polyhedron – formed by the combination of the single-crystal and multiply-twinned parts – have been also observed, corresponding to the different inner morphologies observed by SEM. Electron crystallographic study (Figure 4.33e,f) shows that the mesostructure is composed of two disconnected but interwoven, tetrahedral-connected networks divided by a silica wall grown along a typical diamond minimal surface (D-surface) with a $Pn\bar{3}m$ SG.

These results show that the formation of the hollow mesoporous crystal can be explained by the formation of a 'reverse multiply-twinning'

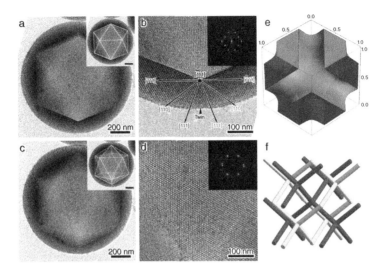

Figure 4.33 HRTEM images of a hollow mesoporous crystal with inner icosahedral shape. (a) TEM image taken from the intersection edge of two triangles with structural model (inset). (b) Magnified TEM image of the twin plane. (c) TEM image taken from the centre of a triangle with structural model (inset). (d) Magnified TEM image of centre. (e) 3D reconstruction of one unit cell along the [111] direction. (f) Schematic representation of the double-diamond bicontinuous structure. Reprinted with permission from [44] Copyright (2011) American Chemical Society.

structure, which is not the icosahedron or decahedron obtained by the packing of tetrahedra sharing {111} facets but rather polyhedral hollows with inner {111} facets formed by the MTP shell. It is worth noting that the MTPs and pentagonal structures are also found in germanium, which has a diamond cubic symmetry, $Fd\bar{3}m$ (which is a subgroup of $Pn\bar{3}m$),[45] the valence bonds of which are flexible enough to overcome the angular misfit. Thus, the reverse MTPs formed by double-diamond structures follow a similar rule to the germanium structure, although the twinning is not formed by the stacking of spherical micelles (like in the FCC crystal) but by the continuous minimal surfaces. Time-course experiments show that the hollow mesoporous crystals were formed by a lamellar-to-cubic shell recrystallisation process. Lamellar-structured vesicles were first formed and then followed a structural transformation into a double-diamond mesostructure. Stacked hexagonal arrays of fusion channels within the surfactant bilayers emerged and then rearranged to form the double-diamond structure with [111]$_\text{cubic}$ perpendicular to the initial bilayer.[46] It is worth noting that the exterior surface was kept as spherical due to the minimum surface free energy effect.

4.7   OTHER TEM TECHNIQUES

4.7.1   STEM and HAADF

A **scanning transmission electron microscope (STEM)** is a type of TEM that can be equipped with additional scanning coils and corresponding detectors. It is different from CTEM in that the electron beam is focused onto a point on a thin specimen and scanned over the sample in a raster. Resolution is basically determined by the size of electron beam on the sample.

As the electron beam scans sequentially point-by-point across the sample, STEM is suited to acquiring information from each point, as in mapping by EDS or EELS. By using STEM and a high-angle annular detector, **annular dark-field imaging (ADF)** and **high-angle annular dark-field imaging (HAADF)** can be observed. As the image contrast of HAADF through phonon excitation (thermal diffuse scattering) is proportional to atomic number Z^n(Z – contrast image, $n \sim 2$) and the image is not affected by the *CTF*, it is possible to form atomic-resolution images. This is in contrast to conventional HRTEM images, in which phase contrast is used and *CTF* plays a great effect. A schematic drawing is shown in Figure 4.34.

Recently important papers on structural study of zeolites using STEM have been published from two independent groups. One is from Spain[47] using mostly STEM-HAADF and another is from Japan[48] using not only HAADF but also aberration-corrected HRTEM.

4.7.2   Electron Tomography

TEM images consist of 2D projected information on 3D structures. Sometimes misleading effects may be caused by artefacts, *e.g.* when locating whether metal particles are dispersed on the surface of the catalytic support or are embedded. In this case, **electron tomography (ET)** can be employed to obtain 3D information by reconstructing the acquired 2D images. This is widely used in medical imaging and will be very useful for non-periodic objects.[49−51] TEM images are collected at incremental degrees of rotation around the centre of the sample and used to reconstruct the 3D image of the target. In principle, the resolution of a reconstructed tomogram is limited by the number of images in and the range of the tile series. In practice, in order to decrease the beam

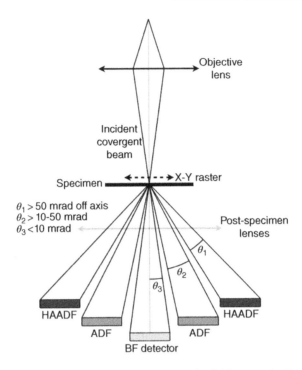

Figure 4.34 Schematic drawing of STEM, bright-field (BF), ADF and HAADF detectors, along with their electron scattering angles.

damage, images are recorded every 1–2° over an angular range ±70° and then back-projected to form a 3D reconstruction. However, a tile may be unreachable due to the thickness of the sample or shadowing from the grid when using CTEM specimen holders. This can lead to artefacts, caused by the missing information (missing wedge). Thus, an alternative approach is to record a second tile series, a 'dual-axis' series, which reduces the missing wedge.[52] In practice, it is usual to rotate the sample by 90° and then record the second tile series.

Bright-field ET is prevalently used for biological samples.[53] It was first applied in materials science in the 1980s in a study of a porous polymer material and later in a study of a porous zeolite.[54,55] However, phase contrast is highly dependent on the *CTF*: when a large defocus is used, the *CTF* changes rapidly and interpretation of the image becomes impossible. This situation become even worse with ET, as large height differences exist within a specimen, sometimes revealing opposite contrast or no contrast at all. STEM HAADF tomography techniques suffer none of these drawbacks: imaging is performed in focus and is not affected by *CTF*, and interpretation of images acquired in this way is considerably more straightforward.

The first example of STEM HAADF tomography was in a study of metallic nanoparticles as heterogeneous catalysts distributed within porous siliceous and carbonaceous support structures by Midgley *et al.*[56] The same authors also showed beautiful images locating randomly arranged metal particles within mesopores.[57] The technique has been used to study the 3D structures of the mesoporous silica MCM-48[58] (Figure 4.35) and platinum supercrystals.[59] The incoherent nature of the dark-field signal leads to directly interpretable images and ET reconstructions, through which the projection of any orientations and slices not available *via* direct imaging techniques can be shown. Very recently, Miao *et al.* observed atomic steps at 3D twin boundaries and the 3D core structure of edge and screw dislocations at atomic resolution in a multiply-twinned platinum nanoparticle, which are not visible in conventional 2D projections, by applying 3D Fourier filtering and ET.[60]

ET coupled with different imaging modes in EM is a powerful tool for the 3D characterisation of a variety of materials at nanometre resolution. However, there is still great scope to develop 3D imaging at the atomic level and to improve time resolution for some beam-sensitive

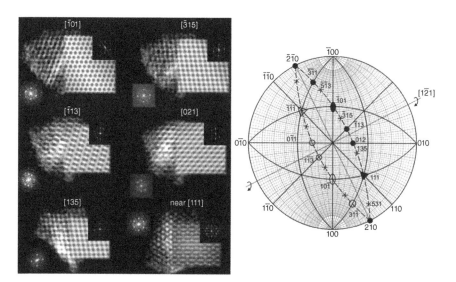

Figure 4.35 Montage of tomographic voxel projections of MCM-48, shown at successive major zone axes as the 3D tomographic reconstruction is rotated about a <112> zone axis, and the stereographic representation of the image series. Reprinted with permission from [58] Copyright (2006) Elsevier Ltd.

samples, such as zeolites and MOFs. Moreover, automation of the technique, including recording of the image and its reconstruction, will be challenging work. Although the need to record numerous images limits the application of ET at the moment, it is clear it will play a key role in materials science in the future.

4.7.3   3D Electron Diffraction

Compared to XRD, ED information is normally incomplete. The Ewald sphere of an electron is almost flat and many reflections are exited simultaneously, but the tilting angle of TEM is relatively small so only few zone axes can be obtained from one crystal by SAED. Even for an experienced scientist, it takes a long time to obtain an ED data set from different zone axes or major diagonals with precise tilting angle in order to get the 3D information on the whole reciprocal space. It was long thought that ED data were not suitable for the solution or refinement of crystal structures, because of the strong dynamic scattering effects.

To overcome these problems, a precession electron diffraction (PED) method was invented by Vincent and Midgley.[61] By rotating the electron beam at several hertz at an angle of 1 $\sim$3° in a conical way around the electron-optical axis, a diffraction pattern is formed by the sum of a continuous set of misaligned diffraction patterns. Under this condition, only a few reflections are in Bragg condition at a given moment (*i.e.* not along zone axes), which greatly reduces the dynamic scattering phenomenon compared with the SAED technique (Figure 4.36a). Using the PED method, it is much easier to collect high-quality ED patterns for reflections in terms of intensity and position, even if the crystal is not perfectly aligned. As PED generates the integrated intensities of the reciprocal-lattice rods (rel-rods for short), the PED patterns can reach higher resolutions than SAED patterns collected from the same crystal.

Recently, an automated diffraction tomography technique has been developed by Kolb *et al.*, using STEM combined with PED.[62] A computer-controlled specimen holder is used and the PED patterns are taken at every 1° by tilting of the specimen holder.[62–66] Figure 4.37 shows projections of the reconstructed 3D ED of semiconductor 6H–SiC. Using raw intensities and assuming easy kinematic approximation, it can determine the full structure *ab initio* by direct methods.[65]

A 3D ED tomography method combined with the tilting of both the electron beam and the specimen holder ws developed by a group at Stockholm University (Hovmöller, Zou, Oleynikov and Zhang) in order

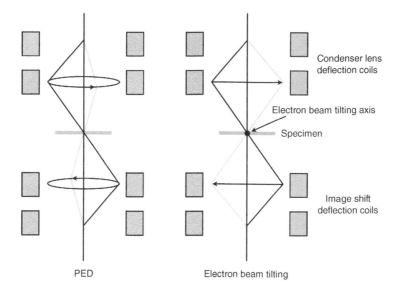

Figure 4.36 Ray diagram of PED (left) and electron beam tilting (right).

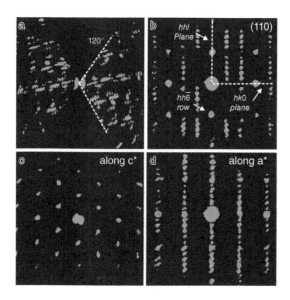

Figure 4.37 Projections of the reconstructed 3D diffraction of 6H–SiC: (a) along the tilt axis; (b) along [110], showing the extinctions along the *hhl*-plane with $l \neq 6N$; (c) along the c^*-axis; (d) along the a^*-axis. Reprinted with permission from [65] Copyright (2012) IOP Publishing Ltd.

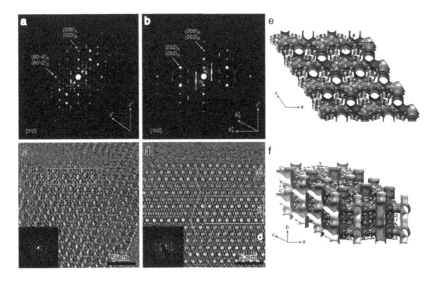

Figure 4.38 Two perpendicular cuts of the reconstructed 3D reciprocal lattice of (a) the $h0l$ slice and (b) the $0kl$ slice. Structure projection reconstructed from 20 HRTEM images along (c) b (c) and (d) a.(e, f) Structural model of the channel system in ITQ-39A. Reprinted with permission from [69] Copyright (2012) Nature Publishing.

to collect 3D ED data on a standard TEM.[67] The tilting axis should not be close to the zone axes. A series of ED patterns is collected *via* beam tilting of the TEM (Figure 4.36b), then the crystal is tilted by a given angle with the tilting of the goniometer and another series of ED patterns is collected. This procedure is repeated until the goniometer tilting limit is reached and the whole set of diffraction data is collected and constructed by computer.

Through using the electron beam tilting method together with the through-focus structure projection reconstruction method,[68] which determines the defocus and astigmatism and corrects for the distortions in each HRTEM image in a through-focus series to retrieve the structure projection, the zeolite ITQ-39 with complex stacking sequence, stacking faults and twinning structures with nanosized domains has been solved by the EC method (Figure 4.38).[69]

4.8 CONCLUSION

We defined **EC** as crystallography that uses electrons as a probe through microscopy and/or diffraction. You are convinced, we hope, that EC is a powerful approach not only to provide crystal structure solutions but also to describe structural deviations or modulation from perfect symmetry.

New atomic crystals and novel nanostructured materials will successively provide interesting new physical properties and functions. Understanding the relationship between structure and function is a key part of materials science, so we need to develop a new approach to solving the structures of crystalline materials at atomic and mesoscopic scale orders, as mentioned above.

We have proposed the use of diffraction-based 3D microscopy,[70] using ED with high coherent illumination, to obtain 3D atomic-scale structures of nanostructured materials and to overcome resolution barriers inherent in HRTEM and tomography. By combining coherent ED patterns with the oversampling phasing method, the actual 3D structure of a nanostructured material will be solved. Following a report on image reconstruction from diffraction patterns using iterative algorithms,[71] atomic-resolution projected images (electron diffractive images) of carbon nanotubes[72,73] and Si[74] have been reported. For electron beam-sensitive crystals, it is quite likely that this kind of experiment can be done at low voltage (30 kV) based on a conventional SEM instrument with a cold FEG.[74]

The combination of intensity measurement of ED patterns by precession or tilting with HRTEM images from nanocrystalline materials will continue to produce more exciting reports on novel nanocrystals.

In situ XRD experiments provide very important information on crystal growth and structural transformation. The advantage of EM lies in its ability to show local structural information, and it will be a useful approach in the study of structures of non-periodic systems or of 'softer' materials and will enable the study of time-evolved crystal structures through the use of 'snap shot' or 'freezing' TEM observation complementary to the *in situ* XRD experiment.

We have shown through some background information and a selection of examples that EM is a very powerful approach to the characterisation of nanostructured materials. We have presented recent progress in EM, together with the basic principles of electron crystallography. As we develop novel materials with different length scales, we should continue to create crystallographic methodologies that can account for defects and modulations, which contain a wealth of information on growth mechanisms and structural transformation. Truly, these defects and modulations are a letter from nature.

ACKNOWLEDGMENT

The authors acknowledge many collaborators who have contributed to the original papers. Financial support from the Swedish Research Council VR, 3DEM-NATUR, EXSELENT, Japan Science and Technology Agency (JST), WCU of Korea (R-31-2008-000-10055-0) and National Natural Science Foundation of China (Grant No. 21201120) is acknowledged.

REFERENCES

[1] E. G. van Putten, D. Akbulut, J. Bertolotti, W. L. Vos, A. Lagendijk and A. P. Mosk, *Phys. Rev. Lett.*, **106**, 193905 (2011).

[2] X. Zou, S. Hovmöller and P. Oleynikov, *Electron Crystallography: Electron Microscopy and Electron Diffraction*, IUCr Text on Crystallography – 16, Oxford University Press, Oxford, 2011.

[3] O. Terasaki, *J. Electron Microscopy*, **43**, 337 (1994).

[4] O Terasaki, *Molecular Sieves*, Vol. 2, Springer-Verlag, 1999, p. 71.

[5] O. Terasaki and T. Ohsuna, *Handbook of Zeolite Science and Technology*, Marcel Dekker, 2003, p. 291.

[6] O. Terasaki, T. Ohsuna, N. Ohnishi and K. Hiraga, *Curr. Opin. Solid State Mater. Sci.*, **2**, 94 (1997).

[7] O. Terasaki and T. Ohsuna, in *Studies in Surface Science and Catalysis 135*, edited by A. Galarneau, F. Di Renzo, F. Fajula and J. Vedrine, Elsevier, 2001, p. 61.

[8] J. M. Thomas, O. Terasaki, P. L. Gai, W. Zhou and J. Gonzalez-Calbet, *Acc. Chem. Res.*, **34**, 583 (2001).

[9] Z. Liu, N. Fujita, K. Miyasaka, L. Han, S. M. Stevens, M. Suga, S. Asahina, B. Slater, C. Xiao, Y. Sakamoto, M. W. Anderson, R. Ryoo and O. Terasaki, *Microscopy*, **62**, 109 (2013).

[10] G. Burns and M. Glazer, *Space Groups for Solid State Scientists*, 2nd edition, Academic Press, 1990.

[11] K. Kanaya and S. Okayama, *J. Phys. D. Appl. Phys.*, **5**, 43 (1972).

[12] S. Che, K. Lund, T. Tatsumi, S. Iijima, S. Joo, R. Ryoo and O. Terasaki, *Angew. Chem., Int. Ed.*, **42**, 2182 (2003).

[13] S. Che, S. Lim, M. Kaneda, H. Yoshitake, O. Terasaki and T. Tatsumi, *J. Am. Chem. Soc.*, **124**, 13962 (2002).

[14] J. Kim, S. Kim and R. Ryoo, *Chem. Commun.*, 259 (1998).

[15] S. Che, Y. Sakamoto, O. Terasaki and T. Tatsumi, *Chem. Mater.*, **13**, 2237 (2001).

[16] S. M. Stevens, K. Jansson, C. Xiao, S. Asahina, M. Klingstedt, D. Grüner, Y. Sakamoto, K. Miyasaka, P. Cubillas, R. Brent, L. Han, S. Che, R. Ryoo, D. Zhao, M. W. Anderson, F. Schüth and O. Terasaki, *JEOL News*, **44**, 17 (2009).

[17] O. Terasaki, T. Ohsuna, Z. Liu, Y. Sakamoto and A. E. Garcia-Bennett, in *Studies in Surface Science and Catalysis 148*, edited by O. Terasaki, Elsevier, 2004, p. 261.

[18] P. M. Arnal, M. Comotti and F. Schüth, *Angew. Chem., Int. Ed.*, **45**, 8224 (2006).

[19] K. Fujiwara, *J. Phys. Soc. Jpn.*, **16**, 2226 (1961).

[20] T. Akaogi, K. Tsuda, M. Terauchi and M. Tanaka, *J. Electron Microsc.*, **53**, 11 (2004).

[21] K. Tsuda, D. Morikawa, Y. Watanabe, S. Ohtani and T. Arima, *Phys. Rev. B*, **81**, 180102 (2010).

[22] J. T. McKeown and J. C. H. Spence, *J. Appl. Phys.*, **106**, 074309 (2009).

[23] Y. Sakamoto, M. Kaneda, O. Terasaki, D. Zhao, J. Kim, G. D. Stucky, H. Shim and R. Ryoo, *Nature*, **408**, 449 (2000).

[24] J. M. Cowley and A. F. Moodie, *Acta Cryst.*, **10**, 609 (1957).

[25] MacTempas,™ www.totalresolution.com (last accessed 17 October 2013); WinHREM™ and MacHREMT,™ www.hremresearch.com (last accessed 17 October 2013).

[26] M. Choi, K. Na, J. Kim, Y. Sakamoto, O. Terasaki and R. Ryoo, *Nature*, **461**, 246 (2009).

[27] K. Lund, N. Muroyama and O. Terasaki, *Micropor. Mesopor. Mat.*, **128**, 71 (2010).

[28] X. Zhang, D. Liu, D. Xu, S. Asahina, K. A. Cychosz, K. V. Agrawal, Y. Al Wahedi, A. Bhan, S. Al Hashimi, O. Terasaki, M. Thommes and M. Tsapatsis, *Science*, **336**, 1684 (2012).

[29] C. T. Kresge, M. E. Leonowicz, W. J. Roth, J. C. Vartuli and J. S. Beck, *Nature*, **359**, 710 (1992).

[30] Z. Liu, O. Terasaki, T. Ohsuna, K. Hiraga, H. J. Shin and R. Ryoo, *Chem. Phys. Chem.*, **2**, 229 (2001).

[31] C. Jin, L. Han and S. Che, *Angew. Chem., Int. Ed.*, **48**, 9268 (2009).

[32] J. M. Cowley, *Acta Cryst.*, **6**, 516 (1953).

[33] S. Kuwabara, *J. Phys. Sc. Jpn.*, **14**, 1205 (1959).

[34] D. L. Dorset and H. A. Hauptman, *Ultramicroscopy*, **1**, 195 (1976).

[35] A. Carlsson, T. Oku, J. O. Bovin, G. Karlsson, Y. Okamoto, N. Ohnishi, O. Terasaki, *Chem. Eur. J.*, **5**, 244 (1999).

[36] D. L. Dorset, C. J. Gilmore, Z. *Kristallogr.*, **226**, 447 (2011).

[37] P. Wagner, O. Terasaki, S. Ritsch, J. G. Nery, S. I. Zones, M. E. Davis and K. J. Hiraga, *Phys. Chem. B*, **103**, 8245 (1999).

[38] T. Ohsuna, Z. Liu, O. Terasaki, K. Hiraga and M. A. Camblor. *J. Phys. Chem. B*, **106**, 5673 (2002).

[39] T. Ohsuna, Y. Sakamoto, O. Terasaki and K. Kuroda, *Solid State Sci.*, **13**, 736 (2011).

[40] K. Miyasaka, L. Han, S. Che and O. Terasaki, *Angew. Chem., Int. Ed.*, **45**, 6516 (2006).

[41] L. Han, Y. Sakamoto, O. Terasaki, Y. Li and S. Che, *J. Mater. Chem.*, **17**, 1216 (2007).

[42] J. Tang, X. Zhou, D. Zhao, G. Q. Lu, J. Zou and C. Yu, *J. Am. Chem. Soc.*, **129**, 9044 (2007).

[43] S. Ino, *J. Phys. Soc. Jpn.*, **21**, 346 (1966).

[44] L. Han, P. Xiong, J. Bai and S. Che, *J. Am. Chem. Soc.*, **133**, 6106 (2011).

[45] S. Mader, *J. Vac. Sci. Technol.*, **8**, 247 (1966).

[46] A. Squires, R. H. Templer, J. M. Seddon, J. Woenckhaus, R. Winter, S. Finet and N. Theyencheri, *Langmuir*, **18**, 7384 (2002).

[47] A. Mayoral, T. Carey, P. A. Anderson, A. Lubk, I. Díaz. *Angew. Chem., Int. Ed.* 50, 11230 (2011).

[48] K. Yoshida, K. Toyoura, K. Matsunaga, A. Nakahra, H. Kurata, Y. H. Ikuhara and Y. Sasaki, *Scientific Reports* 3, 2457, (2013).

[49] J. Frank, *Electron Tomography*, Plenum, New York, 1992.

[50] A. J. Koster, U. Ziese, A. J. Verkleij, A. H. Janssen and K. P. de Jong, *J. Phys. Chem. B*, **104**, 9368 (2000).

[51] G. Prieto, J. Zečević, H. Friedrich, K. P. de Jong and P. E. de Jongh, *Nat. Mater.*, **12**, 34 (2012).

[52] I. Arslan, J. R. Tong and P. A. Midgley, *Ultramicroscopy*, **106**, 994 (2006).

[53] A. J. Koster, R. Grimm, D. Typke, R. Hegerl, A. Stoschek, J. Walz and W. Baumeister, *Struct. Biol.*, **120**, 276 (1997).

[54] R. J. Spontak, M. C. Williams and D. A. Agard, *Polymer*, **29**, 387 (1988).

[55] A. J. Koster, U. Ziese, A. J. Verkleij, A. H. Janssen, J. de Graaf, J. W. Geus and K. P. de Jong, *Stud. Surf. Sci. Catal.*, **130**, 329 (2000)

[56] P. A. Midgley, M. Weyland, J. M. Thomas and B. F. G. Johnson, *Chem. Commun.*, **10**, 907 (2001).

[57] P. A. Midgley and R. E. Dunin-Borkowski, *Nat. Mater.*, **8**, 271 (2009).

[58] T. Yates, J. Thomas, J. Fernandez, O. Terasaki, R. Ryoo and P. Midgley, *Chem. Phys. Lett.*, **418**, 540 (2006).

[59] J. Yamasaki, N. Tanaka, N. Baba, H. Kakibayashi and O. Terasaki, *Philos. Mag.*, **84**, 2819 (2004).

[60] C.-C. Chen, C. Zhu, E. R. White, C.-Y. Chiu, M. C. Scott, B. C. Regan, L. D. Marks, Y. Huang and J. Miao, *Nature*, **496**, 74 (2013).

[61] R. Vincent and P. A. Midgley, *Ultramicroscopy*, **53**, 271 (1994).

[62] U. Kolb, T. Gorelik, C. Kübel, M. T. Otten and D. Hubert, *Ultramicroscopy*, **107**, 507 (2007).

[63] J. Jiang, J. L. Jorda, J. Yu, L. A. Baumes, E. Mugnaioli, M. J. Diaz-Cabanas, U. Kolb, A. Corma, *Science* **333**, 1131 (2011).

[64] E. Mugnaioli, T. Gorelik and U. Kolb, *Ultramicroscopy*, **109**, 758 (2009).

[65] E. Sarakinou, E. Mugnaioli, C. B. Lioutas, N. Vouroutzis, N. Frangis, U. Kolb and S. Nikolopoulos, *Semicond. Sci. Technol.*, **27**, 105003 (2012).

[66] T. E. Gorelik, A. A. Stewart and U. Kolb, *J. Microsc.*, **244**, 325 (2011).

[67] D. Zhang, P. Oleynikov, S. Hovmöller and X. Zou, *Z. Krist.*, **225**, 94 (2010).

[68] W. Wan, S. Hovmöller and X. Zou, *Ultramicroscopy*, **115**, 50 (2012).

[69] T. Willhammar, J. Sun, W. Wan, P. Oleynikov, D. Zhang, X. Zou, M. Moliner, J. Gonzalez, C. Martínez, F. Rey and A. Corma, *Nature Chem.*, **4**, 188 (2012).

[70] J. Miao, T. Ohsuna, O. Terasaki and M. A. O'Keefe, *Phys. Rev. Lett.*, **89**, 155502 (2002).

[71] U. Weierstall, Q. Chen, J. C. H. Spence, M. R. Howells, M. Isaacson and R. R. Panepucci, *Ultramicroscopy*, **90**, 171 (2002).

[72] J. M. Zuo, I. Vartanyants, M. Gao, R. Zhang and L. A. Nagahara, *Science*, **300**, 1419 (2003).

[73] O. Kamimura, Y. Maehara, T. Dobashi, K. Kobayashi, R. Kitaura, H. Shinohara, H. Shioya and K. Gohara, *Appl. Phys. Lett.*, **98**, 174103 (2011).

[74] S. Morishita, J. Yamasaki, K. Nakamura, T. Kato and N. Tanaka, *Appl. Phys. Lett.*, **93**, 183103 (2008).

5

Small-Angle Scattering

Theyencheri Narayanan
European Synchrotron Radiation Facility, Grenoble, France

5.1 INTRODUCTION

Small-angle X-ray and neutron scattering (SAXS and SANS, respectively) are well established methods for probing the nanoscale structure and interactions in materials that are disordered at the atomic scale.[1−7] The scattering contrast in the case of X-rays originates from the spatial fluctuations of the electron density, which systematically varies with the atomic number of constituent elements, while neutron contrast arises from the atomic nuclei without any systematic dependence on the atomic number but displays strong isotope effect. As a result, these techniques provide complementary information in the structural elucidation of multicomponent systems. Both techniques are non-destructive and suitable for *in situ* studies. The amount of structural information obtained from a small-angle scattering (SAS) experiment depends to some extent on the degree of order within the sample. For example, this could be the shape and a few parameters like the radius of gyration and polydispersity in the case of a dilute suspension of particles, while a molecular resolution structural model may be derived with a highly ordered semi-crystalline specimen. Typically, SAS experiments provide ensemble averaged structural information as opposed to the more selective features observed by electron or atomic force microscopies. Therefore, SAS methods and electron and other microscopy techniques

Structure from Diffraction Methods, First Edition. Edited by Duncan W. Bruce,
Dermot O'Hare and Richard I. Walton.
© 2014 John Wiley & Sons, Ltd. Published 2014 by John Wiley & Sons, Ltd.

derive complementary information in the structural elucidation of nanoscale materials. This chapter presents certain unique applications of SANS and SAXS in the study of inorganic materials. The examples presented are illustrative of the basic principles and potential of these techniques and are certainly not meant to be an exhaustive survey of the literature related to inorganic materials.

Both SAXS and SANS techniques are available at all major user facilities (synchrotrons and reactor or spallation sources, respectively).[6,7] The examples presented are obtained at these large-instrument facilities, though the high contrasts provided by the majority of inorganic systems allow very quantitative SAXS experiments even with laboratory instruments. Access to a large experimental facility is mandatory for SANS experiments,[8] while the high brightness offered by modern synchrotron sources permits high-resolution and time-resolved SAXS studies.[9] In general, time-resolved SANS studies are more demanding, due to the relatively low flux of neutrons available, and require many repetition cycles to accumulate sufficient data quality.[10] Moreover, most *in situ* studies require a combination of SANS or SAXS with advanced sample environments that would be readily available at large user facilities.[8,9] Data analysis is an essential step in deriving quantitative structural and kinetic information from a scattering experiment. In this respect, specific program suites available at the user facilities could be an advantage for non-expert users.[11−14]

The importance of SAXS in structural studies of disordered materials was recognised immediately following the success of X-ray diffraction in the investigation of crystalline materials.[15] Guinier and Fournet[1] provide an excellent description of the early development of SAXS and the related bibliography. Many of the theoretical treatments they present are still routinely used for the analysis of SAS data. Glatter and Kratky[2] present the subsequent developments in instrumentation and theoretical modelling of data, and their book is still a source of information for basic theoretical treatments of SAXS. SANS emerged with the development of high-flux research reactor sources in the 1960s and 70s.[16,17,18] With the advent of synchrotron sources, the photon flux became less of a limitation for SAXS and the vast majority of instruments employed the so-called 'pinhole collimation'. Svergun and Feigin[3] describe a unified theoretical framework for SAXS and SANS, in addition to the instrument developments at synchrotron and reactor sources up to the mid 1980s. Further theoretical and experimental advances are provided in a collection of articles in *Modern Aspects of Small-Angle Scattering*, edited by H. Brumberger,[4] while *Neutrons, X-Rays and Light Scattering Methods*

Applied to Soft Condensed Matter, edited by P. Lindner and T. Zemb,[5] gives more up-to-date information on developments up to the late 1990s. Many new developments at modern synchrotron and neutron sources have been covered by more recent books.[6,7] This chapter will primarily give an overview of certain new possibilities for SAXS and SANS at these large instrument facilities. The examples presented are relatively simple and the emphasis is on illustrating the quantitative aspects of SAS and related scattering methods.

5.2 GENERAL PRINCIPLES OF SAS

The basic theoretical formalism of SAS is similar for light, neutrons and X-rays.[5] The important difference is in the interaction of the particular radiation with the scattering medium. Detailed derivations of the expressions are readily available in many textbooks and are therefore omitted in this section.[1−4] In the following, some basic definitions relevant to examples discussed in this chapter are reviewed.

5.2.1 Momentum Transfer

Figure 5.1 depicts the scattering geometry of a typical SAS experiment. A highly collimated and monochromatic X-ray or neutron beam of wavelength λ impinges on a sample and the scattered intensity in the forward direction is recorded by a two-dimensional (2D) detector. The transmitted beam is blocked by a beamstop right in front of the detector. In the case of SAXS, the transmitted primary beam could also be recorded by a point detector embedded in the beamstop,[9],while in the SANS case,

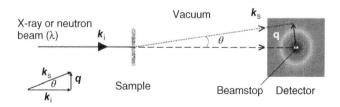

Figure 5.1 Schematic layout of a SAS set-up, depicting the incident, scattered and transmitted beams, the 2D detector and the definition of the scattering vector, q.

transmitted beam intensity is often measured using the same 2D detector with the aid of calibrated attenuators.[10] The entire flight path before and after the sample is in vacuum to avoid absorption and scattering by air. In the experiment, the number of photons or neutrons scattered as a function of the scattering angle, θ, is measured. For a given sample, the quantity that can be compared in different experiments is the number of photons or neutrons scattered into unit solid angle normalised to the incident flux (photons or neutrons per second per unit area). In the case of X-rays, the scattering originates from electrons (Thomson scattering) and at small angles is fully elastic.[19] The magnitudes of the incident and scattered wave vectors are equal, $|k_i| = |k_s| = 2\pi/\lambda$, and the refractive index is close to unity, while cold neutrons used in SANS have comparable energy to typical thermal excitations ($k_B T \approx 25\,\mathrm{meV}$) and consequent inelastic scattering cannot be completely excluded.[8,18] Nevertheless, in steady state SANS experiments usually the elastic component is significantly dominant. Therefore, momentum transfer or scattering vector $q = k_s - k_i$ in both cases, and its magnitude is as follows:

$$q = |q| = \frac{4\pi}{\lambda} \sin(\theta/2) \tag{5.1}$$

This quantity indicates the typical length scales probed by the scattering experiment. In synchrotron SAXS, the q-range covered can be three orders of magnitude, typically $0.006\,\mathrm{nm}^{-1} < q < 6\,\mathrm{nm}^{-1}$ using 1 Å X-ray wavelength, corresponding to real-space dimensions of 1 μm down to 1 nm.[9] Modern SANS instruments also cover similar q-ranges, but using longer wavelengths ($\sim 5\,\text{Å}$) and larger sample–detector distances.[10]

5.2.2 Differential Scattering Cross-Section

In the small angle (neglecting the polarisation factor), the coherent scattered intensity of X-rays from an atom containing Z electrons is given by:

$$I_{\mathrm{coh}} = r_e^2 \sum_1^z f_e^2 = r_e^2 f^2 \tag{5.2}$$

where f is the atomic scattering factor and r_e is the Thomson scattering length, which is equal to the classical electron radius ($2.818 \times 10^{-15}\,\mathrm{m}$).[20] $f \approx Z$, except near the atomic absorption edge, and the product fr_e is the scattering length. The interaction between the incident neutrons and the atomic nucleus is more complex and the corresponding strength is given by a complex quantity, namely the neutron scattering length, b.[18] Table 5.1 lists the scattering lengths for neutrons and X-rays for some representative elements. Note that the scattering length of an electron is roughly equal to the neutron scattering length of sulfur.[6]

The interaction between the incident radiation and the scattering medium within the sample is contained in the differential scattering cross-section ($d\sigma/d\Omega$). In an experiment similar to that depicted in Figure 5.1, the incident intensity per unit area per unit time (I_0) is scattered by a sample and the scattered radiation (photons or neutrons) is acquired by each detector element subtending a solid angle, $\Delta\Omega$, with efficiency ε.[5] The measured scattered intensity is given by:

$$I_S = I_0 \, \varepsilon \, T_r \Delta\Omega \, A_s l_s \frac{d\Sigma}{d\Omega} \tag{5.3}$$

where A_s is the cross-section of the beam, T_r is the sample transmission and $d\Sigma/d\Omega$ is the differential scattering cross-section per unit volume. For X-rays, the transmitted intensity per unit area per second, $I_T = I_0 \exp(-\mu_l l_s)$, where μ_l is the linear absorption coefficient and l_s is the path length of the beam in the sample, which is roughly the sample thickness at low angles, and $T_r = I_T/I_0$. For neutrons, the absorption length is much longer and both scattering and absorption in the sample contribute to the attenuation of the beam. The quantity that can be

Table 5.1 Neutron and X-ray scattering lengths for some representative elements.

Element	Z	Neutron		X-ray scattering length, $r_e f/10^{-15}$ m
		Coherent scattering length, $b/10^{-15}$ m	Incoherent cross-section, $\sigma_{ic}/10^{-28}$ m^2	
H	1	−3.74	80.26	2.82
^{2}H (D)	1	6.67	2.05	2.82
B	5	6.65	0.21	14.1
N	7	9.36	0.5	19.7
O	8	5.80	0.0	22.6
Si	14	4.15	0.004	39.5
S	16	2.85	0.0	45.1
Au	79	7.63	0.43	222.8

directly compared to a model is $d\Sigma/d\Omega$, which contains information on the structure and the interactions in the system over the range of q spanned by the scattering experiment, and it is expressed in units of reciprocal length times reciprocal solid angle ($m^{-1}\,sterad^{-1}$). Therefore, essential steps in reaching a quantitative understanding of the measured intensities are the normalisation of the experimental data to $d\Sigma/d\Omega$, and subtraction of corresponding background contribution. The resulting quantity henceforth will be denoted by $I(q)$ and given simply in units of reciprocal length.

In order to convert measured intensities to $d\Sigma/d\Omega$, the absolute value of the detector efficiency, ε, is needed. This can be measured using a sample of known $d\Sigma/d\Omega$. Of course, the parasitic background of the instrument should be much lower than the scattering from this calibration sample. Over the q-range $1-3\,nm^{-1}$, scattering from a clean liquid, composed of light atoms with known isothermal compressibility, K_T, and density, is an appropriate intensity standard. The SAXS intensity corresponding to the low-q plateau of the liquid structure factor ($q < 3\,nm^{-1}$) is given by $d\Sigma/d\Omega = N^2 n_e^2 r_e^2 k_B T K_T$, where N is the number density of the molecules and n_e is the number of electrons per molecule. Water is a good intensity standard for SAXS and its scattering can be measured by any optimised instrument. For $q < 4\,nm^{-1}$, the intensity is nearly flat after background subtraction with $d\Sigma/d\Omega \approx 1.6 \times 10^{-3}\,mm^{-1}$ at $25\,°C$.[9] Water has been used as a calibration standard for SANS, but the flat signal is primarily contributed by incoherent scattering ($> 99\%$) and involves additional cross-calibration.[10] Therefore, other standards with strong elastic scattering such as glassy carbon, polymer blends are proposed as alternatives.

5.2.3   Non-Interacting Systems

For a dilute system containing N uniform particles per unit volume, the interparticle interactions can be neglected and $I(q)$ mainly depends on the shape and size of particles:

$$I(q) = N\,|F(q)|^2 \tag{5.4}$$

where $F(q)$ is the coherent sum of the scattering amplitudes of the individual scattering centres within a particle, given by the Fourier transform of the scattering length density (electrons or nuclei) distribution.[2,4] In SAS, the scattering length density can be approximated as a continuous function. Therefore, the scattering amplitude of the particle is as follows:

$$F(q) = \int_V \rho(r) \, e^{iqr} \, dV \qquad (5.5)$$

where V is the volume of the particle and ρ is the scattering length density, which is the product of electron density and r_e for X-rays[19] and $\sum b_i / V_M$ for neutrons, with b_i the individual b of constituent atoms[18] and V_M the molar volume. In the case of uniform electron density (or b):

$$\rho = \frac{n_e r_e}{V_M} = \frac{n_e d_M N_A}{M_W} r_e \qquad (5.6)$$

where n_e is the number of electrons in a molecule, d_M is the mass density, N_A is the Avogadro number and M_W is the molar mass. For example, for water ($d_M = 10^3 \, \mathrm{kg\,m^{-3}}$), $\rho \approx 9.4 \times 10^{-4} \, \mathrm{nm^{-2}}$, while for colloidal silica particles ($d_M \approx 2 \times 10^3 \, \mathrm{kg\,m^{-3}}$), $\rho \approx 1.7 \times 10^{-3} \, \mathrm{nm^{-2}}$. When the scattering units are embedded in a medium (e.g. solvent), the relative scattering length density or contrast length density ($\Delta \rho = \rho - \rho_M$) is the relevant parameter that determines the scattering power. For isotropic particles inserting the spatial average of the phase factor in Equation 5.5 reduces the Fourier transform to the 1D form:[2,4]

$$F(q) = 4\pi \int_0^\infty \Delta\rho(r) \, \frac{\sin(qr)}{qr} \, r^2 dr \qquad (5.7)$$

For a uniform spherical particle of radius R_S and volume V_S, Equation 5.7 leads to:

$$|F(q)|^2 = V_S^2 \Delta\rho^2 \left(\frac{3 \left[\sin(qR_S) - qR_S \cos(qR_S) \right]}{(qR_S)^3} \right)^2 = V_S^2 \Delta\rho^2 P(q, R_S)$$

$$(5.8)$$

The shape of the particle is described by the Bessel function inside the brackets and $P(q, R_S)$ is the scattering form factor of a sphere. The product $N \, V_S$ is the volume fraction of the particles, ϕ_S. Table 5.2 lists $P(q, R)$ function for a few shapes that will be used latter in this chapter. A comprehensive list of $P(q, R)$ functions for different particle shapes that frequently occur in SAS can be found in reference [21].

Table 5.2 Form factors of a few commonly observed shapes in scattering from particulate systems. For a more complete list, see reference [21].

Uniform sphere of radius R_S	$P(q, R_S) = \left(\dfrac{3\left[\sin(qR_S) - qR_S \cos(qR_S)\right]}{(qR_S)^3} \right)^2 = F_0^2(qR_S)$
Spherical core–shell of core and shell radii R_1 and R_2	$F^2(q, R_1, R_2) = [V_2 \Delta\rho_2 F_0(qR_2) - V_1 \Delta\rho_1 F_0(qR_1)]^2$, where V_1 and V_2 are the volumes of the inner and outer spheres and $\Delta\rho_1$ and $\Delta\rho_2$ are the contrast between the shell and core and the shell and medium, respectively
Randomly orientated cylinder or circular disc of radius R_C and height H	$P(q, R_C, H) =$ $\displaystyle\int_0^{\pi/2} \left\{ \left[\frac{2J_1(qR_C \sin\varphi)}{qR_C \sin\varphi} \right] \left[\frac{\sin((qH/2)\cos\varphi)}{(qH/2)\cos\varphi} \right] \right\}^2 \sin\varphi \, d\varphi$, where J_1 is the first-order Bessel function and φ is the orientation angle
Ellipsoid of revolution with semi-axes R_1 and R_2	$P(q, R_1, R_2) = \displaystyle\int_0^{\pi/2} F_0^2(q, R) \, \sin\alpha \, d\alpha$, where $F_0(q, R)$ is the sphere function, with $R = \sqrt{(R_1^2 \sin^2\alpha + R_2^2 \cos^2\alpha)}$
Flexible Gaussian chains with radius of gyration R_G	$P(q, R_G) = \dfrac{2[\exp(-q^2\langle R_G^2\rangle) + q^2\langle R_G^2\rangle - 1]}{(q^2\langle R_G^2\rangle)^2}$

5.2.4 Influence of Polydispersity

A distinguishing feature of objects on the nanometre scale as compared to atomic systems is the finite polydispersity in size and other properties. For example, Equation 5.8 has zeros at $qR_S = 4.5, 7.73$ etc., which for real particles appear as minima in the scattered intensity, as depicted in Figure 5.2. For particles with uniform density, these minima can be used to determine the approximate radius. The polydispersity can be described by a size distribution function, $D(R)$, where $\int D(R)dR = 1$ and there is a lower cut-off at $R > 0$. The resulting $I(q)$ is given by[21]:

$$I(q) = N \, \Delta\rho^2 \int_0^\infty D(R) \, V^2(R) \, P(q, R) \, dR \tag{5.9}$$

Analytical expression for Equation 5.9 exists in the case of spherical particles for size distributions including Schulz and Gaussian.[22,23] The polydispersity is given by the ratio of the root mean square

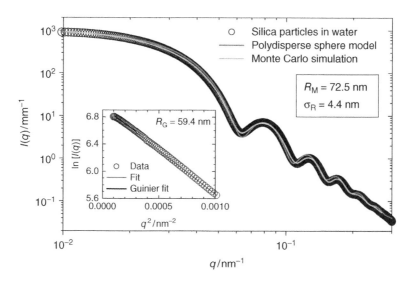

Figure 5.2 SAXS intensity from a dilute suspension of silica colloids after solvent background subtraction. The continuous line is a fit to the polydisperse sphere function given by Equation 5.9, including instrument resolution for a Schultz size distribution with mean radius 72.5 nm and RMS deviation 4.4 nm. For comparison, the scattering curve obtained by Monte Carlo simulation is also shown.[28] The inset depicts the Guinier plot of the low-q region.

(RMS) deviation of $R_S(\sigma_R)$ and the mean value of $R_S(R_M)$. In general, Equation 5.9 can be numerically integrated using an experimentally determined $D(R)$ or a theoretical function to deduce the mean radius and the polydispersity. Moreover, in real systems the distribution may be skewed to larger sizes, which leads to better visibility of the oscillations at high q (see Figure 5.2) for the same mean radius and polydispersity. As a result, the polydispersity will be overestimated if the analysis is based on the first minimum alone. In addition, the first minimum (or several minima) might also be smeared due to instrument resolution that, without correction, leads to an overestimation of polydispersity.

The Schultz distribution is very close to the Gaussian function at low polydispersities and it often represents realistic behaviour. Moreover, it is mathematically tractable, which is an advantage in analytical modelling.[22−25] Many binary and multicomponent systems do not usually form homogeneous spheres, and the core–shell morphology is often observed.[24] In this case, the polydispersity pertains to either the core or the shell, or to the most general-case core and shell. Analytical models exist for polydispersity for the core alone[24] or for both core and shell[25] using a Schultz size distribution. In addition, the density profile of the shell or the interface may not be sharp; in such cases, a smooth density profile, which can be analytically solvable, has been implemented using a Fermi–Dirac-type function.[26] The scattering function of cubes, and of many platonic solids such as icosahedra, dodecahedra etc., is close to that of spheres, as shown by recent models.[27] In order to differentiate shapes using a model, very quantitative data and low polydispersity (which allows observation of higher-order oscillations in $P(q)$) are required. In these cases, complementary electron microscopy images will be helpful in guiding the modelling.

In addition to direct modelling, inversion methods exist to obtain size distribution from $I(q)$ for non-interacting systems.[2,11,13,14] This inversion procedure is mathematically an ill-posed problem and can lead to non-unique solutions or multiple size distributions, depending on the available q-range. The most commonly used approach is based on the indirect Fourier transformation (IFT) method originally proposed by Glatter,[2] which provides stable solutions with appropriate size constraints. Other approaches include the maximum entropy method,[14] Monte Carlo simulation[28] etc. Figure 5.2 gives a comparison of the best-fit scattering curve obtained using the recently proposed Monte Carlo simulation[28] with the same R_M with a Gaussian-like main distribution and some additional weak modes on either side of the peak.

5.2.5 Asymptotic Forms of $I(q)$

For non-interacting particles, irrespective of the shape, the exponential factor in Equation 5.5 can be expanded in terms of the radius of gyration, R_G, at very small q values. This leads to the well-known Guinier law:[1]

$$I(q) = N \, V^2 \Delta\rho^2 \exp\left(-\frac{q^2 R_G^2}{3}\right) \tag{5.10}$$

This approximation is valid only for $qR_G < 1$ (or the leading term in q^2) and is widely used in SAS to determine R_G from the $\ln[I(q)]$ versus q^2 plot. The inset in Figure 5.2 depicts a Guinier plot of the low-q data. Although the Guinier approximation is strictly valid only for $qR_G < 1$ (i.e. $q^2 < 3.0 \times 10^{-4}\,\text{nm}^{-2}$ in Figure 5.2), the linear region extends over a wider q range in the case of spherical particles. In addition, the lowest q range may be affected by the uncertainties in the background subtraction and residual interactions, and these data points should be excluded when determining the R_G. Furthermore, the polydispersity of the system complicates the relationship between R_G and R_M.

In the asymptotic limit, $qR_G \gg 1$, the scattering probes the interface of the particles. In the case of homogenous particles with average surface area S, this leads to Porod behaviour:

$$I(q) = 2\pi \, N \, \Delta\rho^2 S \, q^{-4} \qquad (5.11)$$

signifying a sharp interface between the particle and the medium.[2,4] Moreover, Equation 5.11 provides access to the specific surface S/V ratio.

Power law variation of $I(q)$ is very commonly observed in SAXS from particulate systems composed of both compact and fractal morphologies, as follows:[4]

$$I(q) \propto q^{-p} \begin{cases} p = 4 & \Rightarrow \text{sharp interface} \\ 3 \le p < 4 & \Rightarrow \text{surface fractal} \\ p < 3 & \Rightarrow \text{mass fractal} \\ p \approx 2 & \Rightarrow \text{density fluctuations} \end{cases} \qquad (5.12)$$

The power law exponent, $p > 4$ can be observed in the case of a non-fractal diffuse interface. In practice, extreme care must be exercised when determining the asymptotic power law behaviour from $I(q)$. Improper background subtraction, non-linearity of the detector and so on can undermine an accurate determination of p. In the case of monodisperse particles, this power law region is modulated by the oscillations in Bessel function as shown in Figure 5.2 and the maxima follow the power law.

5.2.6 Multilevel Structures

Many polydisperse systems consist of multiple structural levels, such as primary particles, aggregates and their agglomerates (*e.g.* pyrolitically grown aerosol particles).[29,30] Such systems display structurally limited power law regions with intervening Guinier regions. In this case, the local scattering laws and their crossover can be described by the so-called unified scattering function proposed by Beaucage:[29]

$$I(q) = G \, \exp\left(-\frac{q^2 R_G^2}{3}\right) + B\left\{\frac{\left[erf\left(q \, R_G/\sqrt{6}\right)\right]^3}{q}\right\}^p \qquad (5.13)$$

where $G = N \Delta\rho^2 V^2$, $B = 2\pi N \Delta\rho^2 S$ and *erf* is the error function. This global function does not introduce additional parameters other than those involved in the local scattering laws. For monodisperse spheres, $BR_G^4/G = 1.62$ and analogous ratios can be arrived at for a variety of terminal size distributions.[30] The ratio $BR_G^4/1.62G$ is called the polydispersity index (PDI), which increases from 1 in the case of monodisperse spheres to about 10 for the Debye–Bueche function.[31] For a polydisperse system, R_G is related to mean particle radius, R_M, through the moments of the size distribution.[30] Use of the expression for monodisperse spheres ($R_S^2 = 3/5R_G^2$) leads to severe overestimation of R_M. The unified model provides access to the specific surface (S/V), which is a very important quantity for the characterisation of porous materials.[19] Furthermore, unified model parameters are related to the branch structure of aggregates.[32] Figure 5.3a shows the application of the unified scattering function to a non-fractal silica powder with the parameters indicated in the legend.[30]

Equation 5.13 can be extended to n structural levels by adding the corresponding number of terms for each level. In this case, the power law of the ith level should be cut off at the high-q region by the form factor of the immediate lower level,[29] *i.e.* by multiplying the power law term by $\exp(-q^2 R_G^2/3)$ term of the lower level. Figure 5.3b shows the application of the unified function to model scattering from a blackberry-like two-level structure of polyoxometalate giant inorganic molecular clusters.[33] In aqueous solutions, these polyoxometalate assemblies form shell-like structures (*e.g.* {$Mo_{72}Fe_{30}$}).[34] These self-assembled structures are thought to be metastable intermediates prior to formation of stable crystalline forms.[34] In this case, unified fit provided the structural parameters of the shell, R_{G1}, and the polyoxometalate cluster, R_{G2}, which are among the largest known inorganic molecules. In addition, the power law scattering revealed the surface fractal behaviour of the shell structure with surface fractal dimension, $D_S \simeq 2.7$ ($D_S = 6-p$). The value of R_{G2} is in agreement with the known size of {$Mo_{72}Fe_{30}$} clusters (≈ 2.5 nm). Indeed, the form factor of polyoxometalate clusters shows additional features (oscillations) in addition to the simple power law decay also observed before.[33]

Data analysis tools for multilevel unified scattering functions are readily available in the program suite Irena.[14] While the unified approach is powerful in deducing an array of structural parameters of multilevel systems,[32] it has some pitfalls, as pointed out recently.[35] Allowing the amplitudes G and B to be free in a least-square fitting may lead to undesired artefacts, and extreme care must be exercised

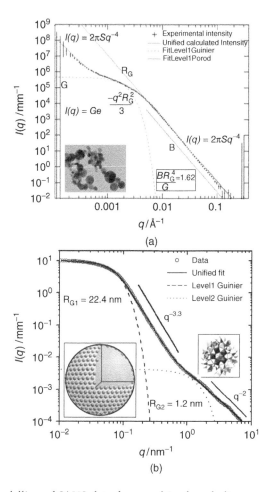

Figure 5.3 Modelling of SAXS data from multiscale polydisperse systems consisting of primary particles and aggregates by the unified function. (a) Non-fractal silica powder with PDI = 23.1 and R_G = 44.2 nm. Inset displays a TEM micrograph, Figure courtesy of G. Beaucage. (b) Aqueous suspension of polyoxometalate assemblies (0.5 mg/ml), consisting of a blackberry shell-like structure of $R_G \approx 22.4$ nm and polyhedral {$Mo_{72}Fe_{30}$} building blocks ($R_G \approx 1.2$ nm). Insets pictorially depict the two structural levels.

when further interpreting the parameters. This problem is partly fixed by the so-called Guinier–Porod model,[35] but then the crossover from the Guinier region to the power law regime occurs at a rather arbitrary value of q. Therefore, comparison with structural parameters deduced by other methods such as electron microscopy is often essential when analysing polydisperse multilevel systems.

5.2.7 Non-Particulate Systems

For non-particulate systems such as porous materials, $I(q)$ can be expressed in terms of the correlation function of scattering length density fluctuations, $\gamma(r)$:[2,4,7,19]

$$I(q) = 4\pi \int_0^\infty \gamma(r) \frac{\sin(qr)}{qr} r^2 dr \tag{5.14}$$

In local fluctuations of scattering length density, $\Delta\rho = \rho - \bar{\rho}$, $\bar{\rho}$ is the average of ρ and

$$\gamma(r) = 4\pi \int_0^\infty \Delta\rho(r') \, \Delta\rho(r' + r) \, r'^2 dr' \tag{5.15}$$

where $\gamma(r)$ varies from $\overline{\Delta\rho^2}$ (mean square fluctuations of scattering length density) to 0 as r increases from 0 to infinity. The pair distribution function $p(r) = r^2\gamma(r)$.[2,4] For an exponentially decaying correlation function $\gamma(r) \sim \exp(-r/\xi)$, where ξ is the correlation length of the fluctuations, Equation 5.14 leads to the Debye–Bueche function:[31]

$$I(q) = \frac{8\pi\xi^3\overline{\Delta\rho^2}}{(1 + q^2\xi^2)^2} \tag{5.16}$$

This expression is widely used to deduce the characteristic size in many disordered materials. Similarly, correlation function decaying as $\gamma(r) \sim 1/r \, \exp(-r/\xi)$ leads to the well-known Ornstein–Zernike structure factor used to describe critical fluctuations near a second-order phase transition:[36]

$$I(q) = \frac{k_B T \overline{\Delta\rho^2} \chi_T}{(1 + q^2\xi^2)} \tag{5.17}$$

where χ_T is the corresponding response function or susceptibility. From the inverse transform of Equation 5.14:

$$\int_0^\infty I(q) \, q^2 dq = 2\pi^2 \, \overline{\Delta\rho^2} = Q_p \tag{5.18}$$

where Q_p is the Porod invariant, proportional to the mean square fluctuations of electron density, which does not change with the shape of the particles.[2,4] For a two-phase system:

$$Q_P = \int_0^\infty I(q) \, q^2 dq = 2\pi^2 \, \Delta\rho^2 \, \phi_1(1 - \phi_1) \tag{5.19}$$

where ϕ_1 is the volume fraction of one of the phases.[4] The invariant is a useful quantity for following phase transformations.[37] It can also be used to normalise the data when the unit of $I(q)$ is uncertain.[30] It is evident that the determination of Q_P requires data over a wide q-range, covering the extrema of the scattering curve.

Q_P and the specific surface given by Equation 5.11 are two powerful quantities for characterising porous and granular materials.[38] However, when dealing with such systems the effective sample thickness is often uncertain. With X-rays, it may be possible to deduce the effective sample thickness from the sample transmission if the precise chemical composition along the beam path is known.[38] This is feasible only for relatively simple systems involving one component. The porosity arises from two contributions: mesopores within the grains and micropores contributed by the intergrain voids. In the general case, a geometric model is required to derive the equivalent grain thickness (*i.e.* the solid and the mesopores) and the effective solid thickness.[38,39] Fortuitously, mesopores and micropores contribute differently to SAS, demarcated by two different Porod regions at high- and low-q regions, respectively. Further differentiation can be made through contrast variation by filling either the mesopores or the micropores with a liquid. With the widely separated size scales of mesopores and macropores, together with different contrasts, it is possible to derive a general expression for the scattered intensity that allows the separation of the specific surfaces of grains. This has been demonstrated for titania, ceria and zirconia mesoporous materials,[38] and more recently for calcium carbonate prepared by different synthesis routes.[39]

5.2.8 Structure Factor of Interactions

When the particulate system is more concentrated, $d\Sigma/d\Omega$ involves an additional term corresponding to the interparticle interactions.[40] This interference term or structure factor, $S(q)$, is a complex function of N and the interaction potential, $U(r)$. In a dilute non-interacting system, $S(q) \approx 1$. For particles with spherical symmetry and narrow size distribution, $I(q)$ can be factorised as given below:[22]

$$I(q) = N\,V^2\Delta\rho^2 P(q)\,S(q) \tag{5.20}$$

$S(q)$ relates scattered intensity to the microstructure through the pair correlation function, $g(r)$, which is related to the probability of finding a particle at a distance r from another particle:[40]

$$S(q) = 1 + 4\pi N \int_0^\infty (g(r) - 1) \frac{\sin(qr)}{qr} r^2 dr \qquad (5.21)$$

Furthermore, in the limit $q \to 0$, $S(0) = Nk_B TK_T$, where k_B is the Boltzmann constant and K_T is the compressibility.[40] Calculation of $g(r)$ involves many-body correlations. The total correlation, $h(r) = g(r) - 1$, is given in terms of the *a priori* unknown direct correlation function, $c(r)$, and indirect correlations by the Ornstein–Zernike (OZ) integral equation.[40] The factorisation of $P(q)$ and $S(q)$ works only at low polydispersities (typically less than 5%) and for spherical particles.[22,40] At higher polydispersities and for anisotropic particles, $P(q)$ and $S(q)$ are strongly coupled. The effective $S(q)$, $S_M(q)$, that can be derived from a scattering experiment is correlated to the scattering amplitude of individual particles.[40] In this case, analytical solutions for $S_M(q)$ exist for relatively simple interaction potentials, such as hard spheres,[41] hard spheres with square-well attraction[42] *etc.* For more complicated interactions, $S_M(q)$ needs to be derived numerically.[40] Figure 5.4a displays the normalised SAXS intensity from an interacting colloidal system of volume fraction ~0.5 composed of sterically stabilised silica particles with a mean radius of 65 nm and a polydispersity of about 8%. The corresponding $S(q)$ is depicted in Figure 5.4b and the $g(r)$ derived from the $S(q)$ model is displayed in Figure 5.4c.

Obtaining real-space information directly from $P(q)$ or $S(q)$ is a mathematically ill-posed problem due to the artefacts introduced by the direct Fourier transform. This problem is circumvented by the IFT method, as mentioned before.[2,43] A more advanced version of this approach, the generalised indirect Fourier transformation (GIFT),[43] allows a model-independent analysis of $P(q)$, but analysis of $S(q)$ still requires a model. Recently, a model-independent IFT approach to deriving $g(r)$ from $S(q)$ that appears to work for different types of system has been proposed.[44]

From the foregoing description, it is clear that data analysis is an essential step in deriving quantitative information from a SAS experiment. A variety of program suites that allow *ab initio* shape analysis for monodisperse systems,[45] shape analysis using analytical models for polydisperse systems,[12,13] determination of size distribution for polydisperse multilevel systems,[14] crystallographic analysis of ordered structures[46] *etc.* are now available. The availability of a large number of instruments at national and international research facilities, together with data-analysis program suites, has made the SAS techniques very accessible to non-expert users.

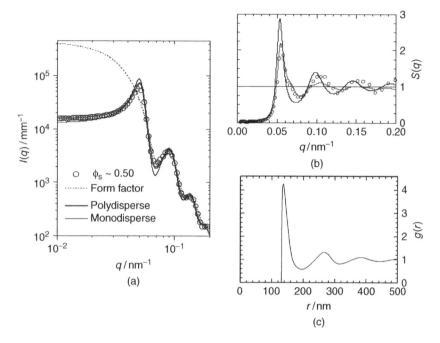

Figure 5.4 (a) Form and structure factors of a colloidal system with hard sphere-like interactions. (b) The effective structure factor, deduced by the division of $I(q)$ by $P(q)$ in (a) and the corresponding monodisperse and polydisperse models. (c) The $g(r)$ derived from the monodisperse model [40].

5.2.9 Highly Ordered Structures

In addition to the fluid-like structures discussed above, many particulate systems organise to form crystalline order.[46] The well-known example is colloidal crystals formed from hard-sphere and charge-stabilised colloids.[47,48] The narrow size distribution of the particles is a prerequisite for crystallisation and the formation of a coherent structure. Usually, spherical systems do not crystallise above a polydispersity of about 6%. SAXS and SANS can elucidate these ordered structures, analogous to crystallography for atomic systems.[15,20] However, synchrotron SAXS has a clear advantage in view of the high q resolution required for these measurements. X-ray or neutron wavelengths are much smaller than the lattice spacing of these crystals. As a result, the Ewald sphere[20] is much larger than the length of the lattice vectors and becomes a plane that intersects many reciprocal lattice vectors simultaneously.[47] Therefore, a large number of diffraction spots can be recorded in each pattern, extending to very high orders. Qualitatively, the positions of Bragg peaks and their relative intensities allow the determination of their crystallographic symmetries. For example, q values of the Bragg peaks are in the ratio $1:2:3:4:5:\ldots$ for a lamellar structure, $\sqrt{3}:2:\sqrt{8}:\sqrt{11}:\sqrt{12}:4:\sqrt{19}:\ldots$ for a face-centred cubic (FCC) lattice, $1:\sqrt{3}:2:\sqrt{7}:3:\sqrt{12}:\sqrt{13}:4:\ldots$ for a hexagonally close-packed layer structure, $1:\sqrt{2}:\sqrt{3}:2:\sqrt{5}:\sqrt{6}:\sqrt{7}:\sqrt{8}:3:\ldots$ for a body-centred cubic (BCC) lattice $etc.$[46]

High-resolution SAXS can reveal further details about the stacking order inside the crystal.[49,50] Figure 5.5 (upper panel) shows the diffraction pattern from a photonic crystal composed of air spheres (spherical voids) of diameter $\sim 600\,nm$ in a titania matrix fabricated by a templated growth.[49] For normal incidence of the X-ray beam, a typical hexagonal pattern can be observed that suggests an FCC structure (beam along the [111] direction). By rotating the crystal along the vertical axis, ω,

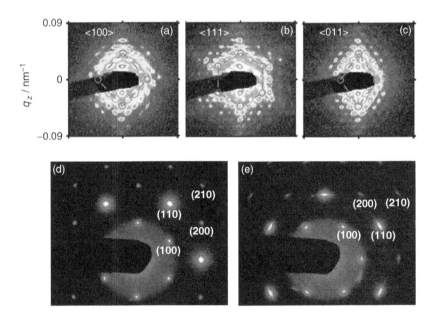

Figure 5.5 SAXS patterns from well-orientated colloidal crystals. Top panel: air spheres in titania with an FCC structure of lattice parameter ca $860\,nm$.[49] Three orientations (a–c), $\omega = -55°$, $0°$ and $35°$, correspond to the $<100>$, $<111>$ and $<011>$ directions, respectively. Reproduced with permission from [49] copyright 2001 American Chemical Society. The lower panel shows the high-resolution diffraction patterns from two colloidal crystals formed from hard-sphere silica colloids for normal incidence (beam along $<111>$), showing the signature of defects in the right image.[50] (d) and (e) Reproduced with permission from [50]. Copyright © IUCr 2006.

other crystallographic planes can be accessed, which confirm the 3D structure of the colloidal crystal. In addition, as the crystal is rotated, stacking faults can be revealed,[48] appearing as Bragg rods connecting the diffraction spots. With very high resolution, such stacking faults can directly be visualised by the azimuthal asymmetry of the Bragg spots.[50] Figure 5.5 (lower panel) shows the diffraction patterns from two different colloidal crystals formed by hard-sphere silica colloids. The left pattern is nearly defect-free, while the right pattern shows stacking faults in the crystal as indicated by the width of the Bragg peaks.[50] These stacking faults need to be eliminated in order to realise the perfect 3D crystal required for photonic applications.[51] High-resolution SAXS provides this information in a non-invasive manner, unlike electron microscopy, which involves specific sample preparation. More quantitatively, these diffraction patterns can be modelled to gain further insight into the periodic structure of such crystalline mesoscale materials.[46]

SANS and SAXS traditionally have been used for the characterisation of ordering in inorganic colloids.[52] A wide range of inorganic colloids have non-spherical shapes, such as platelets,[53] rods[54] and even ribbons.[55] Typical examples are particles of layered materials, which form a variety of liquid crystalline phases; these have been the focus of attention over the last decade.[53−56] They include nickel(II) hydroxide,[53] goethite,[54] vanadium pentoxide,[55] gibbsite,[56] *etc.* These layered material systems have been studied in great detail using high-resolution SAXS, which has led to the identification of many novel liquid crystalline phase transitions as a function of volume fraction and to a deeper understanding of their underlying structures. For example, a magnetic field induced nematic to rectangular columnar phase was observed in goethite nanorod suspensions,[54] while the formation of a novel hexatic phase was revealed in gibbsite suspensions, induced by the polydispersity of platelets.[56] In the latter case, the geometric frustration caused by the size polydispersity of the particles destroyed the long-range translational order and promoted the formation of the hexatic phase, which is characterised by long-range orientational order and short-range translational order.[56] Anisotropic colloidal systems are a much closer realisation of atomic and molecular systems as compared to hard-sphere colloids. Therefore, it is very likely that many new phases will be identified in clay suspensions in the future, and high-resolution SAXS is expected to play a pivotal role in these studies.

Another well-known example of large-scale coherent structures is the vortex lattice in type II superconductors.[57] One of the outstanding applications of SANS has been in the elucidation of vortex lattices.

The length scales associated with vortex lattices are typically much larger than the usual neutron wavelengths, and therefore the diffraction from the corresponding magnetic lattice is best observed in SANS using long sample–detector distances (~ 20 m) and a 2D detector.[58] The classic example is the transition-metal element niobium, in which excess neutron scattering from flux lattice was first demonstrated.[59] Since then, many type II superconductors have been investigated, including high-temperature systems.[58,60,61] Figure 5.6 shows the typical diffraction patterns from vortex lattices for several superconductors.[58] SANS provides information about the vortex lattice structure and alignment,

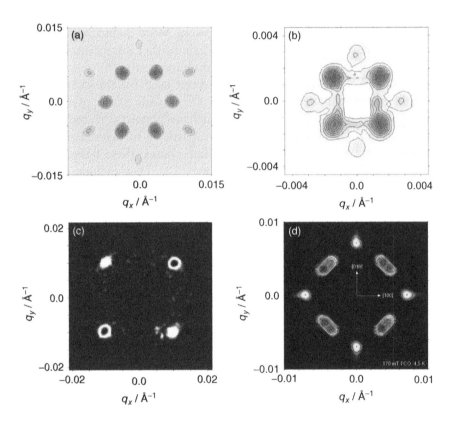

Figure 5.6 Representative SANS patterns of vortex lattices: (a) hexagonal vortex lattice in Nb at 3.2 K and B $\approx$ 0.2 T; (b) square vortex lattice in the unconventional superconductor Sr_2RuO_4 at 50 mK and B $\approx$ 0.025 T; (c) square vortex lattice in the high-temperature superconductor $La_{2-x}Sr_xCuO_4$ at 2 K and B $\approx$ 1.2 T; (d) high-field rhombic vortex lattice in YNi_2B_2C at 4.5 K and B $\approx$ 0.17 T. Reprinted from [58], with permission from Elsevier.

the anisotropy of the superconducting state and the correlation lengths of the vortex lattice positional order.[57] The two characteristic length scales, the magnetic field penetration depth and the superconducting coherence length, which can be determined from the diffraction patterns, are the most important parameters in characterising type II superconductors once their transition temperatures have been established. SANS can also reveal the pinning of a vortex lattice, although the information content in the scattering pattern will consequently be reduced. The shape and orientation of any ordered vortex lattice gives information about the Fermi surface and gap anisotropy. The temperature dependence of the intensity of the diffracted signal is related to the superfluid density and gives an indication of the magnitudes of the superconducting energy gaps, and hence provides information about Cooper pairing. As a result, SANS yields unique information on vortex lattice dynamics and it therefore finds new applications in the study of type II superconductors.[57]

5.3 INSTRUMENTAL SET-UP FOR SAXS

The essential components of a SAXS instrument are a well collimated and monochromatic beam with sufficient photon flux and detectors to measure I_S, I_0 and I_T. In the early days, SAXS experiments were performed using X-ray tubes, and slit collimation was widely employed to compensate for the low intensity of the source.[1,2] Subsequently, rotating anode X-ray generators became available, which provided a higher photon flux, and SAXS instruments more often adopted pinhole collimation.[4] Modern laboratory X-ray tubes, thanks to their advanced design and conditioning optics, provide small beam cross-section and low divergence with flux in excess of 10^8 photons/second. As a result, pinhole collimation has become the standard configuration. The synchrotron is a polychromatic source and therefore requires a monochromator. The desired collimation can be obtained by a combination of focusing elements and slits. The scattered intensity in the forward direction is usually recorded by a 2D detector. The optimum choice of these components and their combination depends on the source properties, desired beam characteristics, available detectors and other resources. In the following, the main components of a synchrotron SAXS instrument are described.

5.3.1   Synchrotron Source

The advent of synchrotron X-ray sources offered new possibilities for SAXS, such as a small beam size and low beam divergence, tunability of λ etc., as well as delivering enhanced photon flux. The quality of the source is expressed in terms of the spectral brilliance or brightness, which is the photon flux per unit phase space volume:[20]

$$B = \frac{\Delta N_{Ph}/\Delta t}{\Delta\Omega_S \,\Delta S\, \Delta E/E} \qquad (5.22)$$

where $\Delta N_{Ph}/\Delta t$ is the number of photons per second, $\Delta\Omega_S$ is the source divergence in $mrad^2$, ΔS is the source area in mm^2 and $\Delta E/E$ is the energy bandwidth in units of 0.1%. The spectral brilliance of modern synchrotron sources is more than 10 orders of magnitude higher than the copper K_α line of a rotating anode source.

In a synchrotron, the charged particles (electrons or positrons) orbit at relativistic speeds (corresponding to energies of several GeV) and the radiation is emitted when these relativistic particles are steered to a curved path. The storage ring consists of bending magnets, which maintain electrons in closed orbits, and straight sections, which accommodate the insertion devices, consisting of periodic magnetic structures. The radiation emitted at the bending magnets is itself a powerful source of X-rays with a continuous spectrum, but the X-ray beams produced by the insertion device (undulator) are much more intense.[20] The periodic magnetic structures of the undulator produce a sinusoidally varying magnetic field in the vertical direction. As a result, the electron beam takes a sinusoidal path in the horizontal direction when passing through the device and the energy lost in the deceleration process is emitted as synchrotron radiation in the tangential direction. In an undulator, the radiation is emitted over a narrow cone, satisfying the condition for interference of emitted radiation from different periods. The coherent addition of the intensities in the far field leads to peaks in intensity at energies corresponding to the fundamental and its harmonics. The peak energy of the undulator spectrum can be varied by changing the strength of the magnetic field, i.e. the magnetic gap (typically 5–20 mm). When the gap is decreased, the spectrum moves to lower energies, but with higher peak intensity. The beam line utilises either the central cone of the radiation of the fundamental or an odd harmonic that is usually more intense and sharper.

The outstanding properties of modern synchrotron sources of the third generation directly stem from the special characteristics of the undulator radiation. The source is extremely bright, because of the small size and low divergence (small cone of emission), and the intensity is peaked at certain energies (due to the interference).[20] By now there are many third-generation sources worldwide, of which the foremost are the European Synchrotron Radiation Facility (ESRF), Grenoble, France, the Advanced Photon Source (APS) at Argonne, IL, USA, the Super Photon Ring – 8 (SPRing8), Harima, Japan and Petra-III, Hamburg, Germany. Fourth-generation X-ray sources based on free-electron lasers are now available (e.g. Linac Coherent Light Source, Stanford, CA, USA), but they are probably too exorbitantly sophisticated for conventional SAXS experiments.

The high brightness of third-generation sources is exploited in different SAXS techniques. For example, the high flux and low divergence of the beam allow real-time studies with high time and q resolution. Typically, millisecond-range time-resolved experiments can be performed even on weakly scattering samples. As a result, structural dynamics and *in situ* processes can be studied with high time resolution.[9] The small source size and high flux further permit focusing of the beam to micrometre and smaller sizes *via* highly demagnifying optics.[62] This is a powerful approach for performing spatially resolved scanning SAXS experiments.[63] Future X-ray sources will have beam divergences and sizes in the range of a few microradians and micrometres, respectively, which will enable micro-SAXS experiments in the standard pinhole configuration. Ultimately, the beam size should be sufficiently greater than the largest structural scale probed to avoid the finite sampling effect.

The above developments equally benefit wide-angle X-ray scattering (WAXS) and surface-sensitive grazing incidence SAXS technique (GISAXS).[64]

5.3.2 X-Ray Optics

In order to take advantage of the high brilliance of the source for a given application, the beam has first to be conditioned using appropriate optical elements.[20,65] The full undulator harmonic is polychromatic and a monochromatic beam is usually obtained by a crystal monochromator. This crystal should be able to withstand the high heat load that results from the absorbed power of the full radiation. The most commonly used monochromator at undulator sources is liquid nitrogen cooled channel-cut or double-crystal silicon-111 ($\Delta\lambda/\lambda \approx 0.014\%$)

used in Bragg geometry (reflection). In addition, curved crystals can be machined, which are suitable for both meridional (in-plane) and sagittal (out-of-plane) focusing of the beam.

A focusing mirror is an essential component of SAXS instrumentation at synchrotron beam lines. First, it serves as an efficient cut-off filter for the higher harmonics reflected by the crystal monochromator.[65] Typically, the sample is placed 30–60 m from the source, and the sample–detector distance can be another 10 m. X-ray mirrors are usually made of a highly polished light material, such as silicon or Zerodur glass ceramic. The reflecting surface is coated (thickness ≈ a few tens of nanometres) by a heavy metal such as rhodium, platinum or gold in order to increase the electron density and thereby increase the critical angle ($\theta_C \propto \lambda \rho^{1/2}$) and reduce the length of the mirror.[65] For low glancing angles (3–4 mrad), the reflectivity is close to 100% at low energies, and it decreases sharply to zero as the energy corresponding to θ_C is crossed, cutting off the high-energy part of the spectrum. Further, the surface roughness and slope errors ideally should be less than 1 Å and 0.1 μrad, respectively. A simple option is a double-focusing elliptical or torroidal mirror fabricated with high precision, which can provide a fixed focal spot. Variable focusing can be achieved by using cylindrically bent mirrors with mechanical benders, which allow the curvature to be tuned. Recently, bimorph mirrors, which are locally deformable, have become available. These devices permit the focus to be varied and aberrations to be minimised.

For SAXS, a straightforward approach is to use a highly parallel beam (of size ≈ 100 μm), which facilitates easy change of the sample-to-detector distance and yields high q resolution. This can be realised by a 1 : 1 focusing or a slightly magnifying optics. An alternative to a parallel beam is to focus the beam on the beamstop, which leads to a slightly higher q resolution, due to the lower effective beam divergence.[4] The beam needs to be refocused with a change in the sample-to-detector distance and the spot size on the sample (scattering volume) varies inadvertently. An alternative to mirror focusing is compound refractive lenses, which are based on the principle that the refractive index of materials is slightly smaller than one ($1-\delta$) at X-ray wavelengths.[63] The focusing is achieved by a set of biconcave parabolic lenses made of beryllium or aluminium. The focal distance can be varied by changing the number of lenses ($f = CR/2n\delta$, where CR is the radius of the curvature and n is the number of lenses). Owing to their low cost and aberration-free focusing, compound refractive lenses are an attractive option.[63]

5.3.3   X-Ray Detectors

A matching detector for the outstanding properties of an undulator source and advanced X-ray optics has been a major bottleneck for SAXS experiments requiring very high time and q resolution.[9] The most commonly implemented SAXS set-up at synchrotron facilities involves pinhole collimation with a 2D detector, which is convenient for recording anisotropic diffraction/scattering patterns and obtaining high-intensity statistics in the case of isotropic scattering. Ideally, the detector should cover a large dynamic range in intensity ($> 10^6$) and q ($q_{MAX}/q_{MIN} \approx 100$). The single-photon signal should be above the noise floor of the detector. Traditionally, gas-filled multiwire proportional chambers (MWPCs) served as the workhorse SAXS detector.[66] However, such detectors have low count-rate capabilities and poor spatial resolution. Recently, the development of pixel-array detectors based on thickly depleted semiconductors has overcome these barriers.[67] In a semiconductor material such as silicon, a given X-ray photon produces as many as an order of magnitude more electron–hole pairs (3.6 eV/pair) than in a gas (~ 25 eV). This, coupled with the high absorption, makes them a very efficient detector. Each pixel has its own charge collection mechanism, amplifier and discriminator, with high count rates ($> 10^6$ counts per second). Advances in microfabrication techniques permit the fabrication of defect-free modules of such pixel-array detectors of reasonable size (many centimetres) and the construction of large-area detectors by the tiling of many such modules (*e.g.* the PILATUS detector from DECTRIS).[67] New-generation pixel-array detectors with individual pixel sizes of the order of 50 μm are under development.

For experiments requiring high spatial resolution, a detector based on large-area CCD sensors is an alternative to the use of pixel-array detectors. The CCD is coupled to a conversion phosphor that converts X-ray photons to visible photons by means of a tapered fibreoptic bundle. Typically, the detecting phosphor is larger than the CCD area and the effective gain (or sensitivity) decreases as the inverse square of the demagnification. With the availability of large CCD arrays (50×50 mm) with high quantum efficiency ($> 60\%$), it is now possible to achieve single-photon sensitivity with this type of detector (*e.g.* FReLoN 4M detector at the ESRF). In general, CCD-based detectors are limited not by the count rate but by the charge capacity of the pixels (several hundred thousand electrons), and they provide good spatial resolution (< 100 μm). The limitations are the time required to readout the CCD, the readout noise and the degradation of quantum efficiency and resolution at low signal levels. A slow-scan CCD must be shielded from ambient light and read out in dark, which is achieved by means of a fast electromechanical beam shutter that opens only during the acquisition period. Furthermore, the phosphor decay should be fast (*ca* a few milliseconds) and the CCD should be cooled (usually thermoelectrically) to lower the dark current. High-performance CCD detectors (split-frame transfer) with frame rates of 100 images per second, pixel array sizes of 1024×1024 and nominal dynamic ranges of 16 bits have recently become available. For isotropic scattering samples, an effective dynamic range of 10^5 can be obtained by azimuthal integration.[9]

Quantitative SAXS experiments also require precise measurements of incident and transmitted intensities. These absolute flux measurements can be made by reverse-biased, highly depleted silicon *p*-*i*-*n* diodes and appropriate photocurrent amplifiers. Avalanche photodiodes are used as point detectors when a high-intensity dynamic ($\sim 10^7$) is required.

5.3.4   SAXS Instrument Layout

Figure 5.7 shows the configuration of a pinhole SAXS instrument at a synchrotron beam line.[9] To avoid absorption and scattering by air, the entire path of the beam (except the sample section) is under vacuum. In practice, several guard slits are used to curtail the low-*q* parasitic scattering from the optical components. The intense parasitic background close to the primary beam has to be blocked by the beamstop to avoid undesirable saturation of the detector. Even for a highly optimised pinhole SAXS camera, the parasitic instrument background dominates

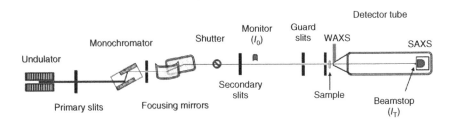

Figure 5.7 Schematic layout of a SAXS beam line at a synchrotron source[9]. The entire beam path except the sample is in vacuum. The SAXS detector is inside a wagon in the evacuated flight tube and the WAXS detector is outside the vacuum. The flux at the sample position can be of the order of 10^{14} photons per second.

over solvent scattering (*e.g.* 1 mm of water, $I(q) \approx 1.6 \times 10^{-3}$) for $q < 0.05\,\text{nm}^{-1}$.[9] For low scattering samples, it is highly desirable to avoid additional windows and place the sample (or the sample cell) in vacuum whenever possible. The measurement of I_T is critical for normalising the measured I_S and can be accomplished by a small detector embedded in the beamstop, as shown in Figure 5.7. Having a WAXS detector allows simultaneous SAXS and WAXS measurement. Typical energy used for SAXS measurements is around 12.4 keV, corresponding to 1 Å X-ray wavelength. Higher energy up to 20 keV is used for strongly absorbing samples. The sample–detector distances may be varied up to 10 m, reaching a $q_{\text{MIN}} \approx 0.006\,\text{nm}^{-1}$(for $\lambda \approx 1$ Å).

5.4   INSTRUMENTAL SET-UP FOR SANS

SANS instruments have similar requirements to those for SAXS, such as a monochromatised and well-collimated neutron beam directed at the sample position and one or more multi-element detectors to record the scattered neutrons in the forward direction. One of the interesting properties of the neutrons is that their velocity is small and the scattered neutrons from the sample arrive at the detector at different times, depending upon their wavelength.[8] This allows time-of-flight (ToF) measurements that can extend the range of the momentum transfer beyond what is defined by the scattering geometry for a fixed wavelength:

$$q = \frac{4\pi}{\lambda} \sin(\theta/2) = \frac{4\pi m_{\text{N}}}{h t_{\text{a}}} L_{\text{a}} \sin(\theta/2) \qquad (5.23)$$

where m_{N} is the neutron mass ($\approx 1.68 \times 10^{-27}$ kg), h is the Planck constant, t_{a} is the arrival time of neutrons at the detector from the source and L_{a} is the corresponding distance or flight path length. Of course, this can also introduce undesired inelastic effects in the SANS measurements, which will be neglected in the following description. SANS experiments can only be performed at large-scale instrument facilities, either at a research nuclear reactor[68] or at an accelerator-based spallation neutron source.[69] The flux of cold neutrons at the sample position is the main constraint for even the most advanced SANS instruments (typically 10^7–10^8 neutrons per square centimetre per second with $\Delta\lambda/\lambda \approx$ 10%).[70] Therefore, the instrumentation must be optimised to maximise the use of available neutron flux. Typically, wavelength spreads of the order of 10–20% and beam sizes of 5–10 mm are used, which make resolution corrections mandatory.

5.4.1   Neutron Sources

As mentioned above, two types of neutron source exist: reactor and spallation sources.[18] For SANS, both sources provide comparable performance;[8,71] the main difference is that reactor sources provide a continuous flux of neutrons, while spallation sources produce neutron pulses with repetition rates of several tens of hertz. The vast majority of neutron sources are based on reactors that use enriched uranium ^{235}U as their fuel. The fission reaction produces fast neutrons with energy of the order of 1–10 MeV, which must be slowed to typical thermal energies (*ca* 25–30 meV) by moderators (water or heavy water).[17,18] At these energies, the neutron velocities (v_N) are in the range of 2300 ms^{-1}, corresponding to a wavelength of 1.6 Å ($\lambda = h/m_N v_N$). However, in SANS studies involving large-scale structures it is desirable to use even longer wavelengths. This is achieved by inserting cooled moderators filled with liquid cryogens (*e.g.* deuterium, methane *etc.*) into the neutron beam path outside the reactor core. These cold sources move the peak of the broad distribution of the wavelength to about 4 – 6 Å. A further cleaning up of short-wavelength neutrons and high-energy radiation is obtained by locating the instrument removed from the reactor source by means of neutron guides. Leading high-neutron flux reactors are the Institut Laue Langevin in Grenoble, France, the National Institute of Standards and Technology, Center for Neutron Research (NCNR) in Gaithersburg, MD, USA, the High Flux Isotope Reactor at Oak Ridge National Laboratory, Oak Ridge, TN, USA, *etc.*[6] The reactor power is limited due to thermal transport through the core and associated safety concerns, setting an upper limit for the thermal neutrons that can be obtained from a reactor source.

Spallation neutron sources are based on high-energy accelerators; typically, protons are accelerated to energies close to 1 GeV and compressed into *ca* 100 ns bunches with repetition rates of 20–50 Hz, which are then directed to a spallation target (*e.g.* tantalum, tungsten, liquid mercury *etc.*).[18,69,71] Each proton that hits the target nuclei knocks out (spalls) several tens of neutrons, which are also bunched with the same repetition rate. These spalled neutrons, in pulsed form, are passed through moderators and then guided to different instruments. Because of the distribution of neutron velocities, the widths of the pulses broaden significantly as they travel along the beam line.[71] As the proton current (number of protons per bunch) is theoretically unlimited, the neutron flux from spallation sources is unbounded, so long as the target can withstand the power density.[18] The ISIS facility near Oxford, UK operates with a proton energy of 0.8 GeV and 50 Hz repetition, reaching a proton current of 0.2 mA. The most advanced Spallation Neutron Source (SNS) at Oak Ridge National Laboratory in Oak Ridge, TN, USA employs a proton current of 2 mA at 1 GeV with a repetition rate of 60 Hz, targeted on liquid mercury, and produces peak neutron fluxes of the order of 10^{17} neutrons per square centimetre per second. A similar neutron flux is expected from the Japan Proton Accelerator Research Complex (J-PARC) facility in Tokai, Japan, which operates with 3 GeV protons at 0.33 mA current. The upcoming European Spallation Source in Lund, Sweden, aims to reach 30 times greater neutron flux using long neutron pulses ($\approx$ 3 ms).

5.4.2 Neutron Optics

The neutron beam from a reactor source requires monochromatisation prior to use in steady-state SANS experiments.[68] Conventional crystal monochromators, as for X-rays, are rarely used in SANS except on ultra small-angle scattering (USAS) instruments.[6] Instead, pyrolytic graphite can be used, which gives a wavelength spread of about 6%.[8] More conventional monochromators are based on velocity selectors that use aircraft turbine-like devices (*e.g.* DORNIER). In this case, the mean wavelength and its bandwidth are controlled by the angular velocity and the blade angle of the turbine, respectively. This provides sufficient flexibility to vary the bandwidth (5–30%). SANS instruments at spallation sources usually operate in ToF mode, and only weak monochromatisation is required (*e.g.* LOQ at ISIS, UK), but the corresponding SANS data reduction is more complex.[69] The ToF SANS measurements at a reactor source can be performed using a mechanical chopper, which gives access to a broad q range but has the disadvantage of throwing away some neutrons.

SANS resolution is also critically determined by the collimation scheme. Collimators are based on a set of pinholes of varying sizes and neutron guides of varying lengths.[72] Neutron guides are based on either glass channels coated with nickel or supermirrors reflecting neutrons above a critical wavelength. The effective collimation or angular divergence can be adjusted by selecting the appropriate length of the collimator and size of the pinholes. More sophisticated collimation systems can be designed on the basis of multiple pinholes, but there are certain technical challenges to be resolved. Focusing is not generally used on SANS instruments but there are proposals (*e.g.* the NG7 instrument at NCNR, USA) to use compound refractive lenses, since the cold neutron refractive index for most materials is only slightly less than unity, as in the case of X-rays.[20] With polarised neutrons, magnetic lenses could be used as either convergent or divergent lenses. The polarisation is usually obtained by means of multilayer mirrors made of magnetic materials[18] and ^{3}He polarised filters act as analysers.

5.4.3   Neutron Detectors

Traditional SANS detectors are based on gas-filled MWPCs filled with ^{3}He as the ionising medium and CF_4 or CH_4-Ar as the quenching gas.[8] Owing to the rarity of ^{3}He, gas-filled detectors have also been designed to use boron trifluoride instead. The neutron absorption by ^{3}He leads to fission reaction releasing one tritium and a proton with 760 keV energy causing primary ionisation of the gas. The anode is made up of a grid of fine wires kept at high voltage (3–5 kV) and the 2D position sensitivity is obtained by two layers of cathode strips on either side of the anode perpendicular to each other (X and Y). The primary proton and tritium nuclei produce a trace of charged atoms and electrons. The electrons move towards the anode, and near the anode electrons are accelerated to produce an avalanche, which also simultaneously induces charges on the cathode strips. The position encoding is realised by introducing a certain fixed delay between successive cathode strips (delay line). The anode signal indicates the arrival of the neutron, and from the subsequent delayed signals from the X and Y grids, the location of the event is computed and histogrammed. Modern SANS detectors have an area of about 1 m^2, with 128 × 128 pixels. However, the total count rate on this type of MWPC is limited to about 100 kHz due to the slow decay of the ionised gas and the dead time effect from the counting electronics.[8,10] As a result, a new generation of SANS detector has been developed using a stack of linear proportional counters (128) in the form of tubes of diameter *ca* 8 mm and length *ca* 1 m, each of which has a central anode wire and independent readout electronics.[10] This has allowed the total count rate over an area of 1 m² to be increased to the order of 2 MHz, instead of 100 kHz.[73] In this case, the dead time correction will be required only for the tube receiving a very high number of neutrons, without influencing the weak signals on other tubes.

The next generation of neutron detectors based on scintillators coupled to light collection (*e.g.* CCD) are in development; these provide not only high spatial resolution but also a higher-intensity dynamic range.[74] However, such integrating detectors will be less suitable for ToF measurements.

5.4.4 SANS Instrument Layout

Figure 5.8 shows the layout of an advanced SANS instrument (D11, ILL, France), including the principal components, velocity selector, collimators and detector tube.[72] The large detector (1 m²) is enclosed in the evacuated flight tube. The beam cross-section on the sample is in the range of 5–10 mm. A beamstop in front of the detector blocks the primary beam of neutrons. The sample–detector distances may be varied up to 40 m. Typical SANS experiments use neutron wavelengths in the range of 5 Å with 10% spread. This allows a nominal $q_{\mathrm{MIN}} \approx 10^{-2}\,\mathrm{nm}^{-1}$ to be reached, though the q resolution is limited by the large beam size, detector pixel element and wavelength spread. Because of the high transmission of neutrons by most materials, it is easier to find low-absorbing and weakly scattering windows.

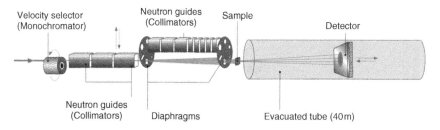

Figure 5.8 Schematic layout of an advanced SANS instrument (D11, ILL, France), illustrating the principal components.[72] The neutron flux at the sample is of the order of 10^8 neutrons per second, with a wavelength spread of 10%. Figure provided by P. Lindner Copyright (2009) ILL.

5.5   ADDITIONAL REQUIREMENTS FOR SAS

5.5.1   Combination with Wide-Angle Scattering

Powder diffraction is a well established technique for the structural elucidation of inorganic materials.[75] A combination of SAS and wide-angle scattering (especially WAXS) is often valuable when investigating hierarchically structured materials. Many semicrystalline materials, such as minerals, hybrid organic–inorganic materials, polymers *etc.*, possess ordering both in the nanoscale and at the molecular level. Precise knowledge of the structure at these different length scales is often important for their practical exploitation. In addition, for a correct interpretation of the SAS data, it may become essential to have some information on the structure at the molecular level. Figure 5.9 depicts the practical implementation of a combined SAXS and WAXS set-up. The q range in WAXS covers typically $6–60 \, nm^{-1}$ and the combined SAXS and WAXS can span over four orders of magnitude in q. In this scheme, the WAXS detector is placed close to the sample (at about 100 mm), which allows the majority of the SAXS pattern to pass unhindered (e.g. the SAXS/WAXS instrument at ID02, ESRF, France). A high-resolution 2D detector is preferred in studies of anisotropic (orientated) samples and in order to improve the intensity statistics for powder and noncrystalline materials. The inset in Figure 5.9 shows a special design of fibreoptically coupled CCD detector for recording simultaneous WAXS. It consists of a large-area CCD sensor and obliquely cut fibreoptics. The wide angle option is not commonly found on SANS instruments, except in a few cases (*e.g.* LOQ at ISIS). Modern SANS instruments also propose

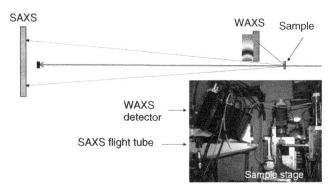

Figure 5.9   Schematic layout of a combined SAXS/WAXS set-up. The SAXS configuration is similar to that in Figure 5.1, to which a WAXS detector is appended. The inset depicts the implementation of a SAXS/WAXS configuration at ID02 Beamline, ESRF, France.

to implement wide-angle detectors simultaneously with small-angle measurements.[70]

For quantitative analysis of WAXS data, additional corrections are necessary, depending on the orientation of the detector, polarisation and the Lorentz factor. The Lorentz term is a kinematic correction resulting from the higher weight of crystallites contributing to intensity at lower Bragg angles.[7,15] For randomly orientated crystallites with the scattering geometry shown in Figure 5.9, the Lorentz factor is roughly $1/[\sin\theta \, \sin(\theta/2)]$, and this correction is usually applied by multiplying $I(q)$ by $s^2[= (q/2\pi)^2]$. The corrected data can be used to evaluate the crystallinity as represented in Figure 5.10. The ratio of the integrated intensities within the crystalline reflections to the total intensity is a measure of the crystallinity within the sample. In the case of neutrons, the incoherent background is an issue at wide angles and usually requires a strong diffracted signal in order to make accurate measurements.

The resolution-corrected full width at half maximum (FWHM), $\Delta\theta_B$, of the Bragg peak is related to the average size of the crystalline domains (l_C) by the Scherrer formula:[15]

$$\Delta\theta_B = K_{1/2}\lambda/[l_C \cos(\theta_B/2)] \text{ or } \Delta q_B = 2\pi K_{1/2}/l_C \qquad (5.24)$$

where $K_{1/2}$ is the Scherrer constant, which depends on the shape of the crystallites ($K_{1/2} \approx 0.94$ and 1.11 for cubic and spherical shapes,

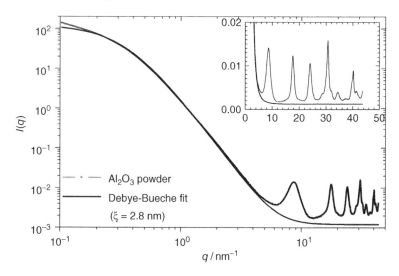

Figure 5.10 SAXS/WAXS pattern from alumina powder with nanometre-sized grains. The SAXS region can be described by the Debye–Bueche function and the low-q region ($q < 0.1 \, \text{nm}^{-1}$) is dominated by scattering from the intergrain voids. The inset shows the WAXS diffraction peaks of nearly equal widths, broadened by the finite size of the grains.

respectively).[76] In general, good agreement can be found between particle sizes derived from the Scherrer relation and SAXS analysis.[77] In Figure 5.10, the Scherrer formula gives $l_C \approx 4.8$ nm, while the Debye–Bueche fit provides $R_M \approx 2.54$ nm or size 5.1 nm (ξ is related to R_M through PDI ≈ 10),[30] which is in reasonable agreement considering that the l_C value is not unique along all directions.

5.5.2 Instrumental Smearing Effects

For SANS, the instrument resolution is an important factor to consider prior to quantitative analysis of the data. This effect originates from both the angular or collimation effect and wavelength spread. The combined resolution is expressed as:[78]

$$\Delta q(\theta) \approx (2\pi/\lambda)\cos(\theta/2)\Delta\theta + q\Delta\lambda/\lambda \qquad (5.25)$$

The collimation effect, which is nearly q-independent, becomes important at low q values, while the influence of wavelength spread is significant at high q ranges and is dependent on the q value. The wavelength smearing term ($q\Delta\lambda/\lambda$) is not significant in SAXS because most often a crystal monochromator is employed, e.g. for Si(111), $\Delta\lambda/\lambda \approx 1.5 \times 10^{-4}$. Therefore, the combined resolution is determined by the beam size and divergence and by the spatial resolution of the detector.[78] When the beam is focused at the detector, smearing due to beam size and divergence becomes less significant and the detector spatial resolution is the dominant contribution. The combined effect of beam size and detector resolution can be obtained by measuring the beam profile directly on the detector with high dynamic range of intensity. The resolution functions are approximately Gaussian and the measured intensity is a convolution of $d\Sigma/d\Omega$ by the combined Gaussian function, $R(q)$:[78]

$$I(q) = \int R(q)\,\frac{d\Sigma}{d\Omega}\,dq \qquad (5.26)$$

In synchrotron SAXS, the deconvolution of the measured $I(q)$ is not often essential when dealing with fluid-like or particulate systems of size ca 100 nm and polydispersity in excess of 5%. However, for larger particles and lower polydispersities, the resolution effect is often manifested by significant smearing of the low q minima in the form factor. In the analysis of crystalline diffraction, the resolution function needs to be incorporated. In general, convoluting the model by the resolution function is numerically more secure when fitting the measured $I(q)$.

5.5.3 Sample Environments

The appropriate selection of sample thickness is crucial for accurate SAS measurements. With inorganic materials, strong absorption is an issue with X-rays, while multiple scattering can become a problem with neutrons.[10] In order to obtain the optimum scattered intensity, the thickness needs to be chosen such that $T_r \approx 1/e (\approx 0.37)$, corresponding to $l_S \approx 1/\mu_l$. However, it may not be possible to use this optimum thickness in all cases, due to multiple scattering at low angles (especially for neutrons) and loss of angular resolution at wide angles. Varying the wavelength can be an option for this optimisation. Liquid samples require a cell, and in this case the absorption and scattering by the empty sample cell should be as low as possible. The high penetration power of neutrons makes the choice of sample cell relatively easy, and optical cells with quartz windows are routinely used.[10]

For X-rays, sample cells with small wall thickness ($\sim$ 10 µm) and made of low-density materials, but which at the same time provide sufficient uniformity and mechanical stability, are mandatory. Thin-walled glass or quartz capillaries are commonly used as containers for SAXS measurements with low-viscosity samples. For high-viscosity samples, cells with thin, flat windows made of mica, Mylar or Kapton are preferred. Mica windows (thickness $\approx$ 10–15 µm) are usually found to be better for low-angle SAXS but their crystalline diffraction is a nuisance in WAXS measurements. For low-scattering samples, accurate background subtraction is delicate because of the variations in scattering of the cell wall, as well as the thickness of the sample at different positions. These uncertainties in background subtraction can be eliminated by using a flow-through cell (*e.g.* a quartz capillary of wall thickness *ca* 10 µm and diameter 2 mm) installed in vacuum. This set-up allows the background and sample scattering to be measured at the same position.

The accessories used to contain and manipulate the sample *in situ* are collectively termed the 'sample environment', which is an essential part of a SAS experiment. For example, a sample environment is needed in static studies to allow a simple change of thermodynamic variables such as temperature, pressure, pH or relative humidity and of external fields such as magnetic and electric fields. In addition, for SANS experiments involving magnetic systems, high magnetic field and cryostats are routinely used. An automatic sample changer is often deployed to enhance the turnover rate in static measurements. Combining scattering techniques with other thermophysical or mechanical probes is sometimes crucial to investigating the transient microstructure of composite materials. Examples include mechanical testing devices used to apply stress or strain on the sample, shearing cells for fluid-like samples and even calorimetry for the monitoring of phase changes.

Precise initiation and synchronisation of a kinetic process in time-resolved experiments is achieved by means of an appropriate sample environment. Examples include stopped-flow rapid-mixing devices, temperature and pressure jumps and high-voltage electric fields. A key advantage of the SAS technique is that it allows a wide variety of sample environments to be combined for *in situ* studies.

5.6 APPLICATION OF SAS METHODS

This section describes some selected applications of SAS methods in the investigation of inorganic materials. Some of the pioneering SAS applications[79] probed phase separation and precipitation in alloys, magnetism in alloys, defects in alloys and semiconductors, among other things. Key results of these early studies include observations of the scaling behaviour of the structure function in phase-separating alloys, the growth of Guinier–Preston zones *etc.*[4,79] Traditionally, SANS has been widely used for a large variety of *in situ* studies, as it has the advantage of high penetration capability, which makes it easier to find low-absorbing windows and heat shield materials for experimental set-ups involving high temperature and pressure.[37] Another example is the study of the growth of semiconductor crystals in a glass matrix, which has revealed a depletion zone[80] and a scaling of the structure factor different from the conventional spinodal decomposition.[81] Many of these solid-state transformation and growth processes occur on slower timescales of hours and therefore the time resolution is not an issue. SAS techniques reveal the formation of well-defined morphologies and underlying growth mechanisms.[80,82]

Both SANS and SAXS have been widely used for the investigation of porous materials, such as mesoporous silicas, activated carbons *etc.*, and the results have provided deeper insights into their nanometre- to micron-scale pore morphologies.[19,83,84] In general, these results are in agreement with other techniques such as gas adsorption and nuclear magnetic resonance (NMR) spectroscopy, but the important difference is that SAS methods probe both open and closed pores, which makes the information obtained unique.[83,84] The contrast provided by the neutrons even allows monitoring of gas adsorption in the pores[85] and investigation of supercritical systems in porous media.[86] *In situ* X-ray scattering techniques have been the key to structural understanding of the formation of nanoporous aluminosilicate and aluminophosphate materials pertinent to catalysis.[87] In the case of more ordered pore structures such as silicate molecular sieve (MCM-41), the contrast variation SANS revealed the channel structure in terms of its straightness and surface roughness.[88]

In the following, some complementary aspects of SAS techniques are illustrated. The methods described are used for bulk or surface studies and extend the range of q, enhance resolution in q and time and allow contrast variation. Third-generation synchrotron sources and high-flux reactors have played key roles in the development of these complementary SAS techniques.

5.6.1 Real-Time and *In Situ* Studies

An important outcome of the developments at large-instrument facilities is the ability to perform time-resolved experiments that can follow chemical processes in real time. Typically, a chemical reaction is induced by rapid mixing of the reactants or a sudden change of temperature or pressure. In addition, many physical processes can be initiated by applying external fields such as electric or magnetic fields. Time-resolved experiments can be performed in two ways: (i) real-time experiments, wherein the scattering is followed continuously as a function of time; and (ii) stroboscopic experiments in which a certain time window is chosen and the experiment is repeated to accumulate the required statistics. In the latter case, the time course is followed by shifting of the time window, relying on the precise synchronisation and temporal reproducibility of the process. The synchronisation of the physical process is achieved by the sample environment. The effective time resolution is determined by the speed of the data acquisition, as well as the ability to synchronise the phenomenon (timescale over which the entire sample behaves like a single entity) with the imposed condition. In principle, real-time studies can be performed by either X-rays or neutrons, but the latter choice is often limited by the available neutron flux when fast kinetics are involved.[9,10] As a result, time-resolved scattering experiments are still developing at neutron sources[6] but have become more mature at synchrotron facilities.[89] Time-resolved SANS experiments with subsecond resolution usually involve multiple repetition cycles in order to accumulate sufficient intensity statistics. Stopped-flow mixing is a widely used technique for triggering reaction kinetics in the millisecond range.[89] A straightforward application is in the study of the nucleation and growth of particles in solution.[90,91] Examples include the synthesis of nanoparticles of silica, calcium carbonate, gold *etc.*

5.6.1.1 *Precipitation of Calcium Carbonate*

The formation of calcium carbonate from supersaturated solutions has been studied for more than a century. In recent decades, it has received renewed interest from the point of view of biomineralisation.[92,93] In particular, amorphous calcium carbonate has emerged as a precursor to the formation of more stable crystalline forms.[92,94] The detailed mechanism of phase transformation leading from the initial amorphous material to the final thermodynamically stable crystalline form (calcite) has been a topic of considerable interest. The reaction can be induced by mixing aqueous solutions of Na_2CO_3 and $CaCl_2$ above pH 10.[92] Time-resolved combined SAXS and WAXS is a good method for probing the underlying mechanism, as illustrated below. Figure 5.11(a) shows the typical time evolution of SAXS intensity following the rapid mixing of the reactants in the stopped-flow device. Each of these time-resolved scattering curves (acquisition time ≈ 20 ms) after subtraction of the water background can be described by a polydisperse sphere scattering function. Figure 5.11(b) shows the typical fits and the inset shows the main parameters derived from this analysis. For a reactant concentration of 4.5 mM, particles are nucleated within 100 milliseconds following the rapid mixing; these grow with time and reach their final size over a few seconds, following Avrami-type kinetics.[95] The analysis further reveals that the number density of particles remains constant throughout the growth, while the polydispersity decreases with time to reach a terminal value. From the deduced number density and mean particle volume, and the absolute scattered SAXS intensity, it turns out that the density of $CaCO_3$ within the particles is roughly 1600 kg m^{-3}, which is significantly lower than the known densities for the stable crystalline forms calcite, aragonite and vaterite.[92] This result, combined with the lack of any crystalline diffraction in the WAXS, directly demonstrates that the particles generated by the reaction of Ca^{2+} and CO_3^{2-} are amorphous.[94]

Figure 5.12 further illustrates that these amorphous $CaCO_3$ particles are metastable and transform to crystalline forms through the

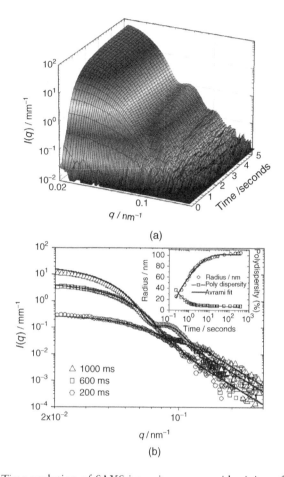

(a)

(b)

Figure 5.11 Time evolution of SAXS intensity upon rapid mixing of initial 9 mM aqueous solutions of Na_2CO_3 and $CaCl_2$ at *ca* pH 10.5. The lower panel shows typical modelling of the time slices in the upper panel. Particles of mean radius *ca* 20 nm are formed within about 200 ms and then grow over 10 seconds to reach a final radius of about 100 nm. Concomitantly, the polydispersity decreases from *ca* 40 *to* 6%. The evolution of mean radius is described by $R(t) = 101.4\{1 - \exp[-(t - 0.15)^{0.5}]\}$.

dissolution and subsequent heterogeneous nucleation of the crystalline modification in the reacting mixture.[94] Typical features of SAXS from crystalline facets[37] and Bragg reflections in WAXS simultaneously appear. The crystalline modifications thus generated can be identified (calcite, vaterite or aragonite, depending on the initial reagent concentration) without any solid–solid transformation. Thus the precipitation of $CaCO_3$ is a prototypical case of the Ostwald rule of

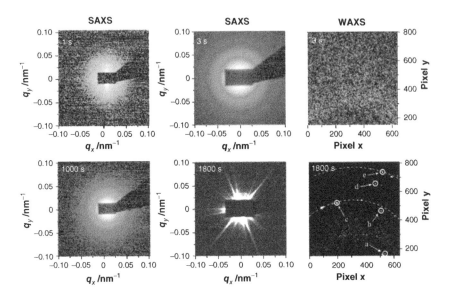

Figure 5.12 Typical SAXS and WAXS patterns during the formation of amorphous CaCO$_3$ particles, their dissolution (after 15 minutes) and the formation of crystalline forms in solution.[94] Typical SAXS patterns from crystalline facets and the simultaneous appearance of Bragg reflections in the WAXS confirm the existence of crystalline modifications after 30 minutes. Reprinted with permission from [94] Copyright (2003) American Chemical Society.

stages. Stopped-flow mixing combined with SAXS/WAXS techniques is therefore well suited to following the mineralisation from aqueous solutions in great detail.[94] Somewhat similar behaviour has been observed in the hydrothermal crystallisation of titania (TiO$_2$). SAXS and WAXS reveal that the anatase form nucleates first and grows to a critical size, and then the rutile form nucleates and grows while the anatase form dissolves.[77]

5.6.1.2 Nucleation and Growth of Gold Nanoparticles

Another interesting example is the nucleation and growth of gold nanoparticles through the reduction of Au(III) to Au(I) or Au(0).[96] This chemical reaction prior to the onset of nucleation can be probed by means of ultraviolet (UV)/visible spectroscopy and the appearance of gold nanoparticles is manifested by a plasmon band around 544 nm. An established synthesis route is *via* the reduction of borohydride salt of gold solubilised in toluene by a cationic surfactant in the presence of an excess of alkyl derivative ligands, either alkylamine or alkanoic acid (both decyl in this case).[96] Time-resolved SAXS allows the nucleation following the initial conversion of Au(III) to be observed. Figure 5.13 shows the typical evolution of the SAXS intensity for the decanoic acid route, while for the decylamine case the plot qualitatively resembles the initial part of growth.[97] The important quantitative difference however is that the nucleation rate and terminal particle size are strongly dependent on the type of ligand used. For a decanoic acid ligand, the nucleation phase is about 1 second, followed by growth with the rate limited by the reaction of the monomers at the interface, resulting in a mean particle radius of about 3.7 nm. On the other hand, for decylamine the nucleation rate is increased by an order of magnitude, thereby quenching the growth through a lack of monomers, yielding

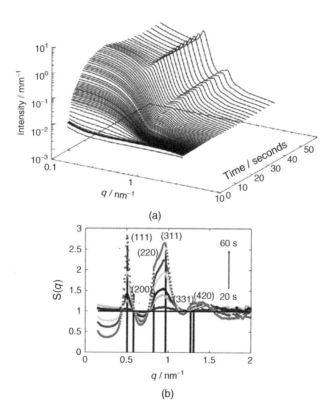

Figure 5.13 (a) Nucleation and growth of gold nanoparticles observed by time-resolved SAXS. (b) Superlattice (FCC) formation at longer timescales. Reprinted with permission from [97] Copyright (2008) American Physical Society.

a particle radius of about 1 nm, with the terminal particle size being reached in about 2 seconds, whereas in the decanoic acid route particle growth continues up to 10 seconds. Simultaneous WAXS data revealed the crystalline nature of gold nanoparticles.[96] More interestingly, at longer times, additional peaks appear in the SAXS pattern despite the low volume fraction of the particles ($\sim 10^{-5}$), corresponding to the formation of an FCC-type superlattice, as shown in the lower panel of Figure 5.13.[97] The growth dynamics was found to be slower than the diffusion-limited process, resulting in a globular structure with an FCC lattice within. Owing to the ligand binding layer on the particles, the gold volume fraction is only 0.33 within the FCC lattice.

An interesting aspect of gold nanoparticles is that van der Waals interactions can be tuned by varying the particle size. The superlattice is formed when particles are large enough to generate sufficient van der Waals attraction to counterbalance the thermal fluctuations. By tuning this interaction, other types of superlattice (*e.g.* BCC) can be formed,[98] enclosing only a low fraction of gold compared to hard-sphere colloids.

5.6.1.3   Formation of Inorganic–Organic Frameworks

An interesting case study is the nucleation and growth of nanocrystals of metal–organic frameworks (MOFs). The 3D frameworks of these hybrid materials are assembled from inorganic units and bridging polytopic organic ligands.[99] A widely investigated MOF is the microporous zeolitic imidazolate framework 8 (ZIF-8). These are found in the form of nanocrystals with a narrow size distribution and can be prepared rapidly by combining methanol solutions of zinc salt and organic ligand at room temperature.[99] In this case, SAXS probes the particle size and distribution, while the crystalline fraction can be quantitatively deduced from WAXS. Monitoring the process *in situ* using combined SAXS/WAXS, as depicted in Figure 5.14, reveals that ZIF-8 nanoparticles start to form after 15 seconds and grow to a size of about 25 nm in 15 minutes, following a monomer addition mechanism, while the Bragg peaks in the WAXS patterns appear after 35 seconds (*i.e.* 7 seconds after the clear emergence of the initial particles). This may be because of the small size of the initial ZIF-8 particles, rather than an amorphous-to-crystalline transformation. Comparison of the evolution of mean particle radius and crystallinity suggests that the late stage of the growth process is governed by Ostwald ripening.

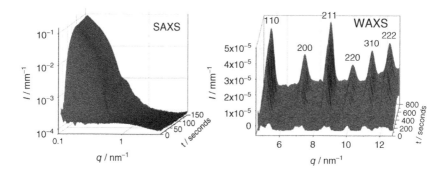

Figure 5.14 Evolution of SAXS and WAXS intensities as a function of time during the formation of the MOF ZIF-8. Reprinted with permission from [99]. Copyright © 2011, WILEY-VCH Verlag GmbH & Co. KGaA.

In the investigation of hybrid inorganic-organic materials, SANS has a specific advantage in terms of its ability to tune the contrast. For example, a recent contrast-matching time-resolved SANS study of the surfactant-templated mesoporous silica clearly illustrated that it is possible to highlight the individual contributions from the silica and surfactant during the growth process.[100] The analysis supported the so-called 'current bun' model of particle growth, which involves adsorption of condensing silica oligomers on to micelles, thereby reducing intermicellar repulsion and promoting aggregation to form initial particle nuclei.[100] Solid mesoporous silica nanoparticles are formed when the intermicellar space is also filled with silica.

5.6.1.4 Pyrolytic Growth of Nanoparticles

A key feature of high-brilliance SAXS is the ability to study extremely dilute and transient systems in gas[101] or plasma states.[102] For example, nanoparticle formation requires conditions far from equilibrium, since particle size decreases with increasing supersaturation in temperature, partial pressure and concentration, as governed by the Gibbs–Thompson equation.[101] Therefore, an ideal route for the formation of nanomaterials is in the gas phase with high temperatures and high concentrations of supersaturated vapours, and the lack of liquid byproducts facilitates easier particle collection from gaseous rather than liquid streams.[103] Pyrolitic processes such as spray flames offer these conditions and lock in the earliest stages of growth with extremely short residence times.[103,104] SAXS has been successfully applied to the study of the nucleation and growth of silica and some metal oxide nanoparticles in scalable flame aerosol reactors.[101,104,105] In this set-up, the height above the burner (HAB) or residence time is a measure of the kinetic time. Therefore, different stages (such as the nucleation and growth of primary particles and their subsequent aggregation and agglomeration) can be mapped by static measurements as a function of HAB.[103,105] The high brightness of the synchrotron source is still required because of the low volume fraction of the particles (ca $10^{-7}-10^{-6}$), together with their small sizes.[104] From the SAS point of view, the gas phase also has the advantage that the background due to the medium is low and the contrast of the particles is very high.

A much higher degree of supersaturation and therefore nanometre sizes range can be realised in spray flames, which has produced an array of functional nanomaterials, such as catalysts, gas sensors, dental and orthopaedic materials and even nutritional supplements.[103] In this case, the height above the nozzle (HAN) is an equivalent residence time that, due to the sonic speed, reaches down into the submillisecond range. This allows the earliest stage of nucleation and growth to be probed.[104] Figure 5.15 depicts a spray flame and corresponding SAXS intensities as a function of HAN during the synthesis of zirconia nanoparticles. Quantitative analysis of the SAXS data using a multilevel unified scattering function provides the mean size, number density, polydispersity

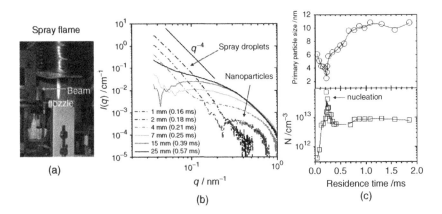

Figure 5.15 (a) Spray flame used for pyrolytic nanomaterial synthesis (temperature $> 2500\,K$, cooling rate $> 10^7\,Ks^{-1}$). (b) Typical SAXS intensity from a spray flame as a function of HAN and equivalent residence time in milliseconds. The Porod scattering at low q for lower HANs corresponds to scattering by spray droplets. (c) Evolution of primary particle size and number density with residence time. Reprinted from [104] with permission from Elsevier.

etc. of primary particles and their aggregates.[30] The dip in particle size corresponds to the highest supersaturation followed by a rapid growth to terminal size by coagulation and sintering, which narrowed the distribution of primary particles within the aggregates. The particle number density, N, shows a maximum located in the nucleation zone. The volume fraction is in the range $10^{-7} - 10^{-6}$ and the size distribution reaches the self-preserving limit within a short residence time of a few milliseconds (kinetic lock-in nanostructures).

In summary, *in situ* SAXS studies and their combination with WAXS have revealed unprecedented details about the nucleation and growth processes in a wide range of inorganic materials. The technique is matured and will remain useful in the future for studies of emerging novel materials of both fundamental and practical importance. In general, the time resolution for SAXS/WAXS is in the millisecond range and is mainly limited by the detector performance for strongly scattering inorganic materials.

5.6.2   Ultra Small-Angle Scattering

Typically the ultra small-angle X-ray and neutron scattering (USAXS and USANS, respectively) range covers sizes from 100 nm to several microns and above.[106] USAS is an alternative to light scattering and microscopy when investigating optically opaque samples.[107] A combination of SAS and USAS allows one to investigate structural scales from 1 nm up to 10 μm, which is very useful in the study of hierarchically organised materials.[108,109] Examples include a wide class of micro- and mesoporous materials, including zeolites, colloidal materials with primary particles, aggregates and agglomerates, ceramic materials *etc.*[108] In addition, USAXS can be used for *in situ* studies of the late stages of particle nucleation and growth.[109]

The experimental scheme is significantly different from the pinhole instruments described before, and involves the Bonse–Hart set-up using a channel-cut crystal monochromator/collimator and analyser.[106,107] The Bragg reflection from a thick crystal has a finite width (Darwin width, ω_D) that depends on the order of reflection[106] and the parasitic tail of the rocking curve can be effectively curtailed by means of well-conditioned channel-cut crystals.[106,107] Figure 5.16 shows a schematic view of a Bonse–Hart instrument. The first crystal (C_1) in the

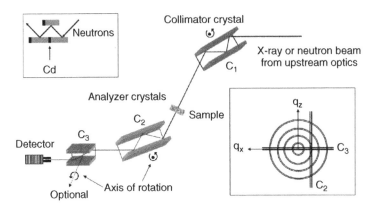

Figure 5.16 Schematic layout of a Bonse–Hart USAS instrument, with C_1 the collimating crystal, C_2 the first analyser and C_3 an optional crossed analyser.[109] The upper insect depicts the Agamalian cut used to suppress the TDS in USANS.[106] The lower inset illustrates the smearing geometry when only one analyser is used to record an isotropic scattering pattern. Reprinted with permission from [109] Copyright (2007) IUCr.

Bragg condition collimates the incident beam on the sample and the second crystal (C_2) is the analyser, which is turned to record the convoluted reflectivity or rocking curve of C_1 and C_2.[9] Typical crystal reflections are Si(111), Si(220) or even higher-order reflections.[106,108,109] The large size and divergence and the wavelength spread of a neutron beam lead to significant loss of flux at the sample position.[106] The beam divergence of an undulator beam is comparable to the width of the rocking curve (*e.g.* Si(111) $\approx$ 22 μrad), and therefore the loss of intensity with multiple reflections is much smaller.[109] Owing to the high penetration power of neutrons, in addition to Bragg reflection, significant thermal diffuse scattering (TDS) arises from the bulk. This scattering background needs to be blocked by inserting a strong absorbing material such as cadmium into the crystal, as shown in the upper inset of Figure 5.16.[106]

In a conventional Bonse–Hart USAS set-up, the measured scattering profiles are smeared in the direction perpendicular to the scan (see the lower inset in Figure 5.16), which requires desmearing correction.[106,108] In synchrotron USAXS, this slit smearing can be avoided by inserting a second analyser (C_3) in the orthogonal direction, which is maintained at the Bragg angle during the scan of C_2.[109] When there is a sample in between, the measured rocking curve is a superposition of the sample scattering and the empty rocking curve. Figure 5.17 shows the normalised rocking curves from a sample and the corresponding empty background. The measured intensity can be directly transformed to

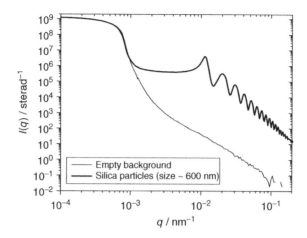

Figure 5.17 Normalised rocking curves measured by a USAXS instrument with crossed analysers.[109] The empty and sample (silica colloidal particles of diameter about 600 nm) rocking curves superimpose at ultra low angles. The peak around $10^{-2} nm^{-1}$ represents $S(q)$ and higher q oscillations depicts $P(q)$. Reprinted with permision from [109] Copyright (2007) IUCr.

$d\Sigma/d\Omega$ from the known widths of the rocking curves without the need for a calibration standard.[108,109] When studying hierarchically ordered porous materials using scattering methods, refraction and multiple scattering can also contribute to the measured intensities at low q region and undermine the correct structural analysis. In USAS, these effects are manifested as a broadening of the rocking curve and thus allow contributions from refraction and multiple scattering to normal scattering to be detected.[109]

One of the issues when studying fractal systems is the lack of access to sufficiently large length scales.[110] This has made it difficult to establish the fractal nature of many natural systems and correctly pin down their fractal dimension. An example is sedimentary rocks, which are formed from a mixture of inorganic and organic debris deposited in an aqueous environment and then buried and compacted over millions of years.[111] Their fluid elusion properties are much more complex than those of mere tightly packed grains. Figure 5.18 presents an example of neutron scattering from a solid rock sample originating from 342.7 m depth in the Urapunga 4 well reserve in Australia.[111] These data were obtained by a combination of SANS and USANS using several facilities, and cover more than three orders in size scale and ten orders in intensity. The analysis based on a fractal model[112] unambiguously established the surface fractal character of this rock, with fractal dimensionality $D_S \approx 2.82$ and

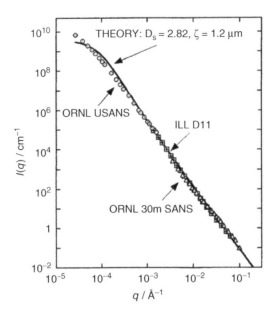

Figure 5.18 Absolute neutron scattered intensity from sedimentary rock sample U116. The continuous line corresponds to a fractal model with fractal dimensionality 2.82 and cut-off length 1.2 μm. This is one of the rare examples of SAS covering such a large dynamic range of intensity. Reprinted with permission from [111] Copyright (1999) American Physical Society.

cut-off length in excess of 1.2 μm. This has allowed a wide variety of rocks from different sources to be characterised as a function of the depth of their origin. Similar surface fractal behaviour has also been observed in low-rank coals (lignite).[110]

Figure 5.19 illustrates another example: USAXS probing the multi-scale structural features of a thermally induced colloidal aggregation in a model system consisting of a sterically stabilised colloidal system (stearyl silica particles in n-dodecane).[113] A transition from hard-sphere repulsive to short-ranged attractive interactions occurs as the temperature is lowered below T_A.[113] In this case, USAXS allows subtle variations in $S(q)$ to be distinguished, corresponding to the transition from repulsive to attractive interactions and the onset of clustering as the interparticle attraction (u) is progressively increased. Modelling of the USAXS curves provides full details of the system, both at the particulate level and in their aggregate state. The morphology of the clusters in this case is characterised by a fractal dimensionality (d_f) of about 2.3 despite the limited size range, and eventually the system transforms to a colloidal gel.[113]

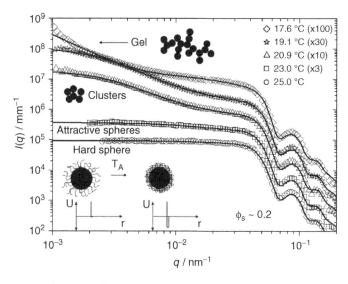

Figure 5.19 Evolution of the USAXS intensity in a short-range interacting colloidal system consisting of stearyl silica particles in *n*-dodecane, illustrating a thermally induced transition from repulsive hard spheres to attractive spheres and the subsequent formation of colloidal clusters and their transformation to a gel.[109,113] Cartoons depict schematically the transformation of the particles and the corresponding interparticle potential and resulting colloidal structure.

Both USAXS and USANS have been employed for non-invasive microstructure characterisation of a variety of materials, such as spray coatings.[108] A combination of SANS and USANS has been used to probe different stages of buckling-driven morphological transitions in the evaporation-induced self-assembly of mixed colloidal particles.[114] This has allowed the condition under which buckling can be avoided to be identified. Another example is the observation of the scaling of the structure factor or lack thereof during the hydration of cement (calcium silicate) by light and heavy water.[115] Rather surprisingly, with light water the mass fractal structure was preserved during the hydration process, while with heavy water the morphology changed from mass fractal to surface fractal and back as the process advanced. USAXS and USANS have been powerful techniques in the characterisation porous materials, especially in allowing the contributions from mesopores and micropores to be separated.[38,39,116] In addition, scattering techniques give access to both open and closed pores in natural materials such as rocks.[116] Despite the significant investments made in the development of USANS and USAXS at large facilities, the techniques remains somewhat under-exploited. Inorganic materials are ideally suited to USAS investigations and hopefully the coming years will attract more applications in this field.

5.6.3 Contrast Variation in SAS

The ability to vary the scattering contrast without altering the physical chemistry of the system under investigation is an important feature widely exploited in SAS studies.[4-6] This allows weak scattering signals otherwise superimposed on a large total intensity to be detected.[117] In the case of neutrons, a large difference in the scattering lengths between different isotopes (*e.g.* hydrogen and deuterium) is key to such experiments (see Table 5.1). For instance, the triumph of SANS in the study of soft matter and biological macromolecules has been significantly contributed by the fact that the scattering length densities of many polymers, surfactants, lipids, proteins, nucleic acids and polysaccharides in aqueous media can be matched by systematically varying the heavy water-to-water ratio.[6] More advanced studies employ partial or complete deuteration of selective components, which provides the ability to distinguish different domains in the scattering objects (*e.g.* particles with core–shell morphology or multicomponent systems).

5.6.3.1 SANS Contrast Variation

The separation of partial scattering intensities has often been crucial for the validation of scattering models when dealing with complex multi-component systems.[117] It is based on at least three different samples: the first with the usual contrast, the second with labelling of a given region and the third at an intermediate contrast.

$$I_1(q) = I_0(q)$$
$$I_2(q) = I_0(q) + \Delta\rho\, I_{0L}(q) + \Delta\rho^2 I_L(q)$$
$$I_3(q) = I_0(q) + \Delta\rho/2\, I_{0L}(q) + \Delta\rho^2/4 I_L(q) \qquad (5.27)$$

The difference $I_1(q) + I_2(q) - 2\,I_3(q)$ yields the partial intensity of the labelled region. While isotope labelling and solvent substitution have been the basis of the classical contrast variation method, the spin-contrast variation is a powerful approach in the study of inorganic materials.[17,18] The underlying principle is the dependence of the scattering amplitude of an atom on the polarisations of nuclear and neutron spins.[118] In this case, the contrast variation is achieved with the same sample and change in polarisation of the incident neutron beam. A high polarisation of protons or deuterons is obtained by dynamic nuclear spin polarisation. However, this is a complex method that involves cooling the sample to low temperatures below 1 K and using high magnetic field and microwave irradiation with frequency below and above the paramagnetic resonance to polarise the nuclei in the desired direction. While such experiments remain very powerful for the amplification of contrast, the measurements are time consuming and require significant effort.[117] Perhaps a more straightforward use of polarised neutrons is in the variation of magnetic contrast or SANS with polarised neutrons (SANSPOL).[119] This method is based on the fact that the contrast of the magnetic component in the scattering object arises from both nuclear and magnetic parts. The sign of the magnetic contrast can be changed by flipping the direction

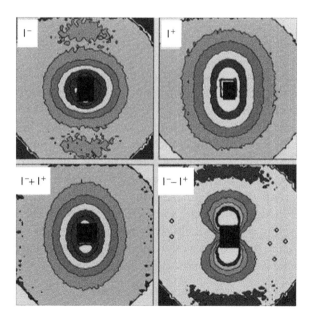

Figure 5.20 SANSPOL patterns obtained for a Co ferrofluid with neutron spins antiparallel [$I^-(q)$] and parallel [$I^+(q)$] to the horizontal field of strength, B, 1.1 T. The arithmetic mean [(I^-) + (I^+)]/2 corresponds to the 2D pattern of non-polarised neutrons. The difference (I^-) − (I^+) yields the interference term with negative values in the centre. Reprinted from [119] with permission from Elsevier.

of polarisation of the incident beam while leaving the non-magnetic contrasts unchanged:[119,120]

$$I^-(q) - I^+(q) = F_M F_N L(x) \sin^2(\alpha) \tag{5.28}$$

where F_M and F_N are magnetic and non-magnetic scattering amplitudes, $L(x) = \cot(x) - 1/x$ (where $x = MB_{Eff}/k_B T$, with M the total magnetic moment and B_{Eff} the effective applied magnetic field), and α is the angle between the applied magnetic field and the scattering vector, q. The information content of such an experiment can be further enhanced by including an additional polarisation analysis of the scattered intensity.[121] Figure 5.20 shows the typical scattering features observed in a SANSPOL experiment using a ferrofluid system. The difference intensities provide the cross-term of the magnetic and non-magnetic scattering amplitudes. One of the most important applications of SANSPOL is in the study of ferrofluids, which are model magnetic liquids with a wide range of practical applications. The fact that the incident polarisation can be flipped relatively quickly allows the dynamics of ferrofluids to be probed stroboscopically.[120]

5.6.3.2   Anomalous SAXS

For X-rays, the atomic scattering factor can be considered a constant ($f \approx Z$) when the incident energy is removed from the atomic absorption edge of any of the constituent elements. Moreover, f varies systematically with the atomic number. As a result, contrast variation involves heavy atom substitution or labelling, which can modify the chemical nature of the system. An alternative approach is based on the fact that near the absorption edge of a constituent element, f becomes a complex function of energy, E, and that not all electrons are involved in the scattering process.[122,123] Above the edge, some electrons absorb the photon, and the corresponding inner-shell electron will be promoted. This absorbed photon will later be emitted as fluorescence at a lower energy.[123] The complex atomic scattering factor is given by:

$$f(E) = f_0 + f'(E) + i\, f''(E) \tag{5.29}$$

where $f_0 \approx Z$ (except for a small relativistic correction at high photon energies).[124] The imaginary part is responsible for absorption, and f' and f'' are related through the Kramers–Kronig dispersion relation.[122,124] The energy dependence of f can be exploited to vary the contrast in SAXS experiments. This contrast-variation SAXS is known as anomalous SAXS (ASAXS).[122,125] For inorganic materials, a large number of constituent elements have an absorption edge in the range 6–24 keV, over which accurate SAXS measurements can be performed.[125] Therefore, ASAXS is a powerful method for deriving the spatial distribution of selective elements. Figure 5.21 shows the typical variation of f' and f'' near the K- and L-edges of certain elements. The anomalous dispersion is stronger near K- and L-III-edges.

The potential of ASAXS was first recognised in the 1980s,[122] and since then it has been widely used in the analysis of nanostructures in metallic alloys, especially to obtain the morphology and size of the precipitates.[125,126] The primary advantage of a synchrotron source in ASAXS is that the energy can be varied continuously. The real and imaginary parts of f (f' and f'') as a function of energy are available for most elements.[124] The K-edge of certain intermediate elements (iron to strontium) and the L-edges of some heavy elements (e.g. gold, platinum and lead) are in a suitable energy range for performing quantitative SAXS.[127,128] The above energy dependence of f near an absorption

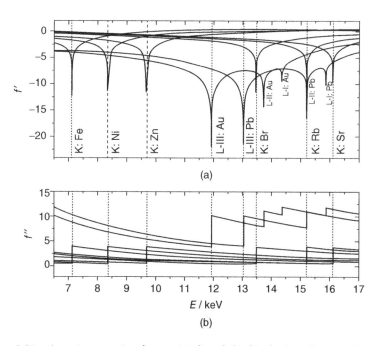

Figure 5.21 Atomic scattering factors (a) f' and (b) f'' of select elements depicting the anomalous effect.[124] Reprinted with permission from [124] Copyright (1992) American Institute of Physics.

edge leads to three terms in the scattered intensity:

$$I(q) = F(q)F^*(q) = F_o^2(q) + 2f' F_o(q)N_R(q) + (f'^2 + f''^2)N_R^2(q) \quad (5.30)$$

where $N_R(q)$ is given by the Fourier transform of the spatial distribution of anomalous or resonant atoms.[127] The first term is a non-resonant intensity, as measured by conventional SAXS far below the absorption edge. The anomalous effect appears as two different terms: the dominant cross-term involving the resonant and non-resonant scattering amplitudes and a smaller self-term resulting from the square of the resonant amplitude.[126,127] The fact that $f'(E)$ decreases near the absorption edge implies that the scattered intensity decreases unless the sign of one of the individual amplitudes is negative. In order to perform ASAXS decomposition, measurements at least three energies below the absorption edge with sufficient differences in f' are required. In reality, measurements are performed at many different energies and then a set of simultaneous equations is solved to extract the three partial intensities with high accuracy.[128]

Figure 5.22 shows an example of the ASAXS decomposition of intensities from selenium-modified ruthenium catalyst on an active carbon black support.[129] Note that the scattered intensity from the ruthenium nanoparticles is several orders of magnitude smaller than the total intensity from the catalyst and the support.[129] ASAXS measurements near the Ru K-edge (22.117 keV) allowed the scattered intensity solely from the ruthenium nanoparticles buried underneath the huge background to be separated.[129] The analysis of this resonant intensity yielded the mean radius and standard deviation of the Ru nanoparticles (1.25 nm and 0.6 nm, respectively). Further ASAXS measurements near the selenium K-edge (12.658 keV) permitted the separation of the partial intensity due to Se clusters.[129] Using these individual size parameters to model the total scattered intensity enabled the determination of the nanostructure of the carbon support itself (15 nm). Therefore, access to several absorption edges is handy when analysing individual nanostructures in multicomponent systems, as also shown for other systems.[130] For example, the role of yttrium and zirconium ions during the sintering of sol-gel-prepared yttria-stabilised zirconia ceramics was probed using *in situ* double ASAXS across the yttrium and zirconium K-edges and the results revealed the dominant role of zirconia in the crystal nucleation and growth.[131]

ASAXS experiments are more demanding than conventional SAXS measurements.[128] Special care is required for the absolute calibration of intensity and energy (as f' strongly depends on E). As the incident

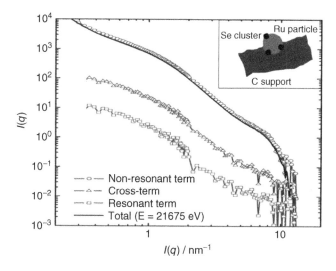

Figure 5.22 Partial scattering intensities for a sample of $Ru_{1-x}Se_x/C$ derived using measurements performed at five energies close to the Ru K-absorption edge. The resonant term corresponds to the partial intensity due to Ru nanoparticles, which is otherwise superimposed on a large intensity from the carbon black support.[129] Reprinted with permission from [129] Copyright (2010) American Chemical Society.

energy is varied, both the detector efficiency and the sample transmission also change. It is important to perform *in situ* calibration of the detector efficiency and energy as this allows one to detect any possible chemical shift when studying complex inorganic systems.[131] In addition, as the absorption edge is approached, resonant inelastic scattering becomes prominent, which gives rise to an excess flat background at the high-q region.[128] This fluorescence background needs to be removed when analysing the data just below and above the absorption edge. In the close vicinity of the absorption edge, the finite energy resolution of the crystal monochromator ($\Delta E/E \approx 0.015\%$ for Si(111)) introduces a smearing of $f'(E)$, and an effective $f'(E)$ must be used in Equation 5.30. The fluorescence contribution can be estimated from the measurements done at two different energies above and below the absorption edge having the same value of $f'(E)$.[128]

One of the key advantages of ASAXS is that it can be applied to functional systems such as catalysts,[129] sintering of ceramics,[131] phase-separating alloys[126] and *in operando* battery cells.[132] The structural information derived is often unique and cannot be accessed by any other methods.

5.6.4 Grazing-Incidence SAS

Grazing-incidence small angle scattering (GISAS) is an important technique for the characterisation of the nanostructure at surfaces, interfaces and in thin films.[133] The feasibility of grazing-incidence SAXS (GISAXS) was first demonstrated in the late 1980s.[134] This technique combines the features of conventional SAS in transmission geometry (*e.g.* collimation, detectors *etc.*) with those of refloctometry (*e.g.* goniometer, sample environments *etc.*).[133] Grazing-incidence experiments can be performed using either X-rays or neutrons, but X-rays have many advantages due to the high brightness provided by the synchrotron sources.[133] The discussion in this section is limited to GISAXS. Figure 5.23 shows the scattering geometry used for a GISAXS experiment, and the scattering vectors along the *x*-, *y*- and *z*-directions are given by:

$$q_{x,y,z} = \frac{2\pi}{\lambda} \left\{ \begin{array}{c} \cos\alpha_f \cos\theta_f - \cos\alpha_i \cos\theta_f \\ \cos\alpha_f \sin\theta_f - \cos\alpha_i \sin\theta_f \\ \sin\alpha_i + \sin\alpha_f \end{array} \right\} \qquad (5.31)$$

The incident beam impinges the sample at a very small angle, α_i, typically below the critical angle, α_c, for specular reflection. The intensity distribution along the q_z-axis provides information on the electron density along the direction perpendicular to the surface. For a given q_z, a section along the q_y-direction elucidates both the morphology and the lateral correlation between the nanostructures present at the surface, as

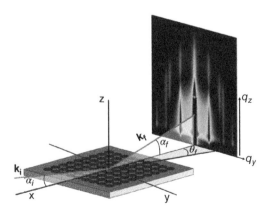

Figure 5.23 Schematic representation of the scattering geometry used for a GISAXS experiment. The definition of the moduli of scattering vectors along the *x*-, *y*- and *z*-directions is given by Equation 5.31. Figure courtesy of O. Konovalov.

in transmission SAXS. A powder ring in the transmission SAXS would appear as an elliptical sector in the $(q_y q_z)$ plane, while Bragg peaks come out elongated in the q_z-direction and become rods in grazing-incidence diffraction.[135] The penetration depth of the X-rays normal to the surface is controlled by the incidence angle, α_i. As α_i increases, the beam penetrates deeper below the surface or interface, which can be used to probe buried structures.

As seen before for transmission SAS, the form factor of an object is calculated from the electron density, assuming Born approximation. However, in the grazing incidence, additional effects such as reflection, refraction *etc.* at the substrate or by the surface microstructure must be included in addition to the scattering. These different effects are incorporated within the framework of distorted-wave Born approximation (DWBA).[136] The calculation of form factors within DWBA for standard shapes can be performed using the computer program written by Lazzari.[137] The calculated DWBA form factor, when combined with the corresponding $S(q)$ of lateral correlations, describes $I(q)$ along the section parallel to the surface. The sample scattering exhibits a maximum at the Yoneda peak defined by $(\alpha_f + \alpha_c)$.[135]

At the grazing incidence, the incident beam probes a large area of the surface due to the long footprint of the beam. This means that GISAS provides statistically averaged nanostructural properties of the surface or interface, as opposed to more local information obtained by imaging techniques such as atomic force microscopy (AFM).[135] For example, the degree of order within monolayers with different deposition techniques can be quantitatively assessed.[138] In addition, the background from the substrate is avoided due to the low penetration depth below the surface. As a result, there has been a rapid development of the technique over the last decade.[133] Since the samples can be investigated as prepared, the technique has been widely used for *in situ* and time-resolved studies. For example, the capability of GISAXS for the *in situ* analysis of particle growth and particle–substrate interactions involving an ultra high-vacuum and high-temperature set-up has been demonstrated for two different cases.[139] In the first case, the formation of Pd nanoparticles on MgO (001), a model system for the growth of metal on oxide surfaces, which is important in the elementary processes of heterogeneous catalysis, was investigated. The process followed three different steps: nucleation, growth and eventually coalescence to form the final particle size.[139] In the second case, the growth of Co nanodots on a herringbone reconstructed Au(111) surface was probed and the optimum conditions for self-organised growth were identified.[139]

Another illustrative example is the evaporation-controlled self-assembly of surfactant-templated silica mesophases on a silicon substrate.[140] This is an alternative route to dip coating by which to obtain mesoporous structures from the same precursor solution and allows thicker films to be formed with tunable morphology, depending upon the evaporation rate and chemical composition (tetraethyl orthosilicate precursor in ethanol/water medium and cetyltrimethylammonium bromide surfactant directing the template structure). *Ex situ* studies showed a 3D hexagonal structure within dip-coated films,[141] while *in situ* GISAXS unravelled the full pathways of the evaporation-controlled self-assembly process. Figure 5.24 shows a transition from micellar solution to 2D hexagonal order, and finally the emergence of a 3D cubic structure during the slow evaporation of the solvent.[140] Varying the incident angle reveals the coexistence of different surfactant mesophases as a function of the film depth due to concentration gradient, with the most ordered phase at the top and the micellar phase at the bottom on the silicon substrate. This points to the importance of the delicate balance between the evaporation rate and silica condensation in the development of the desired structure.[140] This kinetic locking mechanism often plays the pivotal role in developing robust self-assembled structures.

Depending on the evaporation rates, different superstructures can be formed even with colloidal particles. *In situ* GISAXS on evaporating droplets revealed the formation of 2D gold nanocrystal superlattices at the air–water interface for fast evaporation, while 3D superlattices are formed in the slow evaporation case.[142] As in the case of SAXS/WAXS, the combination of GISAXS and grazing-incidence diffraction (GID) is a powerful approach for probing the crystallisation process in films and at interfaces.[143] For example, an investigation of the formation of copper zinc tin sulfide semiconducting thin films from solution precursors by GISAXS/GID allowed for the optimisation of the optical properties pertinent to the photovoltaic application of this material.[143]

Using high-energy X-rays or neutrons, it is also possible to probe nanostructures at buried interfaces (*e.g.* liquid–liquid or liquid–solid interfaces). GISAXS experiments using high-energy X-rays permitted the study of liquid–liquid interfaces suitable for probing novel interfacial reactions.[144] For example, the formation of gold nanoparticles at a toluene–water interface was studied using the reduction reaction, involving an organic precursor of gold kept in a toluene layer and a reducing agent kept in a slightly alkaline water layer. The liquid–liquid

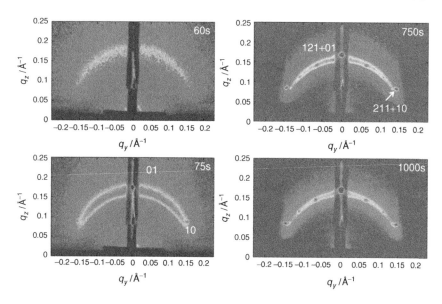

Figure 5.24 Stages in the evaporation-controlled self-assembly of mesoporous silica films on a silicon substrate. The different phases show the wormlike micellar solution, the 2D hexagonal phase, the coexistence of 2D hexagonal and 3D cubic phases and the final 3D cubic phase. Reproduced with permission from [140]. Copyright © 2003, American Chemical Society.

interface facilitated the formation of nanometre-sized particles with low polydispersity. This led to the observation of a monolayer of 'magic clusters'at the water–toluene interface, with each cluster consisting of 13 nanoparticles of about 1.2 nm diameter (similar to Au-55 nanoparticles) coated with an organic layer of about 1.1 nm and an in-plane cluster–cluster separation of 18 nm.[144] With the advancement of source brightness, beam line instrumentation and detectors, GISAXS has become capable of tracking transient processes such as continuous compression of Langmuir films formed by nanoparticles (silver) and of identifying non-equilibrium structures prior to the collapse of the film.[145] Another example is the superstructure formation of colloidal nanorods (CdSe/CdS dot core/rod shell) at the air–water interface. GISAXS has revealed a length-dependent self-organisation of these nanorods.[146]

To summarise, GISAS and GID will continue to thrive as important characterisation tools for inorganic materials at surfaces and interfaces. The exploration of the dynamics of self-assembly processes at interfaces is still in its infancy.

5.7 CONCLUSION

Over the last several decades, SAS and related techniques have found an immense range of applications in the characterisation of nano- and microstructures in inorganic and hybrid materials. This chapter presented a few representative and relatively simple examples to illustrate the different features of SAS, without making an extensive review of the literature on the application of these techniques in the field of inorganic materials. Large-scale facilities such as modern synchrotron, reactor and spallation sources offer new possibilities for the study of the structure and kinetics of materials under processing conditions. Examples include time-resolved SAS, USAS, contrast-variation SAS, microbeam SAXS/WAXS *etc*. Scattering techniques provide statistically averaged nano- and microstructural information that can be directly related to the functional properties of a given material. One of the most attractive features of SAS methods is that they can reveal multiscale structures in real time in a non-invasive manner. Despite the advances in various microscopies (electron, atomic force *etc*.), scattering experiments remain essential for the investigation of liquid samples. The most powerful approach is to combine scattering and imaging methods in a complementary fashion. Not only do these developments take place at large research facilities but also the laboratory SAXS instruments have improved significantly over the last decade.

Advances in detector technology over the last decade have overcome the bottleneck in time-resolved studies to some extent. In order to exploit these developments for the *in situ* characterisation of inorganic materials, more appropriate sample environments are perhaps required. The information content in scattering and diffraction data is usually very high, but often a few qualitative parameters are only deduced. Full extraction of structural information requires more effort in terms of analytical modelling and computer simulations. Many inorganic and hybrid organic–inorganic materials form the backbone of the modern technology. A deeper understanding of their nanostructures and optimisation of their structures are essential to engineering their functional behaviour. In this respect, SAS and related techniques offer many unique capabilities.

ACKNOWLEDGEMENTS

A. R. Rennie (Uppsala) is gratefully acknowledged for constructive comments on the manuscript. The experimental results presented in this chapter involved contributions from many colleagues and collaborators. The ESRF is acknowledged for the financial support and the provision of synchrotron beam time. The editors are thanked for their tenderness with respect to the delay in the submission of the manuscript.

REFERENCES

[1] A. Guinier and G. Fournet, *Small-Angle Scattering of X-rays*, John Wiley, New York, 1955.

[2] O. Glatter and O. Kratky (Eds), *Small-Angle X-Ray Scattering*, Academic Press, London, 1982.

[3] L.A. Feigin and D.I. Svergun, *Structure Analysis by Small-Angle X-Ray and Neutron Scattering*, Plenum Press, New York, 1987.

[4] H. Brumberger (Ed.), *Modern Aspects of Small-Angle Scattering*, Kluwer Academic, Dordrecht, 1995.

[5] P. Lindner and T. Zemb (Eds), *Neutrons, X-Rays and Light: Scattering Methods Applied to Soft Condensed Matter*, Elsevier, Amsterdam, 2002.

[6] T. Imae, T. Kanaya, M. Furusaka and N. Torikai (Eds), *Neutrons in Soft Matter*, John Wiley, New Jersey, 2011.

[7] N. Stribeck, *X-Ray Scattering of Soft Matter*, Springer-Verlag, Berlin-Heidelberg, 2007.

[8] K. Mortensen, in *Neutrons in Soft Matter*, edited by T. Imae, T. Kanaya, M. Furusaka and N. Torikai, John Wiley, New Jersey, 2011, p. 29.

[9] T. Narayanan, in *Soft-Matter Characterization*, edited by R. Borsali and R. Pecora, Springer-Verlag, New York, 2008, p. 899.

[10] I. Grillo, in *Soft-Matter Characterization*, edited by R. Borsali and R. Pecora, Springer-Verlag, New York, 2008, p. 725.

[11] M. V. Petoukhov, D. Franke, A. V. Shkumatov, G. Tria, A. G. Kikhney, M. Gajda, C. Gorba, H. D. T. Mertens, P. V. Konarev and D. I. Svergun, *J. Appl. Cryst.*, **45**, 342 (2012).

[12] S. R. Kline, *J Appl. Cryst.*, **39**, 895 (2006).

[13] J. Kohlbrecher and I. Bressler, https://kur.web.psi.ch/sans1/SANSSoft/sasfit.html (last accessed 20 January 2014).

[14] J. Ilavsky and P. R. Jemian, *J. Appl. Cryst.*, **42**, 347 (2009).

[15] B. E. Warren, *X-Ray Diffraction*, Dover, New York, 1990.

[16] W. Schmatz, T. Springer, J. Schelten and K. Ibel, *J. Appl. Cryst.*, **7**, 96 (1974).

[17] J. W. White and C. G. Windsor, *Rep. Prog. Phys.*, **47**, 707 (1984).

[18] J. Z. Larese, in *Applications of Physical Methods to Inorganic and Bioinorganic Chemistry*, edited by R. A. Scott and C. M. Lukehart, John Wiley, New York, 2007, p. 291.

[19] P. W. Schmidt, in *Modern Aspects of Small-Angle Scattering*, edited by H. Brumberger, Kluwer Academic, Dordrecht, 1995, p. 1.

[20] J. Als-Nielsen and D. McMorrow, *Elements of Modern X-Ray Physics*, John Wiley, New York, 2011.

[21] J. S. Pedersen, in *Neutrons, X-Rays and Light: Scattering Methods Applied to Soft Condensed Matter*, edited by P. Lindner and T. Zemb, Elsevier, Amsterdam, 2002, p. 391.

[22] M. Kotlarchyk and S.-H. Chen, *J. Chem. Phys.*, **79**, 2461 (1983).

[23] E.Y. Sheu, *Phys. Rev. A*, **45**, 2428 (1992).

[24] P. Bartlett and R. H. Ottewill, *J. Chem. Phys.*, **96**, 3306 (1992).

[25] J. Wagner, *J. Appl. Crystallogr.*, **45**, 513 (2012).

[26] T. Foster, *J. Phys. Chem. B*, **115**, 10207 (2011).

[27] X. Li, C. Y. Shew, L. He, F. Meilleur, D. A. A. Myles, E. Liu, Y. Zhang, G. S. Smith, K. W. Herwig, R. Pynna and W.-R. Chen, *J. Appl. Crystallogr.*, **44**, 545 (2011).

[28] B. R. Pauw, J. S. Pedersen, S. Tardif, M. Takata and B. B. Iversen, *J. Appl. Crystallogr.*, **46**, 365 (2013).

[29] G. Beaucage, *J. Appl. Crystallogr.*, **28**, 717 (1995).

[30] G. Beaucage, H. K. Kammler and S. E. Pratsinis, *J. Appl. Crystallogr.*, **37**, 523 (2004).

[31] P. Debye, H. R. Anderson and H. Brumberger, *J. Appl. Phys.*, **28**, 679 (1957).

[32] G. Beaucage, *Phys. Rev. E*, **70**, 031401 (2004).

[33] M. L. Kistler, A. Bhatt, G. Liu, D. Casa and T. Liu, *J. Am. Chem. Soc.*, **129**, 6453 (2007).

[34] S. J. Veen, *Assemblies of Polyoxometalates*, Utrecht University, 2009.

[35] B. Hammouda, *J. Appl. Crystallogr.*, **43**, 1474 (2010). See also B. Hammouda, *J. Appl. Cryst.*, **43**, 716 (2010).

[36] M. E. Fisher and F. J. Burford, *Phys. Rev.*, **156**, 583 (1967).

[37] G. Kostorz, *Z. Kristallogr.*, **218**, 154 (2003).

[38] O. Spalla, S. Lyonnard and F. Testard, *J. Appl. Crystallogr.*, **36**, 338 (2003).

[39] E. A. Chavez Panduro, T. Beuvier, M. Fernandez Martınez, L. Hassani, B. Calvignac, F. Boury and A. Gibaud, *J. Appl. Crystallogr.*, **45**, 881 (2012).

[40] R. Klein and B. D'Aguanno, in *Light Scattering: Principles and Development*, edited by W. Brown, Clarendon Press, Oxford, 1996, p. 30.

[41] A. Vrij, *J. Chem. Phys.*, **71**, 3267 (1979).

[42] D. Gazzillo and A. Giacometti, *J. Chem. Phys.*, **113**, 9837 (2000).

[43] G. Fritz, A. Bergmann and O. Glatter, *J. Chem. Phys.*, **113**, 9733 (2000).

[44] T. Fukasawawa and T. Sato, *Phys. Chem. Chem. Phys.*, **13**, 3187 (2011).

[45] P. V. Konarev, M. V. Petoukhov, V. V. Volkov and D. I. Svergun, *J. Appl. Crystallogr.*, **39**, 277 (2006).

[46] S. Förster, S. Fischer, K. Zielske, C. Schellbach, M. Sztucki, P. Lindner and J. Perlich, *Adv. Colloid Interf. Sci.*, **163**, 53 (2011).

[47] M. Megens, C. M. van Kats, P. Boesecke and W. L. Vos, *Langmuir*, **13**, 6120 (1997).

[48] A. V. Petukhov, I. P. Dolbnya, D. G. A. L. Aarts, G. J. Vroege and H. N. W. Lekkerkerke, *Phys. Rev. Lett.*, **90**, 028304 (2003).

[49] J. E. G. J. Wijnhoven, L. Bechger and W. L. Vos, *Chem. Mater.*, **13**, 4486 (2001).

[50] A. V. Petukhov, J. H. J. Thijssen, D. C. 't Hart, A. Imhof, A. van Blaaderen, I. P. Dolbnya, A. Snigirev, A. Moussaid and I. Snigireva, *J. Appl. Crystallogr.*, **39**, 137 (2006).

[51] J. H. J. Thijssen, A. V. Petukhov, D. C. 't Hart, A. Imhof, C. H. M. van der Werf, R. E. I. Schropp and A. van Blaaderen, *Adv. Mater.*, **18**, 1662 (2006).

[52] J. D. F. Ramsay, *Pure App. Chem.*, **64**, 1709 (1992).

[53] A. B. D. Brown, C. Ferrero, T. Narayanan and A. R. Rennie, *Eur. Phys. J. B*, **11**, 481 (1999).

[54] B. J. Lemaire, P. Davidson, P. Panine and J. P. Jolivet, *Phys. Rev. Lett.*, **93**, 267801 (2004).

[55] O. Pelletier, C. Bourgaux, O. Diat, P. Davidson and J. Livage, *Eur. Phys. J. E*, **2**, 191 (2000).

[56] A. V. Petukhov, D. van der Beek, R. P. A. Dullens, I. P. Dolbnya, G. J. Vroege and H. N. W. Lekkerkerker, *Phys. Rev. Lett.*, **95**, 077801 (2005).

[57] M. R. Eskildsen, E. M. Forgan and H. Kawano-Furukawa, *Rep. Prog. Phys.*, **74**, 124504 (2011).

[58] C. D. Dewhurst and R. Cubitt, *Physica B*, **385–386**, 176 (2006).

[59] D. Cribier, B. Jacrot, L. Madhav Rao and B. Farnoux, *Phys. Lett.*, **9**, 106 (1964).

[60] M. Yethiraj, H. A. Mook, G. D. Wignall, R. Cubitt, E. M. Forgan, D. M. Paul and T. Armstrong, *Phys. Rev. Lett.*, **70**, 857 (1993).

[61] E. M. Forgan, S. J. Levett, P. G. Kealey, R. Cubitt, C. D. Dewhurst and D. Fort, *Phys. Rev. Lett.*, **88**, 167003 (2002).

[62] C. Riekel, *Rep. Prog. Phys.*, **63**, 233 (2000).

[63] A. Snigirev and I. Snigireva, *C. R. Physique*, **9**, 507 (2008).

[64] J. Daillant and M. Alba, *Rep. Prog. Phys.*, **63**, 1725 (2000).

[65] A. K. Freund, in *Complementarity between Neutron and Synchrotron X-Ray Scattering*, edited by A. Furrer, World Scientific, Singapore, 1998, p. 329.

[66] R. A. Lewis, W. I. Helsby, A. O. Jones, C. J. Hall, B. Parker, J. Sheldon, P. Clifford, M. Hillen, I. Sumner, N. S. Fore, R. W. M. Jones and K. M. Roberts, *Nucl. Instr. Meth. Phys. Res., Sect. A*, **392**, 32 (1997).

[67] C. Broennimann, E. F. Eikenberry, B. Henrich, R. Horisberger, G. Huelsen, E. Pohl, B. Schmitt, C. Schulze-Briese, M. Suzuki, T. Tomizaki, H. Toyokawa and A. Wagner, *J. Synchrotron Rad.*, **13**, 120 (2006).

[68] K. Ibel, *J. Appl. Crystallogr.*, **9**, 296 (1976).

[69] R. K. Heenan, J. Penfold and S. M. King, *J. Appl. Crystallogr.*, **30**, 1140 (1997).

[70] C. D. Dewhurst, *Meas. Sci. Technol.*, **19**, 034007 (2008).

[71] T. Otomo, in *Neutrons in Soft Matter*, edited by T. Imae, T. Kanaya, M. Furusaka and N. Torikai, John Wiley, New Jersey, 2011, p. 57.

[72] P. Lindner and R. Schweins, *Neutron News*, **21**, 15 (2010).

[73] P. Van Esch, T. Gahl and B. Guerard, *Nuclear Instrum. Methods Phys. Res. A*, **526**, 493 (2004).

[74] G. Manzin, B. Guerard, F. A. F. Fraga and L. M. S. Margato, *Nuclear Instrum. Methods Phys. Res., Sect. A*, **535**, 102 (2004).

[75] R. E. Dinnebier and S. J. L. Billinge (Eds), *Powder Diffraction: Theory and Practice*, Royal Society of Chemistry, Cambridge, 2008.

[76] A. L. Patterson, *Phys. Rev.*, **56**, 978 (1939).

[77] D. R. Hummer, P. J. Heaney and J. E. Post, *J. Cryst., Growth*, **344**, 51 (2012).

[78] J. S. Pedersen, D. Posselt and K. Mortensen, *J. Appl. Crystallogr.*, **23**, 321 (1990).

[79] G. Kostorz, *J. Appl. Crystallogr.*, **24**, 444 (1991).

[80] V. Degiorgio, G. P. Banfi, G. Righini and A. Rennie, *Appl. Phys. Lett.*, **57**, 2879 (1990).

[81] G. P. Banfi, V. Degiorgio, A. Rennie and J. G. Barker, *Phys. Rev. Lett.*, **69**, 3401 (1992).

[82] W. Bras, S. M. Clark, G. N. Greaves, M. Kunz, W. van Beek and V. Radmilovic, *Cryst., Growth Des.*, **9**, 1297 (2009).

[83] G. Sandi, P. Thiyagarajan, K. A. Carrado and R. E. Winans, *Chem. Mater.*, **11**, 235 (1999).

[84] J. C. Dore, J. B. W. Webber and J. H. Strange, *Colloids and Surfaces A: Physic-ochem. Eng. Aspects*, **241**, 191 (2004).

[85] B. Smarsly, C. Goeltner, M. Antonietti, W. Ruland and E. Hoinkis, *J. Phys. Chem. B*, **105**, 831 (2001).

[86] S. Ciccariello, Y. B. Melnichenko and L. He, *J. Appl. Crystallogr.*, **44**, 43 (2011).

[87] G. Sankar and W. Bras, *Catalysis Today*, **145**, 195 (2009).

[88] K. J. Edler, P. A. Reynolds and J. W. White, *J. Phys. Chem. B*, **102**, 3676 (1998).

[89] P. Panine, S. Finet, T. Weiss and T. Narayanan, *Adv. Colloid Interf. Sci.*, **127**, 9 (2006).

[90] D. Pontoni, T. Narayanan and A. R. Rennie, *Langmuir*, **18**, 56 (2002).

[91] C. J. Gommes, B. Goderis, J.-P., Pirard and S. Blacher, *J. Non-Cryst., Solids*, **353**, 2495 (2007).

[92] J. Bolze, B. Peng, N. Dingenouts, P. Panine, T. Narayanan and M. Ballauff, *Langmuir*, **18**, 8364 (2002).

[93] J. Liu, S. Pancera, V. Boyko, A. Shukla, T. Narayanan and K. Huber, *Langmuir*, **26**, 17405 (2010).

[94] D. Pontoni, J. Bolze, N. Dingenouts, T. Narayanan and M. Ballauff, *J. Phys. Chem. B*, **107**, 5123 (2003).

[95] M. Avrami, *J. Chem. Phys.*, **8**, 212 (1940).

[96] B. Abecassis, F. Testard, O. Spalla and P. Barboux, *Nano Lett.*, **7**, 1723 (2007).

[97] B. Abecassis, F. Testard and O. Spalla, *Phys. Rev. Lett.*, **100**, 115504 (2008).

[98] D. Nykypanchuk, M. M. Maye, D. van der Lelie and O. Gang, *Nature*, **451**, 549 (2008).

[99] J. Cravillon, C. A. Schröder, R. Nayuk, J. Gummel, K. Huber and M. Wiebcke, *Angew. Chem., Int. Ed.*, **50**, 8067 (2011).

[100] M. J. Hollamby, D. Borisova, P. Brown, J. Eastoe, I. Grillo and D. Shchukin, *Langmuir*, **28**, 4425 (2012).

[101] G. Beaucage, H. K. Kammler, R. Strobel, R. Mueller, S. E. Pratsinis and T. Narayanan, *Nature Materials*, **3**, 370 (2004).

[102] E. Jerby, A. Golts, Y. Shamir, S. Wonde, J. B. A. Mitchell, J. L. LeGarrec, T. Narayanan, M. Sztucki, D. Ashkenazi, Z. Barkay and N. Eliaz, *Appl. Phys. Lett.*, **95**, 191501 (2009).

[103] R. Strobel and S. E. Pratsinis, *J. Mater. Chem.*, **17**, 4743 (2007).

[104] T. Narayanan, *Current Opin. Colloid Interf. Sci.*, **14**, 409 (2009).

[105] A. Camenzind, H. Schulz, A. Teleki, G. Beaucage, T. Narayanan and S. E. Pratsinis, *Eur. J. Inorg. Chem.*, **2008**, 911 (**2008**).

[106] M. Agamalian, J. M. Carpenter and W. Treimer, *J. Appl. Crystallogr.*, **43**, 900 (2010).

[107] U. Bonse and M. Hart, *Z. Phys.*, **189**, 151 (1966).

[108] J. Ilavsky, P. R. Jemian, A. J. Allen, F. Zhang, L. E. Levine and G. G. Long, *J. Appl. Crystallogr.*, **42**, 469 (2009).

[109] M. Sztucki and T. Narayanan, *J. Appl. Crystallogr.*, **40**, S459 (2007).

[110] P. W. Schmidt, *J. Appl. Crystallogr.*, **24**, 414 (1991).

[111] A. P. Radlinski, E. Z. Radlinska, M. Agamalian, G. D. Wignall, P. Lindner and O. G. Randl, *Phys. Rev. Lett.*, **82**, 3078 (1999).

[112] J. Teixeira, *J. Appl. Crystallogr.*, **21**, 781 (1988).

[113] M. Sztucki, T. Narayanan, G. Belina, A. Moussaïd, F. Pignon and H. Hoekstra, *Phys. Rev. E*, **74**, 051504 (2006).

[114] D. Sen, J. S. Melo, J. Bahadur, S. Mazumder, S. Bhattacharya, S. F. D'Souza, H. Frielinghaus, G. Goerigk and R. Loidl, *Soft Matter*, 7, 5423 (2011).

[115] S. Mazumder, D. Sen, A. K. Patra, S. A. Khadilkar, R. M. Cursetji, R. Loidl, M. Baron and H. Rauch, *Phys. Rev. Lett.*, 93, 255704 (2004).

[116] C. R. Clarkson, M. Freeman, L. He, M. Agamalian, Y. B. Melnichenko, M. Mastalerz, R. M. Bustin, A. P. Radlinski and T. P. Blach, *Fuel*, 95, 371 (2012).

[117] H. B. Stuhrmann, *J. Phys.: Conf. Ser.*, 351, 012002 (2012).

[118] J. B. Hayter, G. T. Jenkin and J. W. White, *Phys. Rev. Lett.*, 33, 696 (1974).

[119] A. Wiedenmann, *Physica B*, 297, 226 (2001).

[120] A. Wiedenmann, M. Kammel, A. Heinemann and U. Keiderling, *J. Phys.: Condens. Matter*, 18, S2713 (2006).

[121] A. Wiedenmann, *Physica B*, 356, 246 (2005).

[122] H. B. Stuhrmann, *Adv. Polym. Sci.*, 67, 123 (1985).

[123] M. Cianci, J. R. Helliwell, M. Helliwell, V. Kaucic, N. Z. Logar, G. Mali and N. N. Tusar, *Cryst., Rev.*, 11, 245 (2005).

[124] S. Brennan and P. L. Cowan, *Rev. Sci. Instrum.*, 63, 850 (1992).

[125] A. Naudon, in *Modern Aspects of Small-Angle Scattering*, edited by H. Brumberger, Kluwer Academic, Dordrecht, 1995, p. 203.

[126] G. Goerigk, K. Huber, N. Mattern and D. L. Williamson, *Eur. Phys. J. Special Topics*, 208, 259 (2012).

[127] M. Ballauff and A. Jusifi, *Colloid Polym. Sci.*, 284, 1303 (2006).

[128] M. Sztucki, E. Di Cola and T. Narayanan, *J. Appl. Crystallogr.*, 43, 1479 (2010).

[129] S. Haas, G. Zehl, I. Dorbandt, I. Manke, P. Bogdanoff, S. Fiechter and A. Hoell, *J. Phys. Chem. C*, 114, 22375 (2010).

[130] S. Haas, A. Hoell, R. Wurth, C. Rüssel, P. Boesecke and U. Vainio, *Phys. Rev. B*, 81, 184207 (2010).

[131] T. Barnardo, K. Hoydalsvik, R. Winter, C. M. Martin and G. F. Clark, *J. Phys. Chem. C*, 113, 10021 (2009).

[132] A. Braun, S. Seifert, P. Thiyagarajan, S. P. Cramer and E. J. Cairns, *Electrochem. Comm.*, 3, 136 (2001).

[133] G. Renaud, R. Lazzari and F. Leroy, *Surf. Sci. Rep.*, 64, 255 (2009).

[134] J. R. Levine, J. B. Cohen, Y. W. Chung and P. Georgopoulos, *J. Appl. Crystallogr.*, 22, 528 (1989).

[135] P. Mueller-Buschbaum, in *Lecture Notes in Physics: Applications of Synchrotron Light to Scattering and Diffraction in Materials and Life Sciences*, edited by T. Ezquerra, M. C. García-Gutiérrez, A. Nogales and M. A. Gomez, Springer, Berlin-Heidelberg, 776, 2009, p 61.

[136] S. Sinha, E. Sirota, S. Garoff, H. Stanley, *Phys. Rev. B*, 38, 2297 (1988).

[137] R. Lazzari, *J. Appl. Crystallogr.*, 35, 406 (2002).

[138] A. T. Heitsch, R. N. Patel, B. W. Goodfellow, D. M. Smilgies and B. A. Korgel, *J. Phys. Chem. C*, 114, 14427 (2010).

[139] G. Renaud, R. Lazzari, C. Revenant, A. Barbier, M. Noblet, O. Ulrich, F. Leroy, J. Jupille, Y. Borensztein, C. R. Henry, J.-P. Deville, F. Scheurer, J. Mane-Mane and O. Fruchart, *Science*, 300, 1416 (2003).

[140] A. Gibaud, D. Grosso, B. Smarsly, A. Baptiste, J. F. Bardeau, F. Babonneau, D. A. Doshi, Z. Chen, C. J. Brinker and C. Sanchez, *J. Phys. Chem. B*, 107, 6114 (2003).

[141] D. A. Doshi, A. Gibaud, O. V. Goletto, M. Lu, H. Gerung, B. Ocko, S. M. Han and C. J. Brinker, *J. Am. Chem. Soc.*, 125, 11646 (2003).

[142] S. Narayanan, J. Wang and X.-M. Lin, *Phys. Rev. Lett.*, **93**, 135503 (2004).

[143] A. Fischereder, T. Rath, W. Haas, H. Amenitsch, J. Albering, D. Meischler, S. Laris-segger, M. Edler, R. Saf, F. Hofer and G. Trimmel, *Chem. Mater.*, **22**, 3399 (2010).

[144] M. K. Sanyal, V. V. Agrawal, M. K. Bera, K. P. Kalyanikutty, J. Daillant, C. Blot, S. Kubowicz, O. Konovalov and C. N. R. Rao, *J. Phys. Chem. C*, **112**, 1739 (2008).

[145] K. Vegso, P. Siffalovic, E. Majkova, M. Jergel, M. Benkovicova, T. Kocsis, M. Weis, P. Siffalovic, K. Nygård and O. Konovalov, *Langmuir* **28**, 10409 (2012).

[146] F. Pietra, F. T. Rabouw, W. H. Evers, D. V. Byelov, A. V. Petukhov, C. de Mello Donega and D. Vanmaekelbergh, *Nano Lett.*, **12**, 5515 (2012).

Index

Figures are denoted by italic page numbers, **Tables** by bold numbers.

Abbreviations: EC = electron crystallography; ED = electron diffraction; EM = electron microscopy; ET = electron tomography; HRTEM = high-resolution transmission electron microscopy; PDF = pair distribution function; SAED = selected-area electron diffraction; SANS = small-angle neutron scattering; SAS = small-angle scattering; SAXS = small-angle X-ray scattering; SEM = scanning electron microscopy; STEM = scanning transmission electron microscopy; TEM = transmission electron microscopy; XRD = X-ray diffraction

Structure from Diffraction Methods, First Edition. Edited by Duncan W. Bruce, Dermot O'Hare and Richard I. Walton.
© 2014 John Wiley & Sons, Ltd. Published 2014 by John Wiley & Sons, Ltd.